CRC SERIES IN CHROMATOGRAPHY

Gunter Zweig and Joseph Sherma
Editors-in-Chief

GENERAL DATA AND PRINCIPLES

Editors
Gunter Zweig, Ph.D.
U.S. Environmental Protection Agency
Washington, D.C.
Joseph Sherma, Ph.D.
Lafayette College
Easton, Pennsylvania

CARBOHYDRATES

Editor
Shirley C. Churms, Ph.D.
Research Associate
C.S.I.R. Carbohydate Chemistry Research Unit
Department of Organic Chemistry
University of Cape Town, South Africa

DRUGS

Editor
Ram Gupta, Ph.D.
Department of Laboratory Medicine
St. Joseph's Hospital
Hamilton, Ontario, Canada

CATECHOLAMINES TERPENOIDS

Editor
Carmine J. Cosica, Ph.D.
Professor of Biochemistry
St. Louis University School of Medicine
St. Louis, Missouri

STEROIDS

Editor
Joseph C. Touchstone, B.S., M.S., Ph.D.
Professor
School of Medicine
University of Pennsylvania
Philadelphia, Pennsylvania

PESTICIDES

Editors
Joseph Sherma, Ph.D.
Lafayette College
Easton, Pennsylvania
Joanne M. Follweiler, Ph.D.
Lafayette College
Easton, Pennsylvania

LIPIDS AND TECHNICAL LIPID DERIVATIVES

Editor
H. K. Mangold
Executive Director and Professor
Federal Center for Lipid Research
4400 Munster, Germany

HYDROCARBONS

Editors
Willie E. May, Ph.D.
Center for Analytical Chemistry
U.S. Department of Commerce
Washington, D.C.
Walter L. Zielinski, Jr., Ph.D.
Air Program Manager
National Bureau of Standards
Washington, D.C.

INORGANICS

Editor
M. Qureshi, Ph. D.
Professor, Chemistry Section
Zakir Husain College of Engineering and Technology
Aligarh Muslim University
India

PHENOLS AND ORGANIC ACIDS

Editor
Toshihiko Hanai, Ph.D.
University of Montreal
Quebec, Canada

AMINO ACIDS AND AMINES

Editor
Dr. S. Blackburn
Leeds, England

POLYMERS

Editor
Charles G. Smith
Dow Chemical, USA
Midland, Michigan

PLANT PIGMENTS

Editor
Dr. Hand-Peter Kost
Botanisches Institut der Universitat Munchen
Munchen, West Germany

CRC Handbook of Chromatography: Phenols and Organic Acids

Volume I

Editor

Toshihiko Hanai, Ph.D.
Department de Chimie
Universite de Montreal
Montreal, Quebec, Canada

CRC Series in Chromatography

Editors-in-Chief

Gunter Zweig, Ph.D.
School of Public Health
University of California
Berkeley, California

Joseph Sherma, Ph.D.
Professor of Chemistry
Lafayette College
Easton, Pennsylvania

CRC Press, Inc.
Boca Raton, Florida

Library of Congress Cataloging in Publication Data
Main entry under title:

Phenols and organic acids.

(CRC handbook of chromatography)
Bibliography: p.
Includes index.
1. Phenols--Analysis. 2. Acids, Organic--Analysis.
3. Chromatographic analysis. I. Hanai, Toshihiko.
II. Zweig, Gunter. III. Sherma, Joseph. IV. Series.
QD341.P5P38 547'.632 81-19517
ISBN 0-8493-3055-6 AACR2

This book represents information obtained from authentic and highly regarded sources. Reprinted material is quoted with permission, and sources are indicated. A wide variety of references are listed. Every reasonable effort has been made to give reliable data and information, but the author and the publisher cannot assume responsibility for the validity of all materials or for the consequences of their use.

Direct all inquiries to CRC Press, Inc., 2000 Corporate Blvd., N.W., Boca Raton, Florida, 33431.

International Standard Book Number 0-8493-3055-6

Library of Congress Card Number 81-19517
Printed in the United States

CRC HANDBOOK OF CHROMATOGRAPHY

SERIES PREFACE

The present volume in the series of *Handbooks of Chromatography* deals exclusively with the separation by chromatographic techniques of phenols, organic acids, and their derivatives. The volume editor, Dr. Toshihiko Hanai, has succeeded in organizing in an easy-to-locate fashion the most important chromatographic data published during the past 10 years on the chromatographic techniques for phenols and organic acids. The first part of the book is devoted to detailed tables on gas-liquid chromatography (GC); liquid chromatography (mostly high-performance liquid chromatography) (LC); thin-layer chromatography (TLC); and paper chromatography (PC). As in previous volumes of this new series (already published: *DRUGS, CARBOHYDRATES, POLYMERS* and others to come: *LIPIDS & FATTY ACIDS, AMINES & AMINO ACIDS, INORGANICS, HYDROCARBONS, CATECHOLAMINES, PESTICIDES*, and others being planned), each table lists in an easily usable format the experimental conditions of the chromatographic separation (e.g., temperature, packing material or adsorbent, detection technique, etc.) and the respective migration or resolution factors. Shorter descriptive sections of the book deal with the preparation of samples and derivatizations, detection reagents for TLC and PC, and commercial sources for chromatographic supplies. A comprehensive compound index serves as a most useful guide for the ocation of separation techniques for specific phenols or organic acids.

It is inevitable that despite the most careful proof-reading of the manuscript and proofs at every stage of production, some errors might have crept into the final version of the printed book. For this deficiency we beg readers to be most sympathetic but also ask that they communicate either with Dr. Hanai or us so that an errata sheet can be prepared most expeditiously, and that the second printing of the book will have corrected the most obvious errors.

Joseph Sherma
Gunter Zweig
Series Editors

PREFACE

This volume in the *Handbook Series in Chromatography* contains tables and figures for paper, thin-layer, column liquid, and gas chromatography of phenolic compounds and organic acids.

Phenols and organic acids are related to human ecology. The challenge to devise a high-speed qualitative and quantitative analytical system for these compounds is being met by chemists in different fields. The difficulty in analysis of these compounds is due to the existence of many isomers and derivatives. Therefore, analytical research cannot be done without chromatographic techniques.

In the past 8 years, the technical performance of analytical instrumentation has been dramatically improved. However, data from more classical methods were also collected in this volume due to lack of similar, more recent data. These old data can be updated by individual efforts like from thin-layer chromatography to column liquid chromatography or from detection method on paper and/or thin-layer chromatography to post-column reaction detector for column liquid chromatography.

As seen in the data presented here, almost all compounds were subjected to both liquid (column, thin-layer, and paper) and gas chromatography. Rapid development of gas chromatograph as an easily handled instrument has made it a very powerful tool for scientific research. The sensitivity of detectors, the easy operation, and variety of derivatization to form volatile compounds have made gas chromatography a widely acceptable technique to analyze these compounds.

Liquid chromatography is also a suitable separation technique, except for some phenols and esters, and the further development of liquid chromatography to analyze these compounds is predicted. However, one should not ignore data from paper and thin-layer chromatography. The software and detection method are the keys to the operation of modern liquid chromatography. The combination of thin-layer and column liquid chromatography is routinely used by organic chemists. Unfortunately, few data were collected in this volume on liquid chromatography due to lack of experimental detail.

For identification of a structure, a combination of gas chromatograph and mass spectrometer is very powerful, but presently combined liquid chromatography-mass spectrometer has not been readily available due to the complicated handling which will be solved in the near future. Therefore, the selection of identification method should be done on the basis of the individual research purpose.

In this volume, collection of data was basically completed in 1979. A large number of publications in column liquid chromatography and separation of optical isomers have been published in 1980, but the coverage of these subjects will await the second edition of this volume.

In addition, data of fatty acids, cholic acids, and catecholamines are summarized in other volumes of this series. However, some of these data are included in this volume when they were analyzed with phenols or organic acids at the same time. Prostaglandins are excluded. This volume also included current development of column technology, detection methods, and sample preparation method for phenols and organic acids analysis.

Finally, I thank Editors-in-Chief Drs. Gunter Zweig and Joseph Sherma for their skillful help and members of the Advisory Board for their support of this work. I want to express by appreciation to Dr. Joseph Hubert, Universite de Montreal for his personal help. Thanks are also due to Ms. Pamela Woodcock, Amy G. Skallerup, Cathy Walker, and Mr. B. J. Starkoff of CRC Press, for their earnest effort.

I heartily thank the publishers and authors who gave me their permission to use the original data. Thanks are due to everybody who is interested in the analysis of phenols and organic acids. I am sorry that I may have missed some very important data of interest to someone. The editors would greatly appreciate suggestion for improvement and addition from interested readers.

T. Hanai

THE EDITORS-IN-CHIEF

Gunter Zweig, Ph.D., received his undergraduate and graduate training at the University of Maryland, College Park, where he was awarded the Ph.D. in biochemistry in 1952. Two years following his graduation, Dr. Zweig was affiliated with the late R. J. Block, pioneer in paper chromatography of amino acids. Zweig, Block, and Le Strange wrote one of the first books on paper chromatography which was published in 1952 by Academic Press and went into three editions, the last one authored by Gunter Zweig and Dr. Joe Sherma, the co-Editor-in-Chief of this series. *Paper Chromatography* (1952) was also translated into Russian.

From 1953 till 1957, Dr. Zweig was research biochemist at the C. F. Kettering Foundation, Antioch College, Yellow Springs, Ohio, where he pursued research on the path of carbon and sulfur in plants, using the then newly developed techniques of autoradiography and paper chromatography. From 1957 till 1965, Dr. Zweig served as lecturer and chemist, University of California, Davis and worked on analytical methods for pesticide residues, mainly by chromatographic techniques. In 1965, Dr. Zweig became Director of Life Sciences, Syracuse University Research Corporation, New York (research on environmental pollution), and in 1973 he became Chief, Environmental Fate Branch, Environmental Protection Agency (EPA) in Washington, D.C.

During his government career, Dr. Zweig continued his scientific writing and editing. Among his works are (many in collaboration with Dr. Sherma) the now 11-volume series on *Analytical Methods for Pesticides and Plant Growth Regulators* (published by Academic Press); the pesticide book series for CRC Press; co-editor of *Journal of Toxicology and Environmental Health;* co-author of basic review on paper and thin-layer chromatography for *Analytical Chemistry* from 1968 to 1980; co-author of applied chromatography review on pesticide analysis for *Analytical Chemistry,* beginning in 1981.

Among the scientific honors awarded to Dr. Zweig during his distinguished career are the Wiley Award in 1977, Rothschild Fellowship to the Weizmann Institute in 1963/64; the Bronze Medal by the EPA in 1980.

Dr. Zweig has authored or co-authored over 75 scientific papers on diverse subjects in chromatography and biochemistry, besides being the holder of three U.S. patents.

At the present time (1980/81), Dr. Zweig is Visiting Scholar in the School of Public Health, University of California, Berkeley, where he is doing research on farmworker safety as related to pesticide exposure.

Joseph Sherma, Ph.D., received a B.S. in chemistry from Upsala College, East Orange, N.J. in 1955 and a Ph.D. in analytical chemistry from Rutgers University in 1958. His thesis research in ion exchange chromatography was under the direction of the late William Rieman III. Dr. Sherma joined the faculty of Lafayette College in September 1958, and is presently full professor there in charge of two courses in analytical chemistry. At Lafayette he has continued research in chromatography and has additionally worked a total of 12 summers in the field with Harold Strain at the Argonne National Laboratory, Illinois, James Fritz at Iowa State University, Ames, Gunter Zweig at Syracuse University Research Corporation, New York, Joseph Touchstone at the Hospital of the University of Pennsylvania, Philadelphia, Brian Bidlingmeyer at Waters Associates, Framingham, Mass., and Thomas Beesley at Whatman Inc., Clifton, N.J.

Dr. Sherma and Dr. Zweig [who is now with U.S. Environmental Protection Agency (EPA)] co-authored Volumes I and II of the *CRC Handbook of Chromatography,* a book on paper chromatography, and 6 volumes of the series *Analytical Methods for Pesticides and Plant Growth Regulators.* Other books in the pesticide series and further volumes of the *CRC Handbook of Chromatography* are being edited with Dr. Zweig, and Dr. Sherma will co-author the handbook on pesticide chromatography. A book on quantitative TLC (published by Wiley-Interscience, New York) was edited jointly with Dr. Touchstone. Dr. Sherma has been co-author of 7 biennial reviews of liquid chromatography (1968 to 1980) and the 1981 review of pesticide analysis for the journal *Analytical Chemistry*. Dr. Sherma has authored major invited chapters and review papers on chromatography and pesticides in *Chromatographic Reviews* (analysis of fungicides), *Advances in Chromatography* (analysis of fungicides), *Advances in Chromatography* (analysis of nonpesticide pollutants), Heftmann's *Chromatography* (chromatography of pesticides), Race's *Laboratory Medicine* (chromatography in clinical analysis), *Food Analysis: Principles and Techniques* (TLC for food analysis), *Treatise on Analytical Chemistry* (paper and thin layer chromatography), and *CRC Critical Reviews in Analytical Chemistry* (pesticide residue analysis). A general book on thin layer chromatography co-authored by Dr. Sherma is now in press at Marcel Dekker.

Dr. Sherma spent 6 months in 1972 on sabbatical leave at the EPA Perrine Primate Laboratory, Perrine, Fla., with Dr. T. M. Shafik, and two additional summers (1975, 1976) at the U.S. Department of Agriculture (USDA) in Beltsville, Md., with Melvin Getz doing research on pesticide residue analysis methods development. He spent 3 months in 1979 on sabbatical leave with Dr. Touchstone developing clinical analytical methods. A total of more than 200 papers, books, book chapters, and oral presentations concerned with column, paper, and thin layer chromatography of metal ions, plant pigments, and other organic and biological compounds; the chromatographic analysis of pesticides; and the history of chromatography have been authored by Dr. Sherma, many in collaboration with various co-workers and students. His major research area at Lafayette is currently quantitative TLC (densitometry), applied mainly to clinical analysis and pesticide residue determinations.

Dr. Sherma has written an analytical quality control manual for pesticide analysis under contract with the U.S. EPA and has revised this and the EPA Pesticide Analytical Methods Manual under a 4-year contract (EPA) jointly with Dr. M. Beroza of the Association of Official Analytical Chemists (AOAC). Dr. Sherma has also written an instrumental analysis quality assurance manual and other analytical reports for the U.S. Consumer Product Safety Commission, and is currently preparing a manual on the analysis of food additives for the U.S. Food and Drug Administration, both of these projects also in collaboration with Dr. Beroza of the AOAC.

Dr. Sherma taught the first prototype short course on pesticide analysis with Henry Enos of the EPA for the Center for Professional Advancement. He is editor of the Kontes TLC quarterly newsletter and also teaches short courses on TLC for Kontes and the Center for Professional Advancement. He is a consultant for several industrial companies and federal agencies on chemical analysis and chromatography and regularly referees papers for analytical journals and research proposals for government agencies.

Dr. Sherma has received two awards for superior teaching at Lafayette College and the 1979 Distinguished Alumnus Award from Upsala College for outstanding achievements as an educator, researcher, author, and editor. He is a member of the American Chemical Society, Sigma Xi, Phi Lambda Upsilon, Society for Applied Spectroscopy, and the American Institute of Chemists.

THE EDITOR

Dr. Toshihiko Hanai graduated in Biochemistry under Professor Sei Tachibana, Department of Chemistry, Ritsumeikan University, Kyoto, Japan. He joined a research group of Industrial Analytical Chemistry (Faculty of Engineering, Kyoto University, Kyoto, Japan) as a research and teaching assistant. He developed high performance liquid chromatography under the late Professor Wataru Funasaka. He was a member of Chemical Society of Japan, The Japan Society for Analytical Chemistry, and The Research Group of Automatic Liquid Chromatography organized by Professor Hiroyuki Hatano, and taught HPLC outside of the campus as an adviser.

He received his Ph.D. degree in 1974 from Kyoto University. He worked at Northeastern University under Professor Barry L. Karger in 1975 and at Colorado University under Professor Harold F. Walton in 1976. He has worked at Universite de Montreal under Professor Joseph Hulbert since 1977. He is a member of the American Chemical Society and Chemical Institute of Canada.

He has published over 40 papers and is a co-editor of *Experimental HPLC* and *Chromatography, the Separation System* (both in Japanese).

His research interest is controlling liquid chromatography by himself, utilization of chromatography in clinical and environmental analyses, analysis of trace elements in biological fluids by liquid chromatograph coupled with inductive coupled plasma spectrophotometer, and new development of microwave plasma spectrophotometric detector for capillary column gas chromatography.

He enjoys a practical reversed-phase liquid chromatography with gourmandise.

ADVISORY BOARD

TABLE OF CONTENTS

Section I
Chromatographic Data

Gas Chromatography Tables

Section I.I.

GAS CHROMATOGRAPHY TABLES

Table GC-1
PHENOLS

Packing	P1	P1	P1	P2	P3		P3		P3		P4	
Column												
Length (cm)	3000	3000	3000	200	400		400		400		400	
Diameter (mm)	0.5	0.5	0.5	3.3	3		3		3		3	
Material	G	G	G	G	G		G		G		G	
Gas	He	He	He	Ar	N_2		N_2		N_2		N_2	
Flow rate (mℓ/min)	2	2	2	34	25		25		25		25	
Temperature (°C)	150	150	130	140	155		135		125		175	
Detector	FI	FI	FI	FI	FI		FI		FI		FI	
Reference	1	1	1	2	3		3		3		3	
Compound		TMS	TMS									
	Tr	Tr	Tr	Tr	Tr	min	Tr	min	Tr	min	Tr	min
Phenol	1.00	1.00	1.00	1.00	1.00	16.9	1.00	38.3	1.00	47	1.00	28.5
2-Methylphenol	1.55	1.54	1.64	1.11	1.30		1.34		1.34		1.25	
3-Methylphenol	1.78	1.62	1.73	1.78	1.89		1.80		1.85		1.55	
4-Methylphenol	1.78	1.75	1.85	1.50	1.56		1.67		1.72		1.55	
2-Ethylphenol	2.31	2.29	2.51	1.54	2.14		2.15		2.23		1.80	
2,5-Dimethylphenol	2.55	2.44	2.67	1.62	2.19		2.41		2.51		1.94	
3,5-Dimethylphenol	3.05	2.52	2.87	3.53	2.91		3.27		3.45		2.45	
3-Ethylphenol	2.85	2.53	2.89	2.66	2.88		3.15		3.34		2.45	
2,4-Dimethylphenol	2.55	2.58	2.93	1.84	2.18		2.27		2.36		1.94	
4-Ethylphenol	—	2.69	3.03	2.36	2.85		2.93		3.11		2.35	
2,6-Dimethylphenol	2.05	2.78	3.11	1.00	1.54		1.43		1.42		1.34	
2,3-Dimethylphenol	3.29	2.98	3.34	2.54	2.59		2.85		2.98		2.35	
2,4,6-Trimethylphenol	3.50	4.55	5.40	—	—		—		—		—	
2,3,6-Trimethylphenol	4.10	—	—	—	—		—		—		—	
2,3,5-Trimethylphenol	4.98	—	—	—	—		—		—		—	
3,4-Dimethylphenol	—	—	—	3.51	3.10		3.59		3.81		2.81	

Table GC-1 (continued)
PHENOLS

Packing P1 = SE-30 capillary SCOT
P2 = Cadmium chloride-Carbowax® 20M-Chromosorb® P, AW (2:2:96)
P3 = 2.5% Di-(3,3,5-trimethylcyclohexyl)phthalate on Chromosorb® W (80-100 mesh)
P4 = 15% Trimethylolpropane tripelargonate on Celite® 545 (80-100 mesh)

REFERENCES

1. **Rossemyr, L. I.**, Separation and identification of phenols in kerosene using silylation and gas chromatography, *J. Chromatogr.*, 170, 463, 1979.
2. **Frycka, J.**, Gas chromatographic separation of C_{10}-C_{12} naphthalene hydrocarbons and C_6-C_8 monohydric phenols on cadmium chloride modified with Carbowax 20M, *J. Chromatogr.*, 170, 459, 1979.
3. **Husain, S., Kuzelmann, P., and Schildknecht, H.**, Separation of isomeric alkylphenols by high-performance liquid chromatographic and gas-liquid chromatographic techniques, *J. Chromatogr.*, 137, 53, 1977.

Table GC-2
PHENOLS

Packing	P1	P2	P3	P4	P5	P6	P7
Column							
Length (cm)	180	180	180	180	120	1500	4300
Diameter (mm)	3.1	3.1	3.1 od	3.1 od	3.1	0.5	0.4
Material		ss	ss	ss	G	ss	G
Gas	He	N_2	N_2	N_2	He	He	H_2
Flow rate (mℓ/min)	60	20	30	30	60	10 (2/1)[b]	
Temperature (°C)	100	190	150	135	70	115	105
Detector	FI		FI	FI	FI	FI	FI
Reference	1	2	3	3	4	4	5
Compound	Tr	min	Tr	Tr	min[a]	min[a]	min[a]
2,3-Dimethylphenol	1.00	7.75	1.31	2.78	—	—	—
3,4-Dimethylphenol	1.12	8.20	2.18	3.14	—	—	18.0
2,4-Dimethylphenol	0.78	6.40	1.55	2.30	—	—	—
2,6-Dimethylphenol	0.61	5.65	0.51	1.91	—	—	—
2,5-Dimethylphenol	0.78	6.40	1.01	2.30	—	—	—
3,5-Dimethylphenol	—	7.05	1.90	2.71	—	—	—
Phenol	—	3.45	1.00	1.00	1.5	3.2	8.15
2-Methylphenol	—	4.50	0.82	1.37	3.3	5.2	—
3-Methylphenol	—	4.88	1.32	1.61	3.8	5.7	—
4-Methylphenol	—	4.90	1.60	1.58	4.1	6.1	11.0
2-Ethylphenol	—	—	—	—	5.9	7.5	—
3-Ethylphenol	—	—	—	—	7.6	9.2	—
4-Ethylphenol	—	—	—	—	8.7	10.3	17.8
2,4,6-Trimethylphenol	—	—	—	—	—	—	19.0
2,3,5-Trimethylphenol	—	—	—	—	—	—	27.0

[a] Measured on chromatogram.
[b] Split ratio.

Packing P1 = 3% OV-17 on GasChrom® Q (100-120 mesh)
P2 = Tenax GC modified with a solution of 5% polymetaphenylether
P3 = rubidium benzenesulfonate-Carbowax® 20M-ascorbic acid (40:2:2) on Chromosorb® P (60-80 mesh)
P4 = 12% Apiezon L, 0.5% Carbowax® 20M, 1% ascorbic acid on Chromosorb® W (60-80 mesh)
P5 = 3.8% UCW98 on GasChrom® Z (80-100 mesh)
P6 = OV-17 capillary SCOT
P7 = OV-101 capillary

REFERENCES

1. **Smith, R. V., Rosazza, J. P., Engel, K. O., and Humphrey, D. W.,** Gas-liquid and thin-layer chromatographic determination of xylenols in microbial extracts, *J. Chromatogr.*, 106, 235, 1975.
2. **Bartle, K. D., Elstub, J., Novotny, M., and Robinson, R. J.,** Use of a modified Tenax GC column packing for the direct gas chromatographic analysis of phenols in water at the ppm level, *J. Chromatogr.*, 135, 351, 1977.
3. **Bhattacharjee, A. and Bhaumik, A.,** Analysis of low-boiling isomers of phenols by gas chromatography, *J. Chromatogr.*, 136, 328, 1977.
4. **Souter, R. W. and Bishara, R. H.,** Quantitative gas chromatographic analysis on support-coated open tubular capillary columns. I. Analysis of isomeric ethylphenols, *J. Chromatogr.*, 140, 245, 1977.
5. **Malaterre, M., Lohéac, J., Sellier, N., and Guiochon, G.,** Analyse de la fraction monophénolique des condensats de fumée de cigarettes par association de méthodes chromatographiques, *Chromatographia*, 8, 624, 1975.

Table GC-3
PHENOLS AND ORGANIC ACIDS

Packing	P1	P2	P3	P3	P4	P5
Column						
Length (cm)	175	175	150	150	300	125
Diameter (mm)	3	3	3.1	3.1	3	4
Material	G	G	ss	ss	G	G
Gas	N_2	N_2			N_2	He
Flow rate (mℓ/min)	55	37	15	15	50	60
Temperature (°C)	205	220	135	120	170	85-195 (8/min)
Detector	FI	FI	FI	FI	EC	FI
Reference	1	1	2	2	3	4
Compound					Br[b]	AcO[c]
	Tr	Tr	min[a]	min[a]	min[a]	min[a]
Phenol	1.00	1.00	—	—	19.5	2.7
2-Methylphenol	2.31	1.82	—	—	5.7	3.4
3-Methylphenol	2.70	2.03	—	—	24	3.8
4-Methylphenol	2.90	2.14	4.2	—	9	—
4-Ethylphenol	3.90	2.74	—	—	—	—
2-Ethylphenol	5.06	3.37	—	—	—	—
2,6-Dimethylphenol	5.85	3.92	—	—	—	—
2,5-Dimethylphenol	7.38	4.71	—	—	6.5	—
3,5-Dimethylphenol	8.00	5.05	—	—	32.	—
2,3-Dimethylphenol	8.04	5.03	—	—	8.2	—
3,4-Dimethylphenol	9.47	5.86	—	—	13.7	—
Acetic acid	0.21	—	—	—	—	—
Propionic acid	0.31	—	—	—	—	—
Isobutyric acid	0.37	—	—	—	—	—
n-Butyric acid	0.41	—	—	—	—	—
Isopentanoic acid	0.61	—	—	—	—	—
n-Pentanoic acid	0.71	—	—	—	—	—
Isohexanoic acid	1.20	—	—	—	—	—
n-Hexanoic acid	1.43	—	—	—	—	—
n-Heptanoic acid	3.20	—	—	—	—	—
n-Octanoic acid	7.49	—	—	—	—	—
1,2-Dihydroxybenzene	—	—	—	15	—	—
4-Methoxyphenol	—	—	11	19	—	—
3,4-Dihydroxymethylbenzene	—	—	15	—	—	—
2,4-Dichlorophenol	—	—	—	—	—	6.5
1-Hydroxynaphthalene	—	—	—	—	—	11.3

[a] Measured on chromatogram. [b] Converted into bromophenols. [c] Direct acetylation.

Packing P1 = 0.1% SP-1000 on Carbopack C (80-100 mesh)
P2 = P1 + Tenax GC precolumn
P3 = 10% DC-710 on Chromosorb® W, AW, DMCS (80-100 mesh)
P4 = 2% DEGS + 0.5% H_3PO_4 on Chromosorb® W, AW, DMCS (80-100 mesh)
P5 = 1% SP-1240 DA on Supelcoport (100-120 mesh)

REFERENCES

1. **Hoshiba, Y. and Muto, G.**, Gas-liquid chromatographic determination of phenols in air using Tenax-GC and alkaline precolumn, *J. Chromatogr.*, 157, 277, 1978.
2. **Garrett, B. C. and Underwood, A. L.**, Application of gas-liquid chromatography to phenol oxidase studies, *Anal. Biochem.*, 78, 132, 1977.
3. **Hoshiba, Y. and Muto, G.**, Sensitive gas chromatographic determination of phenols as bromophenols using electron capture detection, *J. Chromatogr.*, 179, 105, 1979.
4. **Coutts, R. T., Hargesheimer, E. E., and Pasutto, F. M.**, Gas chromatographic analysis of trace phenols by direct acetylation in aqueous solution, *J. Chromatogr.*, 179, 291, 1979.

Table GC-4
PHENOLS

Packing	P1	P2	P3	P4	P5
Column					
Length (cm)	100	4500	6000	180	4500
Diameter (mm)	2		0.5	2	0.25
Material	ss		ss		ss
Gas			He	He	
Flow rate (mℓ/min)	30	2(1/100)[b]	2(1/25)[b]	25	
Temperature (°C)		140	135	165-195(6/min)	120
Detector	FI	FI	FI	FI	
Reference	1	1	2	2	3

Compound	sec	TMS sec	Tr	min	Tr	min	min[a]
2,6-Di-*tert.*-butyl-4-methylphenol	140	2160	—		—		—
2,6-Di-*tert.*-butylphenol	—	1760	—		—		—
2,6-Di-*sec.*-butylphenol	—	875	914		—		—
2,6-Di-isopropylphenol	270	412	—		—		—
2,3,5,6-Tetramethylphenol	—	806	—		—		—
2,4,6-Trimethylphenol	350	395	4.59		9.53		—
2,6-Dimethylphenol	260	235	2.88		4.58		—
2,5-Di-*tert.*-butylphenol	—	920	—		—		—
2-*tert.*-Butyl-5-methylphenol	550	735	—		—		—
2-*tert.*-Butyl-4-methylphenol	500	466	—		—		—
2-*tert.*-Butylphenol	440	313	—		—		—
2,4,5-Trimethylphenol	710	348	—		—		—
2,3,5-Trimethylphenol	720	339	8.00		11.50		—
2-Propylphenol	—	231	—		—		—
2,5-Dimethylphenol	440	173	4.20		3.15		—
2-Isopropylphenol	410	190	—		—		—
2,4-Dimethylphenol	440	204	4.07		3.13		—
2-Ethylphenol	390	162	3.93		2.58		18.7
2,3-Dimethylphenol	580	232	4.99		4.68		—
2-Methylphenol	340	116	2.65		1.40		12.3
3,5-Di-*tert.*-butylphenol	630	654	—		—		—
4-*tert.*-Pentylphenol	1100	627	—		—		—
4-*sec.*-Butylphenol	850	404	—		—		—
3,5-Dimethylphenol	—	186	5.28		3.90		—
3,4,5-Trimethylphenol	—	449	—		—		—
3-Ethylphenol	640	178	5.16		2.96		—
4-*tert.*-Butylphenol	800	370	—		—		—
4-Propylphenol	—	326	—		—		—
3,4-Dimethylphenol	790	243	6.01		5.14		30.0
4-Ethylphenol	610	202	4.97		3.76		24.5
3-Methylphenol	490	117	3.27		1.57		15.3
4-Methylphenol	480	128	3.16		1.73		14.9
Phenol	371	73	2.14		0.41		9.5
2,3,6-Trimethylphenol	—	—	5.22		10.69		—
Methoxybenzene	—	—	1.00	11.05	1.00	4.40	—
1,2-Dimethylbenzene	—	—	—		—		4.0

[a] Measured on chromatogram.
[b] Split ratio.

Packing P1 = 5% [1,2,3-Tris(cyanoethoxy)propane] on Chromosorb® G,HP (80-100 mesh)
P2 = Apiezon L capillary
P3 = diisodecylphthalate + phosphoric acid capillary
P4 = 0.3% SP-1000 + 0.3% H_3PO_4 on Carbopak A
P5 = 90% didecylphthalate + 10% H_3PO_4, capillary

Table GC-4 (continued)
PHENOLS

REFERENCES

1. **Callmer, K., Edholm, L.-E., and Smith, B. E. F.,** Study of retention behaviour of alkylphenols in straight- and reversed-phase liquid chromatography and application to the analysis of complex phenolic mixtures in conjunction with gas chromatography, *J. Chromatogr.*, 136, 45, 1977.
2. **Brady, R. F., Jr. and Pettitt, B. C.,** Comparison of gas-liquid-solid chromatography with capillary-column gas-liquid chromatography for the analysis of phenols, *J. Chromatogr.*, 93, 375, 1974.
3. **Dietz, W. S.,** Calibration data for gas chromatographic analysis of phenols, *J. Chromatog. Sci.*, 10, 423, 1972.

Table GC-5
PHENOLS

Packing	P1	P1	P1
Column			
Length (cm)	5000	5000	5000
Diameter (mm)	0.25	0.25	0.25
Material	ss	ss	ss
Gas			N_2
Flow rate (mℓ/min)		(1/280)[a]	(1/100)[a]
Temperature (°C)	125	150	130
Detector	FI	FI	FI
Reference	1	1	2
Compound	Tr	Tr	Tr
Phenol	1.00	1.00	1.00
2-Methylphenol	1.20	1.17	1.21
3-Methylphenol	1.69	1.58	1.68
4-Methylphenol	1.57	1.48	1.57
2-Ethylphenol	1.90	1.75	1.91
3-Ethylphenol	2.83	2.51	2.82
4-Ethylphenol	2.66	2.37	2.63
2,3-Dimethylphenol	2.48	2.22	2.43
2,4-Dimethylphenol	1.87	1.75	1.87
2,5-Dimethylphenol	2.02	1.83	2.00
2,6-Dimethylphenol	0.96	1.00	0.96
3,4-Dimethylphenol	3.18	2.80	3.12
3,5-Dimethylphenol	2.83	2.51	2.79
2-*n*-Propylphenol	3.00	2.61	3.02
3-*n*-Propylphenol	4.75	3.97	4.63
4-*n*-Propylphenyl	4.45	3.78	4.48
4-Ethyl-2-methylphenol	3.07	2.66	2.99
2-Ethyl-4-methylphenol	2.83	2.46	2.79
2-Ethyl-5-methylphenol	3.15	2.73	3.09
4-Ethyl-3-methylphenol	5.13	4.24	4.92
3-Ethyl-5-methylphenol	4.60	3.83	4.52
3-Ethyl-6-methylphenol	3.14	2.73	3.14
2,3,4-Trimethylphenol	4.84	4.08	4.74
2,3,5-Trimethylphenol	4.07	3.45	3.94
2,3,6-Trimethylphenol	1.88	1.80	2.08
2,4,6-Trimethylphenol	1.58	1.54	1.61
2,4,5-Trimethylphenol	3.78	3.24	3.63
3,4,5-Trimethylphenol	6.68	5.43	6.39
2-*n*-Butylphenol	5.37	4.43	5.31
3-*n*-Butylphenol	8.54	6.73	8.32
4-*n*-Butylphenol	8.14	6.41	7.85
2-Isopropylphenol	2.65	2.32	2.62
3-Isopropylphenol	3.88	3.24	3.68

Table GC-5 (continued)
PHENOLS

4-Isopropylphenol	3.66	3.13	3.58
2-Methyl-4-*n*-propylphenol	4.98	4.14	4.81
2-Methyl-6-*n*-propylphenol	2.09	1.96	2.12
4-Methyl-2-*n*-propylphenol	4.27	3.58	4.26
2,4-Diethylphenol	4.38	3.66	4.32
2,5-Diethylphenol	5.19	4.24	5.13
3,4-Diethylphenol	7.70	6.08	7.36
2,3,4,6-Tetramethylphenol	3.89	3.36	3.68
2,3,5,6-Tetramethylphenol	3.68	3.36	3.56
2-*sec*.-Butylphenol	3.98	3.36	3.95
4-*sec*.-Butylphenol	4.04	3.36	5.86
2-*tert*.-Butylphenol	4.20	3.50	4.08
4-*tert*.-Butylphenol	5.34	4.42	5.17
3-Methyl-5-isopropylphenol	4.25	3.62	—
2-Isobutylphenol	3.66	3.15	3.93
4-Isobutylphenol	5.92	4.91	5.79
4-Indanol	6.35	5.26	6.07
5-Indanol	8.33	6.59	8.32
4-*tert*.-Butyl-2-methylphenol	5.51	4.42	3.42
2-*tert*.-Butyl-4-methylphenol	5.55	4.43	5.46
4-*tert*.-Butyl-3-methylphenol	12.54	9.62	11.90
2-*tert*.-Butyl-5-methylphenol	6.84	5.38	6.06
2-Ethyl-5-*n*-propylphenol	8.54	6.59	8.44
Pentamethylphenol	9.89	8.15	9.50
3-*n*-Butyl-6-ethylphenol	15.37	11.10	14.73
2-*n*-Pentylphenol	—	—	9.21
2-*n*-Hexylphenol	—	—	16.22
4-*n*-Pentylphenol	—	—	13.80
4-*n*-Hexylphenol	—	—	24.30
2-Ethyl-6-methylphenol	—	—	1.41
3-Methyl-4-*n*-propylphenol	—	—	7.80
3-Methyl-6-propylphenol	—	—	4.92
4-*tert*.-Pentylphenol	—	—	9.72
2-Ethyl-4,5-dimethylphenol	—	—	5.56
3-*tert*.-Butylphenol	—	—	5.40
2-Methyl-6-isopropylphenol	—	—	1.80
3-Methyl-6-isopropylphenol	—	—	4.27
2-Methyl-4-isopropylphenol	—	—	5.25
2,3,4,5-Tetramethylphenol	—	—	7.14
2-*tert*.-Butyl-6-methylphenol	—	—	2.16
2,6-Diisopropylphenol	—	—	2.34
2-Methoxyphenol	—	—	0.57
4-Methoxyphenol	—	—	5.30
3-Methoxyphenol	—	—	6.35
2-Methoxy-4-methylphenol	—	—	0.94
5-Methyl-4-indanol	—	—	5.71

[a] Split ratio.

Packing P1 = 95% tri(2,4-xylenyl)-phosphate + 5% H_3PO_4, capillary

REFERENCES

1. **Macák, J., Buryan, P., and Hrivňák, J.**, Trennung von einwertigen Alkylphenolen durch Gaschromatographie, *J. Chromatogr.*, 89, 309, 1974.
2. **Buryan, P., Macák, J., and Hrivňák, J.**, Separation of monohydric alkylphenols by gas chromatography. II, *J. Chromatogr.*, 137, 425, 1977.

Table GC-6
METHOXY BENZENES

Packing	P1		P2		P3	
Column						
Length (cm)	5000		5000		5000	
Diameter (mm)	0.25		0.25		0.25	
Material	ss		ss		ss	
Gas						
Flow rate (mℓ/min)						
Temperature (°C)	120-160 (1.25/min)		130-230 (2.5/min)		130-180 (1.25/min)	
Detector	FI		FI		FI	
Reference	1		1		1	
Compound	Tr	min	Tr	min	Tr	min
Methoxybenzene	1.00	7.64	1.00	4.86	1.00	5.78
2-Methoxymethylbenzene	1.30		1.23		1.22	
3-Methoxymethylbenzene	1.39		1.28		1.27	
4-Methoxymethylbenzene	1.39		1.28		1.27	
2-Methoxy-1,3-dimethylbenzene	1.50		1.39		1.40	
2-Methoxy-1,5-dimethylbenzene	1.84		1.60		1.63	
2-Methoxy-1,4-dimethylbenzene	1.84		1.60		1.59	
3-Methoxy-1,5-dimethylbenzene	2.01		1.68		1.73	
2-Methoxy-1,6-dimethylbenzene	2.06		1.81		1.79	
3-Methoxy-1,6-dimethylbenzene	2.23		1.90		1.88	
2-Ethylmethoxybenzene	1.64		1.50		1.45	
3-Ethylmethoxybenzene	1.89		1.68		1.62	
4-Ethylmethoxybenzene	1.92		1.68		1.66	
2-Propylmethoxybenzene	2.16		1.91		1.86	
3-Propylmethoxybenzene	2.54		2.15		2.14	
4-Propylmethoxybenzene	2.62		2.20		2.23	
4-Allylmethoxybenzene	2.72		2.25		2.17	
4-Methoxypropylbenzene	3.85		3.07		3.14	
2-Acetylmethoxybenzene	4.35		3.53		2.85	
3-Acetylmethoxybenzene	4.59		3.71		3.08	
4-Acetylmethoxybenzene	5.48		4.34		3.80	
1,2-Dimethoxybenzene	2.46		2.13		1.70	
1,2-Dimethoxy-4-methylbenzene	3.29		2.76		2.30	
4-Ethyl-1,2-dimethoxybenzene	4.12		3.47		3.03	
1,2-dimethoxy-4-propylbenzene	5.11		4.23		4.02	
4-Allyl-1,2-dimethoxybenzene	5.33		4.34		3.94	
1,2-dimethoxy-4-*cis*-propylbenzene	6.21		4.99		4.78	
1,2-dimethoxy-4-*trans*-propylbenzene	7.41		5.45		5.57	
4-Acetyl-1,2-dimethoxybenzene	11.36		6.79		6.48	

Packing P1 = polypropyleneglychol, capillary
P2 = OV-17, capillary
P3 = Apiezon L, capillary

REFERENCES

1. **Weissmann, G.**, Gaschromatographische Trennung von Phenol- und Brenzkatechinmethyläthern, *J. Chromatogr.*, 129, 431, 1976.

Table GC-7
ALKYLPHENOLS

Packing	P1	P1	P2
Column			
Length (cm)	10000	10000	2500
Diameter (mm)	0.25	0.25	0.25
Material			
Gas	N_2	N_2	N_2
Flow rate (mℓ/min)	1	1	1
Temperature (°C)	185		175
Detector	FI	FI	FI
Reference	1	1	1
Compound	acetyl min[a]	acetyl min[a]	acetyl min[a]
4(2′-Hydroxyphenylbenzene)octane	7.4	7.7	—
3(2′-Hydroxyphenylbenzene)octane	8.05	8.0	—
2(2′-Hydroxyphenylbenzene)octane	9.6	9.0	—
5(2′-Hydroxyphenylbenzene)nonane	10.05	9.3	5.7
4(2′-Hydroxyphenylbenzene)nonane	10.5	9.5	5.8
4(4′-Hydroxyphenylbenzene)octane	11.2	9.9	—
3(2′-Hydroxyphenylbenzene)nonane	11.6	10.2	6.1
3(4′-Hydroxyphenylbenzene)octane	12.0	10.6	—
5(2′-Hydroxyphenylbenzene)decane	14.15	11.8	—
2(4′-Hydroxyphenylbenzene)octane	14.15	11.9	—
2(2′-Hydroxyphenylbenzene)nonane	14.15	11.9	6.7
4(2′-Hydroxyphenylbenzene)decane	15.1	12.4	—
5(4′-Hydroxyphenylbenzene)nonane	15.6	12.7	7.3
4(4′-Hydroxyphenylbenzene)nonane	16.1	13	7.4
3(2′-Hydroxyphenylbenzene)decane	16.9	13.5	—
3(4′-Hydroxyphenylbenzene)nonane	17.55	13.9	7.7
6(2′-Hydroxyphenylbenzene)undecane	—	15.3	—
2(2′-Hydroxyphenylbenzene)decane	20.7	15.7	—
2(4′-Hydroxyphenylbenzene)nonane	20.95	15.9	8.5
5(4′-Hydroxyphenylbenzene)decane	22.25	16.7	—
4(4′-Hydroxyphenylbenzene)decane	23.2	17.3	—
3(4′-Hydroxyphenylbenzene)decane	25.35	18.6	—
2(4′-Hydroxyphenylbenzene)decane	30.2	21.6	—
1(2′-Hydroxyphenylbenzene)octane	13.8	—	—
1(3′-Hydroxyphenylbenzene)octane	16.8	—	—
1(4′-Hydroxyphenylbenzene)octane	19.3	—	—
1(2′-Hydroxyphenylbenzene)nonane	21.0	—	—
1(4′-Hydroxyphenylbenzene)nonane	29.0	—	—

[a] Measured on chromatogram.

Packing P1 = DC-550 capillary
P2 = Carbowax® 20M capillary

REFERENCE

1. **Raverdino, V. and Sassetti, P.**, Alkylation of phenol with C_8-C_{10} n-alkenes. Identification and gas chromatographic behaviour of monoalkylphenols, *J. Chromatogr.*, 153, 181, 1978.

Table GC-8
DIHYDROXYBENZENES

Packing	P1	P2
Column		
Length (cm)	240	400
Diameter (mm)	2	4
Material	G	G
Gas	He	He
Flow rate (mℓ/min)		40
Temperature (°C)	160	140
Detector	FI	FI
Reference	1	2

Compound	min[a]	min[a]
1,2-Dihydroxybenzene	2.8	7.2
2,3-Dihydroxymethylbenzene	4.3	8.4
3,4-Dihydroxymethylbenzene	6.0	11.3
3-Ethyl-1,2-dihydroxybenzene	7.7	—
1,3-Dihydroxybenzene	9.3	10.2
2,3-Dihydroxy-1,5-dimethylbenzene	10.8	—
2,6-Dihydroxymethylbenzene	11.3	—
2,4-Dihydroxymethylbenzene	12.3	—
1,4-Dihydroxymethylbenzene	13.8	—
4-Ethyl-1,3-dihydroxybenzene	15.6	18.1
Ethyl-1,4-dihydroxybenzene	16.3	—
4-*tert*.-Butyl-1,2-dihydroxybenzene	22.0	—
2,4-Dihydroxy-*n*-propylbenzene	22.9	19.4
4-*tert*.-Butyl-1,2-dihydroxybenzene	25.3	—
2,4-Dihydroxy-*n*-propylbenzene	30.5	—
1,4-Dihydroxybenzene	—	10.2

[a] Measured on chromatogram.

Packing P1 = 5% PPE on Chromosorb® W, AW, HMDS (0.16-0.20 mm)
P2 = 10% OV-17 on GasChrom® (80-100 mesh)

REFERENCES

1. **Buryan, P. and Macák, J.,** Separation of phenols on a polyphenyl ether with six rings. II, *J. Chromatogr.*, 150, 246, 1978.
2. **Brunnemann, K. D., Lee, H. -C., and Hoffmann, D.,** Chemical studies on tobacco smoke. XLVII. On the quantitative analysis of catechols and their reduction, *Anal. Lett.*, 9, 939, 1976.

Table GC-9
PHENOLS AND ORGANIC ACIDS

Packing	P1	P2	P3
Column			
Length (cm)	325	325	240
Diameter (mm)	2.5	2.5	2
Material	G	G	
Gas	N_2	N_2	N_2
Flow rate (mℓ/min)	24	24	10cm/min
Temperature (°C)	110	75	220
Detector	FI	FI	FI
Reference	1	1	2

Table GC-9 (continued)
PHENOLS AND ORGANIC ACIDS

Compound			TMS		
	Tr	min	Tr	min	min[a]
Phenol	1.00	10.	1.00	10.	1.0
2-Methylphenol	1.20		1.56		1.7
3-Methylphenol	1.68		1.71		2.1
4-Methylphenol	1.56		1.90		2.2
2-Ethylphenol	1.72		2.18		—
3-Ethylphenol	2.60		2.68		—
4-Ethylphenol	2.46		3.16		—
2,3-Dimethylphenol	2.24		3.38		4.3
2,5-Dimethylphenol	1.81		2.36		5.6
2,6-Dimethylphenol	0.91		2.94		8.4
3,4-Dimethylphenol	2.98		3.82		3.7
3,5-Dimethylphenol	2.62		2.79		4.5
2-Methoxyphenol	0.46		3.54		—
3-Methoxyphenol	5.00		5.31		—
4-Methoxyphenol	5.80		5.75		—
2,6-Dimethoxyphenol	3.49		13.40		—
2-Chlorophenol	0.55		2.94		—
3-Chlorophenol	7.00		2.90		—
4-Chlorophenol	6.50		3.54		—
1,2-Dihydroxybenzene	—		3.60		4.1
1,3-Dihydroxybenzene	—		5.72		7.7
1,4-Dihydroxybenzene	—		6.48		6.3
Acenaphthene	2.93		—		—
Benzoic acid	4.50		4.62		—
Phenylacetic acid	7.80		4.25		—
2-Nitrophenol	—		—		2.6
2,4-Dichlorophenol	—		—		4.7
2,4-Dimethylphenol	—		—		5.6
2,4,6-Trimethylphenol	—		—		5.9
2-Methyl-1-nitrophenol	—		—		7.3
o-Vanillin	—		—		10.1
3,4,5-Trimethylphenol	—		—		15.8
1,2,3-Trihydroxybenzene	—		—		16.8
4-Hydroxy-3-methoxybenzaldehyde	—		—		17.8
3-Nitrophenol	—		—		26.5
4-Nitrophenol	—		—		41.1

[a] Measured on chromatogram.

Packing P1 = 0.5% tris-(2,4-xylenyl)phosphate on Chromosorb® W, HP (80-100 mesh)
P2 = 5% tris-(2,4-xylenyl)phosphate on Chromosorb® W, HP (80-100 mesh)
P3 = 0.5% FFAP on Stering FT-G

REFERENCES

1. **Dirmikis, S. M. and Darbre, A.,** Gas-liquid chromatography of simple phenols for urinalysis, *J. Chromatogr.*, 94, 169, 1974.
2. **Di Corcia, A.,** Analysis of phenols by gas-liquid-solid chromatography, *J. Chromatogr.*, 80, 69, 1973.

Table GC-10
PHENOLS

Packing	P1	P2	P2	P3	P4
Column					
Length (cm)	200	170	170	180	45
Diameter (mm)	6	3	3	3.1 o.d.	6.2 o.d.
Material		G	G	ss	ss
Gas	N_2			He	He
Flow rate (mℓ/min)	35			27	
Temperature (°C)	190	170	150	115-230 (16/min)	190-270(10/min)
Detector	FP	FI	FI	FI	FI
Reference	1	2	2	3	3
	diethylphosphate	o-isobutyloxycarbonyl			
Compound	min	min[a]	min[a]	min[a]	min[a]
3-Methyl-6-isopropylphenol	1.9	—	—	—	—
2-Methoxy-4-propylphenol	3.6 & 4.6	3.7	—	—	—
4-Nitrophenol	2.6	—	—	—	—
4-Chlorophenol	1.5	—	5.7	4.6	—
2-Hydroxynaphthalene	6.0	18.0	—	—	9.3
2-Methoxy-4-methylphenol	2.5	—	—	—	—
3,5-Dichlorophenol	—	4.0	—	—	—
2,4-Dichlorophenol	—	5.0	—	—	—
2-Hydroxyacetophenone	—	6.0	—	—	—
2-Nitrophenol	—	8.2	—	—	—
4-Allyl-2-methoxyphenol	—	12.2	—	—	—
1-Hydroxynaphthalene	—	14.5	—	—	—
Phenol	—	—	2.3	2.5	2.9
2,6-Dimethylphenol	—	—	4.3	—	—
3-Methylphenol	—	—	4.3	—	—
2,3-Dimethylphenol	—	—	8.5	—	—
2-Methylphenol	—	—	3.2	3.1	3.6
4-Methylphenol	—	—	4.3	3.4	—
2,5-Dimethylphenol	—	—	5.3	—	—
4-Ethylphenol	—	—	6.7	—	—
2-Methoxyphenol	—	—	8.6	—	—
3,5-Dimethylphenol	—	—	—	4.2	4.8
4-Chloro-3-methylphenol	—	—	—	5.3	6.1
2,4,6-Trichlorophenol	—	—	—	5.9	6.9
Pentachlorophenol	—	—	—	9.6	11.0

[a] Measured on chromatogram.

Packing P1 = 3% OV-1 on Chromosorb® W, (100-120 mesh)
P2 = 5% OV-17 on GasChrom® Q (100-120 mesh)
P3 = 5% OV-17 on Chromosorb® W, AW, DSCS (80-100 mesh)
P4 = Tenax GC

REFERENCES

1. **Heenan, M. P. and McCallum, N. K.,** A new method for the gas chromatographic detection of phenols, *J. Chromatogr. Sci.*, 12, 89, 1974.
2. **Makita, M., Yamamoto, S., Katoh, A., and Takashita, Y.,** Gas chromatography of some simple phenols as their o-isobutyloxycarbonyl derivatives, *J. Chromatogr.*, 147, 456, 1978.
3. **Chriswell, C. D., Chang, R. C., and Fritz, J. S.,** Chromatographic determination of phenols in water, *Anal. Chem.*, 47, 1325, 1975.

Table GC-11
HALOGENATED PHENOLS

Packing	P1		P1		P1		P2	P3
Column								
Length (cm)	2500		2500		2500		160	—
Diameter (mm)	0.35		0.35		0.35		0.35	—
Material	G		G		G		G	—
Gas	He		He		He		Ar-Me	He
Flow rate (mℓ/min)	2-3		2-3		2-3		30	25
Temperature (°C)	150		180		95-180 (3/min)		170	225
Detector	EC		EC		EC		EC	FI
Reference	1		1		1		2	3
	acetate		acetate		acetate		pentachloro	methyl
Compound	Tr	min	Tr	min	Tr	min	min	min
2-Chlorophenol	0.15		0.29		0.24		—	—
3-Chlorophenol	0.16		0.30		0.27		—	—
4-Chlorophenol	0.16		0.33		0.28		—	—
2,6-Dichlorophenol	0.20		0.33		0.34		—	—
2,5-Dichlorophenol	0.21		0.35		0.37		—	—
2,4-Dichlorophenol	0.21		—		0.38		—	—
3,4-Dichlorophenol	0.22		0.35		0.40		—	—
2,3-Dichlorophenol	0.23		0.36		0.43		—	—
3,5-Dichlorophenol	0.24		0.38		0.46		—	—
2,4,6-Trichlorophenol	0.27		0.41		0.49		—	—
2,3,6-Trichlorophenol	0.31		—		0.54		—	—
2,3,5-Trichlorophenol	0.33		—		0.56		—	—
3,4,5-Trichlorophenol	0.40		—		0.67		—	—
2,3,5,6-Tetrachlorophenol	0.50		0.61		0.73		—	—
2,3,4,6-Tetrachlorophenol	0.50		0.61		0.74		—	—
2,3,4,5-Tetrachlorophenol	0.64		0.71		0.83		—	—
Pentachlorophenol	1.00	17.5	1.00	8.0	1.00	25.5	6	—
2-Bromophenol	0.18		0.32		0.32		—	—
3-Bromophenol	0.20		0.33		0.36		—	—
4-Bromophenol	0.20		0.33		0.37		—	—
2,6-Dibromophenol	0.32		0.45		0.55		—	—
2,4-Dibromophenol	0.35		0.48		0.61		—	—
2-Nitrophenol	0.23		0.36		0.46		—	—
3-Nitrophenol	0.29		0.41		0.58		—	—
4-Nitrophenol	0.31		0.42		0.61		—	—
3,4,5,6-Tetrachloro-2-(2′,3′,4′,5′,6′-pentachlorophenoxy)phenol	—		—		—		—	15
2,4,5-Trichlorophenol	0.33		0.46		0.57		—	—
2,3,4-Trichlorophenol	0.38		—		0.64		—	—

Packing P1 = SE-30 capillary
P2 = 1% XF-1105 on Chromosorb® W, AW, DMCS (60-80 mesh)
P3 = 1% SE-30 on Chromosorb® W, AW, DMCS (100-120 mesh)

REFERENCES

1. **Krijgsman, W. and Van de Kamp, C. C.,** Determination of chlorophenols by capillary gas chromatography, *J. Chromatogr.*, 131, 412, 1977.
2. **Gossler, K. and Schaller, K. H.,** Eine gas-chromatographische Method zur quantitativen Bestimmung von pentachlorophenol in Urin und Plasma, *Z. Anal. Chem.*, 290, 111, 1978.
3. **Rappe, C. and Nilsson, C. A.,** An artifact in the gas chromatographic determination of impurities in pentachlorophenol, *J. Chromatogr.*, 67, 247, 1972.

Table GC-12
BIPHENYLS

Packing	P1	P1	P1	P2
Column				
Length (cm)	200	200	200	450
Diameter (mm)	3	3	3	4.2
Material	G	G	G	ss
Gas	N_2	N_2	N_2	N_2
Flow rate (mℓ/min)	40	40	40	40
Temperature (°C)	130-150	110-150 (2/min)	120-150 (2/min)	50-300 (8/min)
Detector	FI	FI	FI	FI
Reference	1	1	1	2
Compound	min[a]	min[a]	min[a]	min
1-Hydroxynaphthalene	5.2	8.7	8.0	—
2,2′-Dihydroxybiphenyl	6.8	11.2	11.0	—
2-Hydroxybiphenyl	7.6	12.0	12.0	25.0
Biphenyl	9.0	13.0	14.0	21.4
2,5-Dihydroxybiphenyl	9.0	14.0	15.8	—
3-Hydroxybiphenyl	14.0	17.3	22.5	—
4-Hydroxybiphenyl	16.9	18.9	25.3	—
3,4-Dihydroxybiphenyl	18.2	19.7	27.1	—
3,3′-Dihydroxybiphenyl	22.0	21.1	29.8	—
4,4′-Dihydroxybiphenyl	29.9	27.1	36.0	—
Benzophenone	—	—	—	28.0

[a] Measured on chromatogram.

Packing P1 = 3% OV-17 on Chromosorb® W, HP (100-120 mesh)
P2 = 1% SE-30 on Chromosorb® Q (80-100 mesh)

REFERENCES

1. **Davis, P. J., Jamieson, L. K., and Smith, R. V.,** Determination of biphenyl and eight biphenyls in microbial extracts by gas chromatography and thin-layer chromatography, *Anal. Chem.*, 50, 736, 1978.
2. **Beernaert, H.,** Determination of biphenyl and o-phenylphenol in citrus fruits by gas chromatography, *Anal. Chem.*, 77, 331, 1973.

Table GC-13
PHENOLIC COMPOUNDS

Packing	P1	P1	P1	P2	P3	P4	P5	P6
Column								
Length (cm)	150	150	150	150	150	150	150	150
Diameter (mm)	4.7	4.7	4.7	4.7	4.7	4.7	4.7	4.7
Material					G			
Gas	N_2	N_2	N_2	N_2	N_2	N_2	N_2	N_2
Flow rate (mℓ/min)	45	45	45	45	45	45	45	45
Temperature (°C)	220	235	250	220	230	200	200	220
Detector	FI	FI	FI	FI	FI	FI	FI	FI
Reference	1	1	1	1	2	3	4	4
Compound	min	min	min	min	min	min	min	min
Cardanol	19.2	12.4	10.0	19.6	—	—	—	—
Cardol	47.0	27.8	16.8	48.6	—	—	—	—
2-Methylcardol	52.6	—	—	—	—	—	—	—

Table GC-13 (continued)
PHENOLIC COMPOUNDS

Compound	min	min	min	min	min	min	min	min
Methyl anacardate	36.2	22.4	13.4	36.2	19.9	52.5	—	—
Dimethylanacardate	—	—	—	—	20.8	86.5	53.0	25.7
Cardanol (15:0)	21.0	—	—	—	—	—	—	—
Cardol (15:0)	52.0	—	—	—	—	—	—	—
2-Methylcardol (15:0)	39.0	—	—	—	—	—	—	—
Cardanol methylether (15:0)	17.6	—	—	—	—	14.0	11.6	10.3
Cardol dimethylether (15:0)	33.4	—	—	—	—	—	32.6	20.1
2-Methylcardol dimethylether (15:0)	38.8	—	—	—	—	—	30.0	23.3
Methylanacardate *O*-methylether (15:0)	41.4	—	—	—	—	—	—	—
Cardanol acetate	22.8	—	—	—	11.7	—	—	—
Cardol diacetate	59.0	—	—	—	33.7	—	—	—
2-Methylcardol diacetate	70.4	—	—	—	37.4	—	—	—
Methylanacardate *O*-acetate (15:0)	55.2	—	—	—	—	—	—	—
Cardanol acetate (15:0)	25.0	—	—	—	—	—	—	—
Cardol diacetate (15:0)	65.2	—	—	—	—	—	—	—
2-Methylcardol diacetate	74.8	—	—	—	37.4	—	—	—
Cardanol methylether (15:1)	—	—	—	—	—	15.5	—	—
Cardanol methylether (15:2)	—	—	—	—	—	18.9	—	—
Cardanol methylether (15:3)	—	—	—	—	—	23.0	—	—
Cardol dimethylether (15:1)	—	—	—	—	—	44.2	—	—
Cardol dimethylether (15:2)	—	—	—	—	—	53.5	—	—
Cardol dimethylether (15:3)	—	—	—	—	—	66.3	—	—
2-Methylcardol dimethylether (15:1)	—	—	—	—	—	41.0	—	—
2-Methylcardol dimethylether (15:2)	—	—	—	—	—	50.2	—	—
2-Methylcardol dimethylether (15:3)	—	—	—	—	—	61.3	—	—
Methylanacardate (15:1)	—	—	—	—	—	57.3	—	—
Methylanacardate (15:2)	—	—	—	—	—	72.0	—	—
Methylanacardate (15:3)	—	—	—	—	—	90.9	—	—
Cardanol (15:1)	—	—	—	—	—	92.1	—	—
Cardanol (15:2)	—	—	—	—	—	112.2	—	—
Cardanol (15:3)	—	—	—	—	—	137.2	—	—
Cardanol methylether	—	—	—	—	8.0	—	—	—
Cardanol dimethylether	—	—	—	—	16.1	—	—	—
2-Methylcardol dimethylether	—	—	—	—	18.4	—	—	—

Packing P1 = 3% SE-30 on diatomite C, AW, silanized
P2 = 3% OV-17 on diatomite C, AW, silanized (80-100 mesh)
P3 = 3% Dexsil® on diatomite C, AW, silanized
P4 = 2% PEGA on Celite® AW, silanized (80-100 mesh)
P5 = 2% PEGA on diatomite C, AW, silanized
P6 = 5% SE-52 on diatomite C, AW, silanized

REFERENCES

1. **Tyman, J. H. P.,** Long-chain phenols. V. Gas chromatographic analysis of cashew nut-shell liquid (Anacardium occidentale), *J. Chromatogr.*, 111, 285, 1975.
2. **Tyman, J. H. P.,** Long-chain phenols. XII. Compositional studies: the polymeric material in unsaturated phenols of Anacardium occidentale, *J. Chromatogr.*, 156, 255, 1978.
3. **Tyman, J. H. P.,** Long-chain phenols. IV. Quantitative determination of the olefinic composition of the component phenols in cashew nut-shell liquid, *J. Chromatogr.*, 111, 277, 1975.
4. **Tyman, J. H. P.,** Long-chain phenols. IX. Compositional studies on the unsaturated phenols of Anacardium occidentale (cashew nut-shell liquid), *J. Chromatogr.*, 138, 97, 1977.

Table GC-14
PHENOLS

	P1	P2
Packing	P1	P2
Column		
Length (cm)	200	100
Diameter (mm)		3
Material	G	ss
Gas		N_2
Flow rate (mℓ/min)		25
Temperature (°C)	150-280 (8/min)	220-250 (10/min)
Detector	(MS)	
Reference	1	2
Compound	min[a]	min
Phenol	1.2	—
2-Methylolphenol	2.8	—
4-Methylolphenol	3.5	—
2,6-Dimethylolphenol	5.9	—
2,4-Dimethylolphenol	6.2	—
2,2′-Dihydroxydiphenylmethane	7.9	—
2,4,6-Trimethylolphenol	7.9	—
2,4′-Dihydroxydiphenylmethane	8.7	—
4,4′-Dihydroxydiphenylmethane	9.4	—
Dihydrodehydrodiconiferylalcohol	—	12.5
Dihydro-di-1-hydroxy-2-methoxy-4-propylbenzene	—	3.5
trans-2,4′-Dihydroxy-3,3′-dimethoxy-5-(3-hydroxypropyl)-stilbene	—	15.9
2,4′-Dihydroxystilbene	—	2.7
3-Hydroxymethylstyryl arylether	—	3.3

[a] Measured on chromatogram.

Packing P1 = 3% OV-17 on Chromosorb® W, AW, DMCS (100-120 mesh)
P2 = 2% OV-1 on Chromosorb® G, AW, DMCS (80-100 mesh)

REFERENCES

1. Lindner, W., Trennung und Identifizierung von Phenol-Formaldehyde Reaktionsprodukten mittles Gas-Chromatographie-Massenspektroskopie, *J. Chromatogr.*, 151, 406, 1978.
2. Miksche, G. E., Über das Verhalten des Lignins bei der Alkalikochung. VI. Zum Abbau von *p*-Hydroxy-phenylcumaranstrukturen durch Alkali, *Acta Chem. Scand.*, 26, 3269, 1972.

Table GC-15
PHENOLIC COMPOUNDS

Packing	P1	P2	P3	P4
Column				
Length (cm)	180	90	270	180
Diameter (mm)	3.1	3.1	2	6.2
Material	ss	ss	G	G
Gas	N_2	N_2	N_2	
Flow rate (mℓ/min)			20	
Temperature (°C)	200	190	70-250 (4/min)	185
Detector	FI	FI	FI	
Reference	1	1	2	3
Compound	Tr	Tr	min[a]	min
(6) Gingeryl methylether acetate	1.0	—	—	—
(6) Gingeryl diacetate	1.4	1.6	—	—
(8) Gingeryl diacetate	2.8	3.6	—	—
(10) Gingeryl diacetate	5.4	7.8	—	—
(6) Shogaol	1.0	1.0	—	—
(8) Shogaol	1.9	2.0	—	—
(10) Shogaol	3.7	3.9	—	—
(6) Paradol	0.9	0.7	—	—
(8) Paradol	1.7	1.6	—	—
(10) Paradol	3.5	3.3	—	—
(6) Paradyl monoacetate	1.1	1.4	—	—
(8) Paradyl monoacetate	2.2	—	—	—
2-Phenylisopropanol	—	—	14.4	—
2-Phenyl-*n*-propanol	—	—	18.8	—
1,3-Dihydroxy-5-pentylbenzene	—	—	—	4.6

[a] Measured on a chromatogram.

Packing P1 = 1.6% SE-30 on Chromosorb® W, AW (80-100 mesh)
P2 = 3% OV-1 on Chromosorb® W, AW (80-100 mesh)
P3 = 5% OV-17 on Chromosorb® W, HP (100-120 mesh)
P4 = 5% OV-7 on Chromosorb® W

REFERENCES

1. **Connell, D. W. and McLachlan, R.,** Natural pungent compounds IV. Examination of the gingerols, shagaols, paradols and related compounds by thin-layer and gas chromatography, *J. Chromatogr.*, 67, 29, 1972.
2. **Goenechea, S., Olek, K., and Wardenbach, P.,** Simultaneous gas chromatographic determination of the cumene metabolites 2-phenylpropanol-1 and 2-phenylpropanol-2 in urine, *J. Cbromatogr.*, 154, 282, 1978.
3. **Bailey, K.,** Formation of olivetol during gas chromatography of cannabinoids, *J. Chromatogr.*, 160, 288, 1978.

Table GC-16
PHENOLIC COMPOUNDS

Packing	P1	P1	P2	P2	P3	P3
Column						
Length (cm)	180	180	180	180	180	180
Diameter (mm)	3	3	3	3	2	2
Material	G	G	G	G	G	G
Gas					N_2	N_2
Flow (m*l*/min)	45	45	45	45	65	65
Temperature (°C)	200	200	220	210	260	280
Detector	FI	FI	FI	FI	FI	FI
Reference	1	1	1	1	2	2
Compound	TMS min	methyl min	TMS min	methyl min	methyl min[a]	butyl min[a]
Diphenylhydantoin	3.6	2.7	4.3	5.2	—	—
2-Hydroxyphenyl-diphenylhydantoin	7.5	4.6	7.5	9.0	—	—
3-Hydroxyphenyl-diphenylhydantoin	7.8	5.2	8.0	9.6	—	—
4-Hydroxyphenyl-diphenylhydantoin	10.2	6.5	10.0	12.2	—	—
Dihydrodiol-diphenylhydantoin	11.8	—	8.4	—	—	—
3,4′-Dihydroxyphenyl-diphenylhydantoin	20.7	12.6	16.5	25.6	—	—
4,4′-Dihydroxyphenyl-diphenylhydantoin	27.0	15.5	21.4	33.5	—	—
2,4-Dihydroxyphenyl-diphenylhydantoin	17.0	15.2	13.0	21.5	—	—
3,4-Dihydroxyphenyl-diphenylhydantoin	15.0	9.8	12.0	20.0	—	—
New metabolite	15.0	9.8	12.0	20.1	—	—
Phenyltoin	—	—	—	—	2.0	4.5
1,3-Dimethyl-5-(3′-methoxyphenyl)-5-phenylhydantoin	—	—	—	—	4.0	—
1,3-Dimethyl-5-(4′-methoxyphenyl)-5-phenylhydantoin	—	—	—	—	4.8	—
5-(4′-Hydroxyphenyl)-5-phenylhydantoin	—	—	—	—	7.0	—
5-(3,4-Dihydroxyphenyl)-5-phenylhydantoin	—	—	—	—	—	5.1
5-(4-Hydroxy-3-methoxyphenyl)-5-phenylhydantoin	—	—	—	—	—	5.9
5-(4-Hydroxy-3-methoxyphenyl)-5-phenylhydantoin tributylether	—	—	—	—	—	7.5
5-(4-Hydroxy-3-methoxyphenyl)-5-phenylhydantoin dibutylether	—	—	—	—	—	8.3
5-(3,4-Dihydroxyphenyl)-5-phenylhydantoin tetrabutylether	—	—	—	—	—	9.5
5-(3,4-Dihydroxyphenyl)-5-phenylhydantoin tributylether	—	—	—	—	—	10.8

[a] Measured on chromatogram.

Packing P1 = 3% OV-1 on GasChrom® Q (100-120 mesh)
P2 = 3% OV-17 on GasChrom® Q (100-120 mesh)
P3 = 5% methylphenylsilicone (75% phenyl) on diatomite HP, AW, DMCS (100-120 mesh)

REFERENCES

1. **Chang, T., Okerholm, R. A., and Glazko, A. J.,** Identification of 5-(3,4-dihydroxyphenyl)-5-phenylhydantoin: a metabolite of 5,5-diphenylhydantoin (dilantin) in rat urine, *Anal. Lett.,* 5, 195, 1972.
2. **Midha, K. K., Hindmarsh, K. W., McGilveray, I. J., and Cooper, J. J.,** Identification of urinary catechol and methylated catechol metabolites of phenyltoin in humans, monkeys, and dogs by GLC and GLC-mass spectrometry, *J. Pharm. Sci.,* 66, 1596, 1977.

Table GC-17
PHENOLIC COMPOUNDS

Packing	P1	P1
Column		
Length (cm)	300	300
Diameter (mm)	2	2
Material	G	G
Gas	N_2	N_2
Flow rate (mℓ/min)	25	25
Temperature (°C)	100-260 (4/min)	
Detector	FI	FI
Reference	1	1
Compound	TMS min[a]	TMS min[a]
Phenol	2.0	—
3-Methylphenol	4.6	—
1,2-Dihydroxybenzene	5.6	—
2-Hydroxybenzoic acid	6.6	—
Cinnamic acid	7.2	—
1,3,5-Trihydroxybenzene	7.5	—
4-Hydroxy-3-methoxybenzaldehyde	7.8	—
4-Hydroxybenzoic acid	8.5	—
4-Hydroxy-3,5-dimethoxybenzaldehyde	9.7	—
4-Hydroxy-3-methoxybenzoic acid	10.3	—
2,6-Dihydroxybenzoic acid	10.5	—
2,5-Dihydroxybenzoic acid	10.8	—
3,4-Dihydroxybenzoic acid	11.1	—
7-Hydroxycoumarin	11.4	—
3,4,5-Trihydroxy-1-cyclohexane-1-carboxylic acid	12.0	—
4-Hydroxy-3,5-dimethoxybenzoic acid	12.3	—
1,2,3,4-Tetrahydroxycyclohexanecarboxylic acid	12.6	—
2-Hydroxycinnamic acid	13.0	—
4-Hydroxycinnamic acid	13.3	—
3,4,5-Trihydroxybenzoic acid	14.0	—
2,8-Dihydroxycoumarin	14.4	—
7-Hydroxy-6-methoxycoumarin	14.9	—
4-Hydroxy-3-methoxycinnamic acid	15.4	—
3,4-Dihydroxycinnamic acid	15.8	—
4-Hydroxy-3,5-dimethoxycinnamic acid	19.0	—
Epicatechin	24.4	4.9
Catechin	24.7	5.1
3′,5,7-Trihydroxy-4′-methoxyflavanone	25.3	5.6
Epicatechin gallate	25.6	—
Cyanadin	26.3	7.7
34′57-Tetrahydroxyflavone	27.3	8.6
Chlorogenic acid	27.9	8.9
Neochlorogenic acid	28.3	—
Isocholorogenic acid	28.6	—
3,3′,4′,5,7-Pentahydroxyflavone	29.4	9.7
3,3′,4′,5,5′,7-Hexahydroxyflavone	30.3	10.5
Ellagic acid	31.2	—
2-Phenylchromone	—	2.0
Rutin	—	2.4
Naringin	—	2.7
Hesperidin	—	3.0
Naringenin	—	4.5
Epicatechin + leucocyanidin	—	4.9
Taxifolin	—	7.0
Delphindin	—	7.3

Table GC-17 (continued)
PHENOLIC COMPOUNDS

Compound	TMS min[a]	TMS min[a]
Kaempferol-3-rhamnoglucoside	—	8.6
Pelargonidin	—	9.4
Quercitrin	—	9.7
Myricitrin	—	10.5
Luteolin-7-glucoside	—	11.
Malvidin	—	11.5
Delphinin	—	32.8
Vitexin	—	33.2
Cyanin	—	33.7
Pelargonin	—	34.2
Peonin	—	40.9
Malvin	—	47.7
Oenin	—	58.8

[a] Measured on chromatogram.

Packing P1 = 3% SE 52 on Chromosorb® W, AW.DMCS (80-100 mesh).

REFERENCE

1. **Drawert, F. and Leupold, G.,** Gas-chromatographische Analyse der Trimethysilyl-Derivate phenolischer Verbindungen, *Chromatographia*, 9, 605, 1976.

Table GC-18
PHENOLS

Packing	P1
Column	
Length (cm)	180
Diameter (mm)	6.2
Material	Cu
Gas	He
Flow rate (mℓ/min)	75
Temperature (°C)	122
Detector	TC
Reference	1

Compound	min
2-Methylphenol	4.1
2-Methylphenol-ethylether	1.1
2-Methylphenol-*n*-propylether	1.7
2-Methylphenol-*n*-butylether	2.8
2-Methylphenol-*n*-pentylether	4.8
2-Methylphenol-isopropylether	1.0
2-Methylphenol-*sec.*-butylether	1.5
3-Methylphenol-ethylether	5.8
3-Methylphenol-*n*-propylether	1.4
3-Methylphenol-*n*-butylether	2.0
3-Methylphenol-*n*-pentylether	5.9
3-Methylphenol-isopropylether	1.2
3-Methylphenol-*sec.*-butylether	1.8
4-Methylphenol-ethylether	5.9
4-Methylphenol-*n*-propylether	1.6

Table GC-18 (continued)
PHENOLS

Compound	min
4-Methylphenol-*n*-butylether	2.4
4-Methylphenol-*n*-pentylether	7.2
4-Methylphenol-isopropylether	1.4
4-Methylphenol-*sec.*-butylether	2.1

Packing P1 = 7% *p*-((*p*-methoxybenzylidene)-amino)phenylbenzoate on Chromosorb® W, AW, DMCS (60-100 mesh)

REFERENCE

1. **Cook. L. E. and Spangelo, R. C.**, Separation of monosubstituted phenol isomers using liquid crystals, *Anal. Chem.*, 46, 122, 1974.

Table GC-19
HALOGENATED PHENOLS

Packing	P1	P1	P2
Column			
Length (cm)	270	270	150
Diameter (mm)	2	2	2
Material	G	G	
Gas	N_2	N_2	N_2
Flow rate (mℓ/min)	35	35	15-20
Temperature (°C)	80-170 (8/min)	80-170 (8/min)	100
Detector	EC	EC	EC
Reference	1	1	2

Compound	heptafluorobutyl- min	min	min
1,4-Dichlorobenzene	—	1.7	—
1,2-Dichlorobenzene	—	2.0	—
1,2,4-Trichlorobenzene	—	3.1	—
1,2,3-Trichlorobenzene	—	3.8	—
Phenol	4.2	—	—
1,2,4,5-Tetrachlorobenzene	—	4.9	—
2-Methylphenol	5.2	—	—
2,4,6-Trichloromethoxybenzene	—	5.7	—
1,2,3,4-Tetrachlorobenzene	—	6.0	—
4-Chlorophenol	6.4	—	—
2-Chlorophenol	6.7	—	—
Pentachlorobenzene	—	7.3	—
2-Bromophenol	7.75	—	—
4-Bromophenol	7.75	—	—
2,5-Dichlorophenol	8.1	—	—
2,3,4,6-Tetrachloromethoxybenzene	—	8.3	—
2-*sec.*-Butylphenol	8.35	—	—
2,6-Dichlorophenol	8.5	—	—
2,4,5-Trichloromethoxybenzene	—	8.6	—
4-*tert.*-Butylphenol	8.8	—	—
2,4,6-Trichlorophenol	9.0	—	—
Hexachlorobenzene	—	9.4	—
2,4,5-Trichlorophenol	9.5	—	—
2,4-Dibromophenol	10.4	—	—
Pentachloromethoxybenzene	—	10.6	—

Table GC-19 (continued)
HALOGENATED PHENOLS

Compound	heptafluorobutyl- min	min	min
1,3-Dihydroxybenzene	11.1	—	—
2,3,4,5-Tetrachloromethoxybenzene	—	11.2	—
1,2-Dihydroxybenzene	11.5	—	—
4-Nitrophenol	11.6	—	—
2-Hydroxybiphenyl	12.1	—	—
2,6-Dichlorophenoxyacetic acid methylester	—	12.5	—
2,4-Dichlorophenoxyacetic acid methylester	—	13.6	—
tert.-Butyl-1,2-dihydroxybenzene	13.8	—	—
2,4,5-Trichlorophenoxypropionic acid methylester	—	14.0	—
2,4,5-Trichlorophenoxyacetic acid methylester	—	15.2	—
3-Methyl-6-isopropylphenol monochloroacetate	—	—	3.1
3-Methyl-6-isopropylphenol monofluoroacetate	—	—	1.2
3-Methyl-6-isopropylphenol pentafluorobenzylether	—	—	5.8

Packing P1 = not noticed
P2 = 1% SE-52 on Diatoport S

REFERENCES

1. **Lamparski, L. L. and Nestrick, T. J.,** Determination of trace phenols in water by gas chromatographic analysis of heptafluorobutyryl derivatives, *J. Chromatogr.*, 156, 143, 1978.
2. **McCallum, N. K. and Armstrong, R. J.,** The derivatization of phenols for gas chromatography using electron capture detection, *J. Chromatogr.*, 78, 303, 1973.

Table GC-20
PHTHALIC ACIDS

Packing	P1
Column	
Length (cm)	240
Diameter (mm)	1.8
Material	G
Gas	N_2
Flow rate (mℓ/min)	20
Temperature (°C)	160-234 (3/min)
Detector	FI
Reference	1

Compound	Tr	min[a]
1,2-Benzenedicarboxylic acid	0.18	—
1,2-Benzenedicarboxylic acid dimethylester	0.24	—
1,2-Benzenedicarboxylic acid diethylester	0.34	—
1,2-Benzenedicarboxylic acid diisopropylester	0.39	—
1,2-Benzenedicarboxylic acid dipropylester	0.55	6.0
1,2-Benzenedicarboxylic acid diisobutylester	0.69	—
1,2-Benzenedicarboxylic acid dibutylester	0.85	—
1,2-Benzenedicarboxylic acid dimethylethylester	1.13	—
1,2-Benzenedicarboxylic acid di-*n*-pentylester	1.20	—
1,2-Benzenedicarboxylic acid diethylethylester	1.25	—
1,2-Benzenedicarboxylic acid di-*n*-heptylester	1.58	17.0
1,2-Benzenedicarboxylic acid butylbenzylester	1.71	—
1,2-Benzenedicarboxylic acid di-2′,2′,4′-trimethyl pentylester	1.77	—
1,2-Benzenedicarboxylic acid hexylheptylester	1.80	—
1,2-Benzenedicarboxylic acid butylnonylester	1.80	—
1,2-Benzenedicarboxylic acid dibutylethylester	1.93	—
1,2-Benzenedicarboxylic acid di-2′-ethylhexylester	1.93	—
1,2-Benzenedicarboxylic acid di-*n*-hexylester	2.01	—
1,2-Benzenedicarboxylic acid dicyclohexylester	2.02	21.5
1,2-Benzenedicarboxylic acid di-3′,5′,5′-trimethylhexylester	2.27	—
1,2-Benzenedicarboxylic acid diphenylester	2.31	—
1,2-Benzenedicarboxylic acid di-*n*-octylester	2.36	—
1,2-Benzenedicarboxylic acid di-3′,5′,5′-trimethylcyclohexylester	2.55	—
1,2-Benzenedicarboxylic acid di-*n*-nonylester	2.76	—
1,2-Benzenedicarboxylic acid di-*n*-decylester	3.31	35.0
1,2-Benzenedicarboxylic acid diphenylethylester	4.80	51.0
1,2-Benzenedicarboxylic acid di-*n*-dodecylester	5.40	57.0
Methylarachidate	1.00	10.6

[a] Measured on chromatogram.

Packing P1 = 1% QF-1 on Chromosorb® Q (100-200 mesh)

REFERENCE

1. **Bloom, P. J.,** Application des chromatographies sur courche mince et gaz-liquide à l'analyse qualitative et quantitative des esters de l'acide phthalique, *J. Chromatogr.*, 72, 35, 1972.

Table GC-21
BENZENE CARBOXYLIC ACIDS

Packing	P1
Column	
Length (cm)	100
Diameter (mm)	3
Material	ss
Gas	Ar
Flow rate (mℓ/min)	60
Temperature (°C)	190
Detector	FI
Reference	1

Compound	methyl Tr (1,2-benzenedicarboxylic acid)
Benzoic acid	0.40
2-Hydroxybenzoic acid	0.50
3-Hydroxybenzoic acid	0.90
4-Hydroxybenzoic acid	0.97
2-Methoxybenzoic acid	0.70
3-Methoxybenzoic acid	0.70
4-Methoxybenzoic acid	0.75
1,2-Benzenedicarboxylic acid	1.00
3-Hydroxy-1,2-benzenedicarboxylic acid	1.30
4-Hydroxy-1,2-benzenedicarboxylic acid	3.00
3-Methoxy-1,2-benzenedicarboxylic acid	2.00
4-Methoxy-1,2-benzenedicarboxylic acid	2.20
1,3-Benzenedicarboxylic acid	1.18
2-Hydroxy-1,3-benzenedicarboxylic acid	2.57
4-Hydroxy-1,3-benzenedicarboxylic acid	1.70
5-Hydroxy-1,3-benzenedicarboxylic acid	3.80
2-Methoxy-1,3-benzenedicarboxylic acid	2.05
4-Methoxy-1,3-benzenedicarboxylic acid	2.75
5-Methoxy-1,3-benzenedicarboxylic acid	2.56
1,4-Benzenedicarboxylic acid	1.18
Hydroxy-1,4-benzenedicarboxylic acid	1.70
Methoxy-1,4-benzenedicarboxylic acid	2.15
4-Hydroxy-5-methyl-1,3-benzenedicarboxylic acid	2.15
4-Methoxy-5-methyl-1,3-benzenedicarboxylic acid	3.40
1,2,4-Benzenetricarboxylic acid	3.40
3-Hydroxy-1,2,4-benzenetricarboxylic acid	8.10
3-Methoxy-1,2,4-benzenetricarboxylic acid	6.70
5-Hydroxy-1,2,4-benzenetricarboxylic acid	5.00
5-Methoxy-1,2,4-benzenetricarboxylic acid	7.30
1,3,5-Benzenetricarboxylic acid	4.48
Hydroxy-1,3,5-benzenetricarboxylic acid	7.20
Methoxy-1,3,5-benzenetricarboxylic acid	6.30
1,2,3,4-Benzenetetracarboxylic acid	10.80
5-Hydroxy-1,2,3,4-benzenetetracarboxylic acid	16.20
5-Methoxy-1,2,3,4-benzenetetracarboxylic acid	25.30

Packing P1 = 5% SE-30 on Chromosorb® W, DMCS

REFERENCE

1. **Maximov, O. B. and Shapovalov, V. E.,** Detection of methoxyaromatic acids in the oxidation products of coals and humic acid by gas-liquid chromatography, *J. Chromatogr.*, 66, 29, 1972.

Table GC-22
ORGANIC ACIDS

Packing	P1	P1	P2	P3		P3		P4	P5
Column									
Length (cm)	180	180	360	180		180		180	120
Diameter (mm)	3	3	3.1	5		5		3	6 o.d.
Material	G	G							ss
Gas	Ar	Ar	N_2	N_2		N_2		N_2	
Flow rate (mℓ/min)	80	80	30					32	30
Temperature (°C)	50-240(2/min)	70-240(2/min)	140-160	130		180		80-210(3/min)	90-150(4/min)
Detector	FI	FI	FI	FI		FI		FI	FI
Reference	1	1	2	3		3		4	5
	TMS	TMS	TMS	TMS		TMS		TMS	TMS
Compound	min	min	min[a]	Tr	min	Tr	min	min[a]	min
Pyruvic acid MO	4.5	8.4	—	—	—	—	—	8.1	—
Oxalic acid	7.6	12.4	6.5	—	—	—	—	—	—
Succinic acid	14.2	19.5	13.7	0.51	—	—	—	14	—
Fumaric acid	16.5	21.8	—	0.62	—	—	—	16	—
Oxalacetic acid MO	20.2	25.6	—	1.11	—	—	—	23.5	—
Malic acid	21.6	27.0	—	1.43	—	—	—	22	10.0
2-Oxoglutaric acid MO	25.2	30.6	—	—	—	—	—	—	—
cis-Aconitic acid	34.1	39.6	—	—	—	1.07	—	32	—
Citric acid	36.3	41.9	—	—	—	1.57	—	34	19.5
Isocitric acid	36.4	—	—	—	—	1.45	16	—	—
Malonic	—	—	9.5	—	—	—	—	—	—
3,3-Dimethylglutaric acid		—	—	1.00	—	—	—	—	—
2-Oxoglutaric acid	—	—	—	2.08	29	—	—	27	—
Tricarbullylic acid	—	—	—	—	—	1.00	—	—	—
Lactic acid	—	—	—	—	—	—	—	5.7	—
Glyoxalic acid	—	—	—	—	—	—	—	7.5	—
Glutaric acid	—	—	—	—	—	—	—	—	5.7
Quinic acid	—	—	—	—	—	—	—	—	20.8
Ascorbic acid	—	—	—	—	—	—	—	—	22.2

Table GC-22 (continued)
ORGANIC ACIDS

[a] Measured on chromatogram.

Packing P1 = 3% Dexsil®-300 on Supelcoport (100-120 mesh)
P2 = 5% SE-30 on Chromosorb® G, AW, DMCS (60-80 mesh)
P3 = 5% SE-30 on GasChrom® P
P4 = 12% DC-560 on Chromosorb® W, AW, DMCS
P5 = 10% OV-17 on GasChrom® Q (80-100 mesh)

REFERENCES

1. **Langenbeck, U. and Seegmiller, J. E.**, Separation of Krebs cycle intermediates on a low-bleed gas chromatographic column. Temperature dependence of methylene unit values, *J. Chromatogr.*, 78, 420, 1973.
2. **Nicolai, H., von, and Zilliken, F.**, Gaschromatographische Bestimmung von Oxalsäure, Malonsäure und Bernsteinsäure aus biologischem Material, *J. Chromatogr.*, 92, 431, 1974.
3. **Nulton, C. P., Naworal, J. D., Campbell, I. M., and Grotzinger, E. W.**, A combined radiogas chromatograph/mass spectrometer detects intermediates in mycophenoic acid biosynthesis, *Anal. Biochem.*, 75, 219, 1976.
4. **Rosenqvist, H., Kallio, H., and Nurmikko, V.**, Gas chromatographic analysis of citric acid cycle and related compounds from Escherichia coli as their trimethylsilyl derivatives, *Anal. Biochem.*, 46, 224, 1972.
5. **Ryan, J. J.**, Chemical composition of Canadian apple juice, *J. Assoc. Anal. Chem.*, 55, 1104, 1972.

Table GC-23
ORGANIC ACIDS

Packing	P1	P2
Column		
Length (cm)	183	9500
Diameter (mm)	3	0.5
Material	G	G
Gas	N_2	He
Flow rate (mℓ/min)	30-40	2-4
Temperature (°C)	110-280 (5/min)	80-240
Detector	FI	MS
Reference	1 & 1′	2
Compound	TMS min	TMS min[a]
Glyoxylic acid MO	3.0	—
Pyruvic acid MO	4.3	—
Glyoxylic acid EO	4.4	—
L-Lactic acid	4.7	8.0
Glycolic acid	5.0	—
Pyruvic acid	5.2	—
Pyruvic acid EO	5.8	—
Oxalic acid	6.6	—
Methylmalonic acid	10.0	—
Glyoxylic acid	10.3	—
Benzoic acid	10.7	—
Succinic acid	12.9	21.5
Fumaric acid	14.1	—
Oxaloacetic acid MO	17.1	—
Oxaloacetic acid EO	18.0	—
L-Malic acid	18.1	—
Glyoxylic acid BO	20.4	—
2-Oxoglutaric acid EO	20.8	—
2-Oxoglutaric acid	21.0	—
Hippuric acid	25.2	—
Citric acid	26.0	—
2-Oxoglutaric acid	26.3	—
4-Hydroxyphenylpyruvic acid EO	27.7	—
Undecanoic acid	28.8	—
Oxaloacetic acid BO	29.4	—
2-Oxaloacetic acid	31.6	—
Glyoxylic acid DiPED	35.7	—
4-Hydroxyphenylpyruvic acid BO	37.2	—
2-Hydroxy-4-methylpentanoic acid	—	17.5
Phenylacetic acid	—	22.5
2,6-Di-*tert.*-butylcresol	—	37.0
β-Phenylacetic acid	—	40.0
4-Hydroxyphenylacetic acid	—	45.0
4-Hydroxyphenylpropionic acid	—	51.7
Tetradecanoic acid	—	52.8
Hexadecanoic acid	—	63.0
9-Octadecenoic acid	—	72.5
Octadecanoic acd	—	73.2
Cholesterol	—	136.0

[a] Measured on chromatogram.

Note: MO: methoxime TMS; EO; ethoxime TMS; BO: *O*-benzyloxime.
DiPED: *N,N′*-diphenylethylenediamine derivatives-TMS.

Table GC-23 (continued)
ORGANIC ACIDS

Packing P1 = 10% OV-101 on HP Chromosorb® W, AW. DMCS (80-100 mesh)
P2 = OV-17 capillary, split ratio 3:1 - 20:1

REFERENCES

1. **Chalmers, R. S. and Watts, R. W. E.,** Derivatives for the identification and quantitative determination of some keto- and aldo-carboxylic acid by gas-liquid chromatography, *Analyst (London),* 97, 951, 1972.

1′. **Chalmers, R. A. and Watts, R. W. E.,** The quantitative extraction and gas-liquid chromatographic determination of organic acids in urine, *Analyst (London),* 97, 958, 1972.

2. **Ward, M. E., Politzer, I. R., Laseter, J. L., and Alam, S. Q.,** Gas chromatographic mass spectrometric evaluation of free organic acids in human saliva, *Biomed. Mass Spectrom.,* 3, 77, 1976.

Table GC-24
ORGANIC ACIDS

Packing	P1		P2	P3	P4	P5	P6	P7
Column								
Length (cm)	183		180	95	180		200	150
Diameter (mm)	6.3		6.2 o.d.	3.5	3.1 o.d.		2	3.1
Material			G		ss		ss	G
Gas	He		He		N_2		N_2	
Flow rate (mℓ/min)	55		40		40		30	
Temperature (°C)	145-240(6/min)		200	140-180(10/min)	105	190-250	120-150	110-180(10/min)
Detector	FI		N		FI		FI	FI
Reference	1		2	3	4	5	6	7
	n-butyl		*n*-butyl	methyl	methyl	benzyl	benzyl	ethyl
Compound	Tr	min	min	min	min[a]	min	min[a]	min
Malonic acid	0.17	—	—	—	—	—	—	—
Succinic acid	0.26	—	—	—	—	—	—	—
Fumaric acid	0.27	—	—	1.1	—	—	—	—
Cinnamic acid	0.32	—	—	—	—	—	—	—
Malic acid	0.41	—	—	—	—	—	—	—
2-Hydroxybenzoic acid	0.44	—	—	—	—	—	—	—
N-methyliminodiacetic acid	0.50	—	—	—	—	—	—	—
Iminodiacetic acid	0.53	—	—	—	—	—	—	—
Tartaric acid	0.55	—	—	—	—	—	—	—
4-Hydroxy-3-methoxybenzoic acid	0.60	—	—	—	—	—	—	—
Tetradecanoic acid	0.70	—	—	—	—	—	—	—
Shikimic acid	0.83	—	—	—	—	—	—	—
Hexadecanoic acid	0.94	—	—	—	—	—	—	—
1,2-Benzenedicarboxylic acid	1.00	13.3	—	—	—	—	—	—
Hydroxycinnamic acid	1.07	—	—	—	—	—	—	—
4-Hydroxy-3,5-dimethoxybenzoic acid	1.12	—	—	—	—	—	—	—
9,10-Octadecadienoic acid	1.14	—	—	—	—	—	—	—
9,12,15-Octadecatrienoic acid	1.16	—	—	—	—	—	—	—
Octadecanoic acid	1.17	—	—	—	—	—	—	—
4-Hydroxy-3-methoxycinnamic acid	1.23	—	—	—	—	—	—	—

Table GC-24 (continued)
ORGANIC ACIDS

Compound	*n*-butyl Tr	*n*-butyl min	*n*-butyl min	methyl min	methyl min[a]	benzyl min	benzyl min[a]	ethyl min
Citric acid	1.29	—	—	—	—	—	—	—
Eicosanoic acid	1.31	—	—	—	—	—	—	—
Nitriloacetic acid	1.36	—	—	—	—	—	—	—
4-Hydroxy-3,5-dimethoxycinnamic acid	1.46	—	—	—	—	—	—	—
Docosanoic acid	1.50	—	—	—	—	—	—	—
Nitrilotriacetic acid	—	—	4.4	—	—	—	—	—
Oxalic acid	—	—	—	0.8	5	29.0	—	8
Itaconic acid	—	—	—	1.9	—	—	—	—
Maleic acid	—	—	—	2.1	—	—	—	—
Adipic acid	—	—	—	3.0	—	—	—	—
Azelacnic acid	—	—	—	5.5	—	—	—	—
Sepacic acid	—	—	—	6.3	—	—	—	—
1,4-Benzenedicarboxylic acid	—	—	—	7.2	—	—	—	—
1,3-Benzenedicarboxylic acid	—	—	—	7.8	—	—	—	—
1,2-Benzenedicarboxylic acid	—	—	—	8.6	—	—	—	—
Formic acid	—	—	—	—	—	6.1	8.5	—
Acetic acid	—	—	—	—	—	6.8	10.5	—
Propionic acid	—	—	—	—	—	8.4	13.5	—
Lactic acid	—	—	—	—	—	730	38.0	—
Butyric acid	—	—	—	—	—	—	19.0	—
n-Hexanoic acid	—	—	—	—	—	—	32.0	—
Glycolic acid	—	—	—	—	—	—	42.0	—

[a] Measured on chromatogram.

Packing P1 = 3% OV-210 on Chromosorb® W, AW (60-80 mesh)
P2 = 3% OV-1 on Chromosorb® W, HP (80-100 mesh)
P3 = 10% SP-216-PS on Supelcoport (100-120 mesh)
P4 = 20% diethyleneglycolsuccinate + 5% isophthlic acid on Chromosorb® W, AW
P5 = 10% Carbowax® 20M on Chromosorb® P, DMCS (80-100 mesh)
P6 = 3% butane-1,4-diolsuccinate polyester on Chromosorb® W (120-140 mesh)

P7 = 15% DEGS on Chromosorb® W, AW (80-100 mesh)

REFERENCES

1. **Larson, R. A., Weston, J. C., and Howell, S. M.**, Quantitative gas chromatographic determination of nitrilotriacetic acid in the presence of other carboxylic acids, *J. Chromatogr.*, 111, 43, 1975.
2. **Williams, D. T., Benoit, F., Muzika, K., and O'Grady, R.**, Gas chromatographic determination of nitrilotriacetic acid using a nitrogen-selective detector, *J. Chromatogr.*, 136, 423, 1977.
3. **Mlejnek, O. and Cvečková, L.**, Evaluation of new phases for the gas chromatography of dibasic acids, *J. Chromatogr.*, 82, 377, 1973.
4. **Mee, J. M. L. and Stanley, R. W.**, A rapid gas-liquid chromatographic method for determining oxalic acid in biological materials, *J. Chromatogr.*, 76, 242, 1973.
5. **Triebig, G., Schaller, K. H., and Gossler, K.**, Eine einfache zuverlässige gas chromatographische Bestinmung von Ameisensäure in urine, *Z. Anal. Chem.*, 290, 114, 1978.
6. **Bethge, P. O. and Lindstrom, K.**, Determination of organic acids of low relative molecular mass (C_1 to C_4) in dilute aqueous solution, *Analyst (London)*, 99, 137, 1974.
7. **Charransol, G., Barthelemy, C., and Desgrez, P.**, Rapid determination of urinary oxalic acid by gas-liquid chromatography without extraction, *J. Chromatogr.*, 145, 452, 1978.

Table GC-25
ORGANIC ACIDS

Packing	P1	P1	P1	P1	P1	P2	P3	P4
Column								
Length (cm)	152	152	152	152	152	3000	200	200
Diameter (mm)	4.5	4.5	4.5	4.5	4.5	0.35	3.1	2
Material	G	G	G	G	G	G	ss	
Gas	N_2	N_2	N_2	N_2	N_2	N_2	N_2	N_2
Flow rate (mℓ/min)	60	60	60	60	60	1.4	15	23
Temperature (°C)	130	175	190	215	275	70—220(5/min)	150	80—185(5/min)
Detector	FI	FI	FI	FI	FI	FI	FI	FI
Reference	1	1	1	1	1	2	3, 4	5

Table GC-25 (continued)
ORGANIC ACIDS

Compound	benzyl		benzyl		benzyl		benzyl		benzyl		TMS	Me	TMS
	Tr	min	Tr	min	Tr	min	Tr	min	Tr	min	min[a]	min	min
Glyoxylic acid	0.34		—		—		—		—		—	—	—
Glycolic acid	0.60		—		—		—		—		—	—	—
Lactic acid	0.63		—		—		—		—		—	—	—
Isovaleric acid	0.72		—		—		—		—		—	—	—
Crotonic acid	0.93		—		—		—		—		—	—	—
Thioglycolic acid	1.00	7.2	—		—		—		—		—	—	—
4-Hydroxybutyric acid	1.15		—		—		—		—		—	—	—
Trichloroacetic acid	—		0.29		—		—		—		—	—	—
Indoleacetic acid	—		0.33		1.59		—		—		—	—	—
Benzoic acid	—		0.64		—		—		—		11.9	—	—
Nicotinic acid	—		0.78		—		—		—		13.7	—	—
Phenylacetic acid	—		0.78		—		—		—		—	—	—
2-Methylbenzoic acid	—		0.89		—		—			21.3	—	—	
2-Hydroxybenzoic acid	—		1.00	7.45	—		—			18.7	—	—	
Mandelic acid	—		1.15		—		—		—		—	11.8	—
Phenylpropionic acid	—		1.24		—		—		—		—	—	—
6-Aminohexanoic acid	—		1.33		—		—		—		—	—	—
Quinaldic acid	—		1.33		—		—		—		—	—	—
2-Nitrobenzoic acid	—		1.70		—		—		—		—	—	—
2-Acetoxybenzoic acid	—		1.93		—		—		—		—	—	—
4-Nitrobenzoic acid	—		2.60		—		—		—		—	—	—
Fumaric acid	—		—		0.82		—		—		—	—	8.57
Maleic acid	—		—		0.88		—		—		—	—	7.34
Succinic acid					1.00	12.3	1.00	7.7	—		—	—	7.65
Citraconic acid	—		—		1.00		—		—		—	—	—
Citramalic acid	—		—		1.12		—		—		—	—	11.66
Malic acid	—		—		1.33		—		—		—	—	12.14
3,3-Dimethylglutaric acid			—		1.47		—		—		—	—	—
Mesaconic acid	—		—		1.54		—		—		—	—	—
Tartaric acid	—		—		1.81		—		—		—	—	15.74

1,2-Benzenedicarboxylic acid			—		3.22	—	0.59		—	—	—
Oxalic acid	—		—		—	0.50	—		—	—	—
Malonic acid	—	—		—	0.66	—		—	—	5.32	
Glutaric acid	—	—		—	1.36	—		—	—	9.74	
Adipic acid	—		—		—	1.85	—		—	—	—
Octadecanoic acid	—		—		—	—	0.66		—	—	—
Abietic acid	—		—		—	—	1.00		7.1	—	—
Eicosanoic acid	—		—		—	—	1.00		—	—	—
Docosanoic acid	—		—		—	—	2.08		—	—	—
Citric acid	—		—		—	—	2.24		—	—	19.40
Isocitric acid	—		—		—	—	2.41		—	—	—
4-Aminobenzoic acid	—		—		—	—	—		25.6	—	—
Phenylglyoxylic acid	—		—		—	—	—		—	7.0	—
Methylmalonic acid	—		—		—	—	—		—	—	5.59
trans-Aconitic acid	—		—		—	—	—		—	—	17.60

[a] Measured on chromatogram.

Packing P1 = 3% SE-30 on Chromosorb® G, AW, DMCS (80-100 mesh)
P2 = OV-17, capillary
P3 = 5% ethylene glycol diadipate on GasChrom® P (100-140 mesh)
P4 = 3% SE-30

REFERENCES

1. **Corina, D. L.**, Calibration data for the gas chromatographic determination carboxylic acids as their benzyl esters, *J. Chromatogr.*, 87, 254, 1973.
2. **Christophersen, A. S., Rasmussen, K. E., and Tønnesen, F.**, Glass capillary column gas chromatography of carboxylic acids after flash-heater trimethylsilylation, *J. Chromatogr.*, 179, 87, 1979.
3. **Schaller, K.-H., Gossler, K., Bost, H.-P., and Valentin, H.**, Gaschromatographische Methodes zur Bestimmung von Styrol im Blut sowie von Mandelsäure und Phenylglyosilsäure in Urin, *Arbeitsmed. Sozialmed. Präventivmed.*, 11(Heft 1), 24, 1976.
4. **Zschiesche, W., Gossler, K., Triebig, G., und Schaller, K. H.**, Zur Überwachung beruflich styrolexponierter Personen mit modernen Analysenmethoden. Möglichkeiten und Grenzen des Biological Monitoring, Arbeitsmedizinische Probleme des Dienstleistungsgewerbes, Arbeitsmedizinisches Kolloquium, Frankfurt, Mai 1978, 355.
5. **Stumpf, D. K. and Burris, R. H.**, A Micromethod for the purification and quantification of organic acids of the tricarboxylic acid cycle in plant tissues, *Anal. Biochem.*, 95, 311, 1979.

Table GC-26
DICARBOXYLIC ACIDS

Packing	P1	P1	P2
Column			
Length (cm)	30	30	50
Diameter (mm)	0.5	0.5	0.5
Material	ss	ss	ss
Gas	N_2	N_2	N_2
Flow rate (mℓ/min)			
Temperature (°C)	120	100	130
Detector	FI	FI	FI
Reference	1	1	1

Compound	min	min	min
Dimethylmalonate	1.5	3.4	2.3
Dimethylsuccinate	2.6	6.3	4.1
Dimethylglutarate	5.0	12.3	7.8
Dimethyladipate	9.3	26	15
Dimethylpimelate	17	48	24
Dimethylsuberate	32	93	48
Dimethylazelate	58	183	85
Dimethylsebacate	108	348	150
Methylethylmalonate	2.1	5.0	2.7
Methylethylsuccinate	3.8	9.5	5.2
Methylethylglutarate	6.9	18	9.8
Methylethyladipate	13	35	18
Methylethylpimelate	25	70	32
Methylethylsuberate	45	133	60
Methylethylazelate	81	260	110
Methylethylsebacate	155	—	200
Diethylmalonate	2.8	7.1	3.6
Diethylsuccinate	5.3	14	7.1
Diethylglutarate	9.8	27	14
Diethyladipate	19	52	25
Diethylpimelate	35	101	48
Diethylsuberate	63	190	82
Diethylazelate	115	382	150
Diethylsebacate	212	—	260
Dipropylmalonate	7.9	22	11.5
Dipropylsuccinate	16	44	24
Dipropylglutarate	29	84	40
Dipropyladipate	55	174	71
Dipropylpimelate	98	327	129
Dipropylsuberate	180	—	246
Dipropylazelate	355	—	402
Dipropylsebacate	640	—	800

Packing P1 = Ucon Oil LB-550X, capillary
P2 = Apiezon M, capillary

From Strauss, C. R., *J. Chromatogr.*, 87, 576, 1973. With permission.

Table GC-27
HYDROXY ACIDS

Packing	P1	P2
Column		
Length (cm)	300	200
Diameter (mm)	2	2
Material		
Gas	N_2	N_2
Flow rate (mℓ/min)	40	40
Temperature (°C)	120 (1/min)	150 (2/min)
Detector	FI	FI
Reference	1	1

Compound	TMS Tr	Acetate Tr
Citric acid	1.83	2.13
Erythronic acid	0.78	1.00
Erythronolactone	0.86	1.00
2-Galaturonic acid	2.26	2.31
3-Galaturonic acid	2.39	2.31
Isocitric acid	1.90	2.13
Lactic acid	1.94	2.13
Malic acid	0.78	0.48
Ribonic acid	1.37	2.01
Ribonolactone	1.45	2.01
Glucaric acid	2.19	3.22
Glucaric acid 1,4-lactone	2.33	3.22
Mesotartaric acid	1.00	1.00

Packing P1 = 10% OV-17 on GasChrom® Q (80-100 mesh)
P2 = 3% OV-225 on Varaport 30 (100-120 mesh)

From Kringstad, R., *J. Chromatogr.*, 144, 209, 1977. With permission.

Table GC-28
ORGANIC ACIDS

Packing	P1		P1		P2	P3
Column						
Length (cm)	180		180		240	240
Diameter (mm)	5		5		3.1	2
Material					ss	ss
Gas	N_2		N_2		N_2	N_2
Flow rate (mℓ/min)					29	35
Temperature (°C)	130		180		75-220(2/min)	60-170(2.5&5/min)
Detector	FI		FI		FI	FI
Reference	1		1		2	2
Compound	TMS		TMS		TMS	
	Tr	min	Tr	min	min[a]	min[a]
Succinic acid	0.51		—		30.0	—
Fumaric acid	0.62		—		—	—
3,3-Dimethylglutaric acid	1.00		—		—	—
Oxalacetic acid	1.11		—		—	—
Malic acid	1.43		—		—	—
2-Oxoglutaric acid	2.08	29	—		—	—
Tricarballylic acid	—		1.00		—	—
cis-Aconitic acid	—		1.07		—	—
Citric acid	—		1.57	16	—	—
Isocitric acid	—		1.45		—	—
Lactic acid	—		—		10.0	—
3-Hydroxybutyric acid	—		—		18.5	—
Urea	—		—		26.0	—
2-Ethoxyethoxyacetic acid	—		—		27.0	33.0
4-Phenylbutyric acid	—		—		43.0	—
4-Hydroxyphenylacetic acid	—		—		51.5	—
Acetic acid	—		—		—	9.0
Propionic acid	—		—		—	12.0
Isobutyric acid	—		—		—	13.0
Butyric acid	—		—		—	14.5

n-Valeric acid	—	—	—	17.5
Benzoic acid	—	—	—	35.0

[a] Measured on chromatogram.

Packing P1 = 5% SE-30 on GasChrom® P
P2 = 5% GESE-52 on Chromosorb® W, AW, DMCS (100-120 mesh)
P3 = 5% Neopentylglycoladipate + 2% H_3PO_4 on Chromosorb® W, AW, DMCS (100-120 mesh)

REFERENCES

1. **Pinnelli, A. and Colombo, A.**, Gas chromatographic separation of Krebs-cycle metabolites, *J. Chromatogr.*, 118, 236, 1976.
2. **Kamerling, J. P., Duran, M., Bruinvis, L., Ketting, D., Wadman, S. K., De Groot, C. J., and Hommes, F. A.**, (2-Ethoxyethoxy) acetic acid: an unusual compound found in the gas chromatographic analysis of urinary organic acids, *Clin. Chim. Acta*, 77, 397, 1977.

Table 29
ORGANIC ACIDS

Packing	P1	P1	P1	P2	P2
Column					
Length (cm)	17	17	17	2500	2500
Diameter (mm)				0.3	0.3
Material	G	G	G	G	G
Gas				N_2	N_2
Flow rate (mℓ/min)	25	25	25	1	1
Temperature (°C)	90	90	90	150	200
Detector	FI	FI	FI	FI	FI
Reference	1	1	1	2	2
Compound	TMS min	pyrazoline min	cyclopropane min	*O*-acetylated min[a]	methyl min[a]
trans-Aconitic acid	2.4	7.4	3.0	—	—
R-Lactic acid	—	—	—	15	—
S-Lactic acid	—	—	—	13	—
R-Glyceric acid	—	—	—	—	14
S-Glyceric acid	—	—	—	—	13

[a] Measured on chromatogram.

Packing P1 = 4% SE-30 on Chromosorb® W, AW, DMCS (80-100 mesh)
P2 = SP-1000 capillary

REFERENCES

1. **Dilli, S. and Szyszka, J.,** Gas chromatography of the pyrazoline derivatives of trimethyl aconitate, *J. Chromatogr.*, 132, 148, 1977.
2. **Kamerling, J. P., Gerwig, G. J., Vliegenthart, J. F. G., Duran, M., Ketting, D., and Wadman, S. K.,** Determination of the configurations of lactic and glyceric acids from human serum and urine by capillary gas-liquid chromatography, *J. Chromatogr.*, 143, 117, 1977.

Table 30
ORGANIC ACIDS

Packing	P1		P1		P2
Column					
Length (cm)					180
Diameter (mm)					3.1 o.d.
Material					ss
Gas	N_2		N_2		
Flow rate (mℓ/min)	30		30		
Temperature (°C)	220		220-260(4/min)		100-200(4/min)
Detector	EC		FI		FI
Reference	1		1		2
	methylester of 2,4-dinitrophenylhydrazone				methyl
Compound	Tr	min	Tr	min	min[a]
Glyoxylic acid	0.63		0.63		—
Pyruvic acid	0.82		0.82		—
2-Oxobutanoic acid	1.00		1.00		—
2-Oxo-3-methylbutanoic acid	1.06		1.06		—
2-Oxo-3-methylvaleric acid	1.31		1.31		—
2-Oxo-4-methylvaleric acid	1.32		1.32		—
Oxalacetic acid	2.04		1.84		—
2-Oxoglutaric acid	2.91	32.5	2.13	24	—
cis-2-Hydroxycyclohexanecarboxylic acid	—		—		10.5
trans-2-Hydroxycyclohexanecarboxylic acid	—		—		12.5
cis-3-Hydroxycyclohexanecarboxylic acid	—		—		16.2
trans-3-Hydroxycyclohexanecarboxylic acid	—		—		17.0

[a] Measured on chromatogram.

Packing P1 = 4% SE-30 on Chromosorb® W, AW, TMCS (80-100 mesh)
P2 = 2.5% SE-30 on Chromosorb® G (100-120 mesh)

REFERENCES

1. **Kallio, H. and Linko, R. R.,** Gas-liquid chromatographic analysis of 2,4-dinitrophenylhydrazones of keto acid methylesters, *J. Chromatogr.*, 76, 229, 1973.
2. **Kaneda, T.,** Separation of cis and trans isomers of hydroxycyclohexanecarboxylic acids by gas-liquid chromatography, *J. Chromatogr.*, 92, 191, 1974.

Table 31
KETO ACIDS

Packing	P1	P2
Column		
Length (cm)	180	180
Diameter (mm)	3	3
Material	G	G
Gas	N_2	N_2
Flow rate (m*l*/min)	60	60
Temperature (°C)	50-200 (1, 2, 4/min)	50-180 (2/min)
Detector	FI	FI
Reference	1	2

	O-trimethylsilylquinoxalinol	
Compound	min[a]	min[a]
2-Hydroxyisovaleric acid	3.8	—
Pyruvic acid	17.2	21.0
2-Oxoisovaleric acid	21.0	25.0
2-Oxo-*n*-valeric acid	23.3	27.0
2-Oxo-3-methyl-*n*-valeric acid	23.9	27.5
2-Oxoisohexanoic acid	24.3	28.5
Hexadecanoic acid	38.7	—
2-Oxoglutaric acid	38.7	44.0
2-Hydroxy-3-methyl-*n*-valeric acid	5.4	—
2-Hydroxyisohexanoic acid	5.4	—
2-Oxobutanoic acid	—	23.5
2-Oxooctanoic acid	34.2	38.0
2-Oxo-4-(methylthio)butyric acid	—	39.5
Phenylpyruvic acid	—	44.5
2-Oxoadipic acid	—	48.0

[a] Measured on chromatogram.

Packing P1 = 3% OV-1
P2 = 3% OV-1 on GasChrom® Q(100-200 mesh)

REFERENCES

1. **Langenbeck, U., Hoinowski, A., Mantel, K., and Möhring, H-U.,** Quantitative gas chromatography and single-ion detection of aliphatic α-keto acids from urine as their O-trimethylsilyquinoxalinol derivatives, *J. Chromatogr.,* 143, 39, 1977.
2. **Langenbeck, U., Möhring, H.-U., and Dieckmann, K.-P.,** Gas chromatography of α-keto acids as their *O*-trimethylsilylquinoxalinol derivatives, *J. Chromatogr.,* 115, 65, 1975.

Table GC-32
ORGANIC ACIDS

Packing	P1	P2
Column		
Length (cm)	180	180
Diameter (mm)	3.1	
Material		G
Gas		N_2
Flow rate (mℓ/min)		60
Temperature (°C)	70-285(4/min)	70-160(2/min)
Detector	MS	MS
Reference	1	2
Compound	TMS Tr	*O*-trimethylsilylquinoxalinol min[a]
Glycolic acid	1131	—
3-Hydroxyisobutyric acid	1206	—
Pyruvic acid	1224	6.4
2-Methylphenol	1235	—
3-Hydroxyisovaleric acid	1244	—
Urea	1369	—
Methylfumaric acid	1472	—
Adipic acid	1606	—
Pyroglutamic acid	1676	—
5-Hydroxymethylfuroic acid	1690	—
2-Oxoglutaric acid	1728	25
3-Hydroxyphenylacetic acid	1735	—
4-Hydroxyphenylacetic acid	1766	10
2,5-Furandicarboxylic acid	1806	—
trans-Aconitic acid	1858	—
Citric acid	1888	—
3-(3-Hydroxyphenyl)-hydracrylic acid	1943	—
Hippuric acid	2118	15.9
3-Indoleacetic acid	2191	—
Octadecanoic acid	2288	—
3-Hydroxyhippuric acid	2380	—
4-Hydroxyhippuric acid	2459	—
Phenylacetic acid	—	2.0
Mandelic acid	—	5.3
2-Hydroxyphenylacetic acid	—	7.8
Phenyllactic acid	—	8.4
2-Oxovaleric acid	—	10.6
Phenylpyruvic acid	—	25.6
4-Hydroxyphenylpyruvic acid	—	37.2

[a] Measured on chromatogram.

Packing P1 = 10% OV-17 on GasChrom® Q (100-120 mesh)
P2 = 3% OV-1

REFERENCES

1. **Anderson, P. J., Fitch, W. L., and Halpern, B.,** Rapid and simplified extraction procedure for gas chromatographic mass spectrometric profiling of urinary organic acids, *J. Chromatogr.*, 146, 481, 1978.
2. **Langenbeck, U., Mench-Hoinowski, A., Dieckman, K.-P., Möhring, H.-U., and Petersen, M.,** *O*-Trimethylsilylquinoxalinol derivatives of aromatic α-keto acids. Mass spectra and quantitative gas chromatography, *J. Chromatogr.*, 145, 185, 1978.

Table GC-33
AROMATIC ACIDS

Packing	P1	P1	P2	P2	P3	P3	P4	P4	P5
Column									
Length (cm)	100	100	100	100	100	100	100	100	200
Diameter (mm)	6 o.d.	6 o.d.	6 o.d.	6 o.d.	6 o.d.	6 o.d.	6 o.d.	6 o.d.	2
Material	G	G	G	G	G	G	G	G	ss
Gas	N_2	N_2	N_2	N_2	N_2	N_2	N_2	N_2	N_2
Flow rate (mℓ/min)	30	30	30	30	30	30	30	30	25
Temperature (°C)	195	174	195	174	195	174	195	174	130
Detector	FI	FI	FI	FI	FI	FI	FI	FI	FI
Reference	1	1	1	1	1	1	1	1	2
Compound	ru	ru	ru	ru	ru	ru	ru	ru	Tr
Benzene	685	703	773	750	715	748	732	746	—
Methoxybenzene	957	—	1067	—	989	—	1006	—	—
Methylbenzoate	1093	1119	1349	1223	1282	1214	1296	1219	0.84
Methyl-2-methoxybenzoate	1451	—	1686	—	1577	—	1581	—	—
Methyl-3-methoxybenzoate	1419	—	1622	—	1555	—	1545	—	—
Methyl-4-methoxybenzoate	1482	—	1687	—	1606	—	1650	—	—
Toluene	—	820	—	858	—	856	—	855	—
Methyl-2-methylbenzoate	—	1203	—	1299	—	1294	—	1293	1.14
Methyl-3-methylbenzoate	—	1230	—	1336	—	1331	—	1330	1.38
Methyl-4-methylbenzoate	—	1238	—	1343	—	1342	—	1340	1.46
Acetophenone	—	—	—	—	—	—	—	—	1.00

Packing P1 = 10% Apiezon L on silylated Celite® (80-100 mesh)
P2 = 10% 4-dodecylphenyl-2-methoxybenzoate on silylated Celite® (80-100 mesh)
P3 = 10% 4-dodecylphenyl-3-methoxybenzoate on silylated Celite® (80-100 mesh)
P4 = 10% 4-dodecylphenyl-4-methoxybenzoate on silylated Celite® (80-100 mesh)
P5 = 10% polyethyleneglycol 1500 on Chromosorb® W (60-80 mesh)

REFERENCES

1. **Vernon, P., Sharples, W. E., and Kyffin, T. W.,** The steric factor in the gas-liquid chromatographic retention of toluic and methoxybenzoic esters, *J. Chromatogr.*, 111, 117, 1975.
2. **Flek, J. and Šedivec, V.,** Determination or toxic substances and their metabolites in biological fluids by gas chromatography. VII. Toluric or toluic acids in urine, *Collect. Czech. Chem. Commun.*, 38, 1754, 1973.

From Vernon, *J. Chromatogr.*, 111, 117, 1975. With permission.

Table GC-34
AROMATIC ACIDS

Packing	P1	P2	P3		P3		P4	P4
Column								
Length (cm)	130	183	200		200		150	150
Diameter (mm)	2	2	3		3		6 o.d.	6 o.d.
Material		G					G	G
Gas		He	N_2		N_2		N_2	N_2
Flow rate (mℓ/min)	38	20	50		50		60	60
Temperature (°C)	230	100-200(6/min)	140		160		186	228
Detector	FI	FI	FI		FI		EC	EC
Reference	1	2	3		3		4	4
		TMS	methyl		methyl		flophemesyl	
Compound	min[a]	min	Tr	min[a]	Tr	min[a]	min	min
Benzoic acid	3.5	12.2	—	—	—	—	6.6	—
2-Methylbenzoic acid	6.6	—	—	—	—	—	—	—
3-Methylbenzoic acid	8.6	—	—	—	—	—	—	—
2-Hydroxybenzoic acid	9.5	—	—	—	—	—	—	—
4-Methylbenzoic acid	10.0	15.4	—	—	—	—	—	7.5
2-Chlorobenzoic acid	12.0	—	1.00	17	—	—	—	—
3-Chlorobenzoic acid	13.0	—	0.65	11	—	—	—	—
4-Chlorobenzoic acid	14.5	16.1	0.57	9	—	—	—	—
3-Hydroxybenzoic acid	25.5	—	—	—	—	—	—	—
4-Hydroxybenzoic acid	32.5	—	—	—	—	—	—	—

Table GC-34 (continued)
AROMATIC ACIDS

		TMS	methyl		methyl		flophemesyl	
Compound	min[a]	min	Tr	min[a]	Tr	min[a]	min	min
2-Nitrobenzoic acid	36.0	—	—	—	—	—	—	—
Methylbenzoylformate	—	16.4	—	—	—	—	—	—
Methyl-4-methylbenzoylformate	—	19.5	—	—	—	—	—	—
Methyl-4-chlorobenzoylformate	—	20.0	—	—	—	—	—	—
2-Bromobenzoic acid	—	—	—	—	1.00	14.5	—	—
3-Bromobenzoic acid	—	—	—	—	0.69	10.0	—	—
4-Bromobenzoic acid	—	—	—	—	0.60	8.5	—	—
2-Iodobenzoic acid	—	—	—	—	1.00	27.0	—	—
3-Iodobenzoic acid	—	—	—	—	0.71	18.5	—	—
4-Iodobenzoic acid	—	—	—	—	0.62	17.0	—	—
Phenol	—	—	—	—	—	—	4.1	—
4-Methylphenol	—	—	—	—	—	—	5.2	—
4-Bromophenol	—	—	—	—	—	—	6.45	—
4-Chlorophenol	—	—	—	—	—	—	6.55	—
2,4-Dichlorobenzoic acid	—	—	1.80	30	—	—	—	—

[a] Measured on chromatogram.

Packing P1 = 0.3% FFAP on FT-G (Graphitized sterling FT)
P2 = 10% OV-17 on Chromosorb® WHP (80-100 mesh)
P3 = 5% bentone, 5% trimeric acid on Chromosorb® W, AW
P4 = 10% SE-30 on Chromosorb® P, AW, DMCS

REFERENCES

1. **Di Corcia, A.**, Gas-liquid/solid chromatography of free acids, *Anal. Chem.*, 45, 492, 1973.
2. **White, E. R., Sutton, B. M., Blank, J. E., Moeckel, E., and Zarembo, J. E.**, Application of ozonolysis and gas chromatography for structure elucidation of substituted vulpinic acids, *Anal. Chem.*, 44, 1582, 1972.
3. **Felder, E., Tiepolo, U., and Mengassini, A.**, Gas chromatographic analysis of mono-halobenzoic acid isomers, *J. Chromatogr.*, 82, 390, 1973.
4. **Francis, A. J., Morgan, E. D., and Poole, C. F.**, Flophemesyl derivatives of alcohols, phenols, amines and carboxylic acids and their use in gas chromatography with electron-capture detection, *J. Chromatogr.*, 161, 111, 1978.

Table GC-35
CINNAMIC ACIDS

Packing	P1	P2	P3
Column			
Length (cm)	275	152	300
Diameter (mm)	4	4	2
Material	G	G	G
Gas	Ar	Ar	N_2
Flow rate (mℓ/min)	55	50	40
Temperature (°C)	195	175	210
Detector	FI	FI	FI
Reference	1	1	2

Compound	TMS min	TMS min	TMS sec
cis-Hydroxycinnamic acid	12.5	5.1	—
trans-Hydroxycinnamic acid	19.2	9.1	—
cis-3,4-Dihydroxycinnamic acid	22.4	11.6	256
trans-3,4-Dihydroxycinnamic acid	32.6	22.4	407
cis-4-Hydroxy-3-methoxycinnamic acid	23.7	8.4	206
trans-4-Hydroxy-3-methoxycinnamic acid	40.8	16.7	336
cis-4-Hydroxy-3,5-dimethoxycinnamic acid	48.3	14.5	304
trans-4-Hydroxy-3,5-dimethoxycinnamic acid	86.7	31.0	522

Packing P1 = 5% OV-25 on Diatomite CQ (80-100 mesh)
P2 = 3% SE-30 on Diatomite CQ (80-100 mesh)
P3 = 3.5% OV-1 on Chromosorb® W (80-100 mesh)

REFERENCES

1. **Hartley, R. D. and Jones, E. C.,** Effect of ultraviolet light on substituted cinnamic acids and the estimation of their cis and trans isomers by gas chromatography, *J. Chromatogr.*, 107, 213, 1975.
2. **Fenton, T. W., Mueller, M. M., and Clandinin, D. R.,** Isomerization of some cinnamic acid derivatives, *J. Chromatogr.*, 152, 517, 1978.

Table GC-36
PHENOLS AND ORGANIC ACIDS

Packing	P1	P2	P3	P4	P5	P6	P6	P6	P7	P8
Column										
Length (cm)	360	180	182	2000	200	2800	2800	5000	2800	—
Diameter (mm)	2	3.1	2	0.32	4	0.38	0.38	0.55	0.38	
Material		G	G		G	G	G	G	G	
Gas			He	He	N_2	He	He	He	He	
Flow rate (mℓ/min)			22	1.8	20	0.55	0.55	0.55	0.55	
Temperature (°C)	60-260	80-280	80-300	70-240	150-350	70-260	70-260	70-260	70-260	
Gradient (°C/min)	4	4	4	3	6	5	2	2	2	
Detector	FI	MS	FI		FI	FI	MS	MS	MS	
Reference	1	2	3	4	5	6	6	6	6	6
Compound	TMS min	TMS Tr	TMS min	TMS min[a]	methyl Tr	TMS ri[b]	TMS ri[b]	TMS ri[b]	TMS ri[b]	TMS ri[b]
2-Hydroxyisobutanoic acid	10.4	—	—	—	—	—	—	—	—	—
Lactic acid	11.3	1094	8.50	5.3	—	1082	1082	1082	1095	1107
Glycolic acid	12.5	1132	9.75	—	—	1092	1100	1100	1105	1138
Phenol	—	1133	—	—	—	—	—	—	—	—
2-Hydroxybutanoic acid	—	—	10.95	—	—	—	—	—	—	—
3-Hydroxyisobutanoic acid	—	1205	1239	—	—	—	—	—	—	—
3-Hydroxybutanoic acid	14.9	—	1239	—	—	1175	—	1181	1173	1206
Pyruvic acid	15.3	1222	10.95	—	—	—	—	—	—	—
Methylphenol	16.0	1239	—	—	—	—	—	—	—	—
2-Methylphenol	—	—	—	—	—	1133	1134	—	—	—
3-Methylphenol	—	—	—	—	—	1144	1141	—	—	—
4-Methylphenol	—	—	—	—	—	1154	1152	1153	1156	—
1-Octanol	—	—	—	—	—	1186	1185	1187	1187	—
cis-Glyceraldehyde	—	—	14.53	—	—	—	—	—	—	—
trans-Glyceraldehyde	—	—	15.33	—	—	—	—	—	—	—
Dihydroxyacetone	—	—	16.89	—	—	—	—	—	—	—
3-Hydroxyisovaleric acid	—	1243	—	—	—	—	1231	1226	1215	1234
Oxalic acid	16.2	—	—	7.9	—	—	—	—	—	—
2-Methylmalonic acid	—	—	—	—	—	1240	1243	1239	—	1299

Glycerol	17.3	—	—	—	—	—	—	—	—	—
Benzoic acid	—	—	—	14.2	—	1260	1258	1251	1235	1372
Phosphoric acid	20.7	1353	18.40	17.6	—	1295	1305	1302	1263	1360
Urea	—	1366	18.44	15.6	—	—	—	—	—	—
4-Deoxyerythronic acid	21.2	1383	—	—	—	—	1372	1367	1364	1377
Phenylacetic acid	—	—	—	16.8	—	—	—	—	—	—
1,2-Dihydroxybenzene	—	—	—	—	—	1323	1323	1322	—	—
Succinic acid	22.0	—	20.12	19.3	—	1340	1335	1334	1313	1408
Glycine	—	1368	—	—	—	1360	1358	1358	1347	1372
Fumaric acid	—	—	—	—	—	1365	1368	—	1357	1416
Deoxyerythronic acid	22.9	—	—	—	—	—	—	—	—	—
Glutaric acid	23.8	—	—	—	—	—	—	—	—	—
4-Deoxyerythronic acid	—	—	—	—	—	—	1383	1371	1372	1380
Threono-1,4-lactone	—	—	—	—	—	—	1388	1380	1366	1573
3-Deoxy-2,4-dioxo-tetrahydrofuran	—	1449	—	—	—	—	1441	1443	1433	1459
Erythrono-1,4-lactone	—	—	—	—	—	—	1456	1453	1413	—
2-Deoxy-2,4-dioxotetrahydrofuran	—	1465	—	—	—	—	1466	1465	1456	1475
Erythronic acid	26.9	1557	—	—	—	—	1587	1587	1584	1571
4-Hydroxycyclohexanecarboxylic acid	—	1565	—	—	—	—	—	—	—	—
Malic acid	—	—	24.99	—	—	—	—	—	—	—
Threonic acid	27.7	1587	—	—	—	—	1606	1603	1600	1598
Adipic acid	28.9	1602	—	—	—	1530	1528	1529	1505	1611
3-Methyladipic acid	29.3	—	—	—	—	—	—	—	—	—
2-Hydroxyglutaric acid	29.3	—	27.90	—	—	—	—	—	—	—
2-Deoxyerythropentonic acid	—	1663	—	—	—	—	—	—	—	—
4-Hydroxymethylfuroic acid	—	1689	—	—	—	—	—	—	—	—
3-Hydroxy-3-methylglutaric acid	30.1	—	—	—	—	1625	1606	1631	1621	1735
1,4-Arabinolactone	—	1710	—	—	—	—	—	—	—	—
2-Hydroxyphenylacetic acid	—	—	—	30.2	—	1583	1581	—	1561	1691
3-Hydroxybenzoic acid	30.7	—	—	—	—	—	—	—	—	—
Pyroglutamic acid	30.7	—	28.74	—	—	—	—	—	—	—
Phenyllactic acid	—	—	—	31.1	—	—	—	—	—	—
Arabonic acid	31.1	—	—	—	—	—	—	—	—	—
4-Hydroxybenzoic acid	32.0	—	—	—	—	—	—	—	—	—
Lauric acid	—	—	29.52	—	—	—	—	—	—	—
2-Oxoglutaric acid	—	—	30.38	—	—	—	—	—	—	—
3-Hydroxyphenylacetic acid	32.0	1733	—	31.9	—	1625	1624	1629	1602	1731

Table GC-36 (continued)
PHENOLS AND ORGANIC ACIDS

Arabono-1,4-lactone	—	—	—	—	—	—	1665	1663	1641	—
Tartaric acid	—	—	—	—	—	1685	1686	1689	1705	1710
2-Deoxypentanoic acid	—	—	—	—	—	—	1703	1703	1700	1653
1,4-Xylonolactone	—	1744	—	—	—	—	—	—	—	—
Ribono-1,4-lactone	—	—	—	—	—	1710	1714	1717	1690	1776
Pentonic acid	—	—	—	—	—	—	1807	1812	1814	1759
Pentonic acid	—	—	—	—	—	—	1829	1829	1831	1788
4-Hydroxyphenylacetic acid	33.0	1765	—	33.1	—	1660	1660	1660	1631	1683
β-Glycerophosphoric acid	34.8	—	—	—	—	—	—	—	—	—
Phenylpyruvic acid	—	—	—	35.6	—	—	—	—	—	—
3-Hydroxyphenylpropionic acid	31.1	1846	—	—	—	—	—	—	—	—
4-Hydroxy-3-methoxybenzoic acid	34.8	—	—	37.7	—	—	—	—	—	—
Aconitic acid	—	1854	—	—	—	—	—	—	—	—
1,4-Benzenedicarboxylic acid	34.8	—	—	—	—	—	—	—	—	—
β-1,3-Dihydroxybenzene	34.8	—	—	—	—	—	—	—	—	—
α-Glycerophosphoric acid	35.0	—	—	—	—	—	—	—	—	—
Citric acid	35.8	1886	34.66	41.0	—	1860	1858	1864	1858	1889
4-Hydroxy-3-methoxyphenylacetic acid	36.2	—	—	37.8	—	—	—	—	—	—
Furoylglycine	—	1891	—	—	—	—	—	—	—	—
Myristic acid	—	—	34.96	—	—	—	—	—	—	—
Isocitric acid	—	—	35.10	41.2	—	—	—	—	—	—
Quinic acid	—	—	—	—	—	1906	1908	1907	1912	1860
3(3-Hydroxyphenyl)hydracrylic acid	36.2	1935	—	—	1.12	—	—	—	—	—
Hexono-1,5-lactone	—	—	—	—	—	—	1907	1905	1891	—
Hexono-1,5-lactone	—	—	—	—	—	—	1921	1920	1905	—
1,4-Gluconolactone	—	1949	37.02	—	—	—	1938	1940	—	—
1,5-Gluconolactone	—	1949	—	—	—	1930	1931	1930	1915	—
3,4-diMethoxybenzoic acid	36.9	—	—	—	—	—	—	—	—	—
4-Hydroxy-3-methoxymandelic acid	37.6	—	—	42.5	1.06	—	—	—	—	—
Hippuric acid di TMS	—	1989	—	—	—	—	—	—	—	—
4-Hydroxyphenyllactic acid	—	—	37.45	—	—	—	—	—	—	—

Hexono-1,4-lactone	—	—	—	—	—	—	1967	1966	—	—
Ascorbic acid	—	—	—	—	—	2000	1995	1997	2095	2033
4-Hydroxyphenylhydracrylic acid	38.1	—	—	—	—	—	—	—	—	—
3,4-Dihydroxyphenylpropionic acid	—	2046	—	—	—	—	—	—	—	—
Parmitic acid	—	—	39.98	—	1.24	—	—	—	—	—
Hexanoic acid	—	—	—	—	—	—	2008	2010	2031	1934
Hexuronic acid	39.8	—	—	—	—	—	—	2024	—	2060
Hexuronic acid	—	—	—	—	—	—	2035	2034	—	—
4-Hydroxymandelic acid	—	—	—	39.2	—	—	—	—	—	—
Hippuric acid	42.0	2111	—	39.6	1.00	1870	1871	1873	1807	2110
4-Hydroxy-3-methoxycinnamic acid	42.0	—	—	—	—	—	—	—	—	—
3,4-Dihydroxyphenylacetic acid	—	—	—	40.3	—	—	—	—	—	—
Gluconic acid	—	—	—	—	—	2050	2049	2052	—	1975
Indoleacetic acid	44.0	—	—	—	1.15	—	—	—	—	—
3-Hydroxyhippuric acid	—	—	—	—	1.51	—	—	—	—	—
3,4-Dihydroxycinnamic acid	44.9	—	—	—	—	—	—	—	—	—
Dimethylphenobarbital	—	—	—	—	1.22	—	—	—	—	—
Caffeine	—	—	—	—	1.32	—	—	—	—	—
Hexonic acid	—	—	—	—	—	—	2068	2069	2060	—
Hexuronic acid	—	—	—	—	—	—	2092	2097	2095	—
Uric acid	46.2	—	44.17	—	—	—	—	—	—	—
5-Hydroxyindoleacetic acid	48.7	—	—	52.1	—	—	—	—	—	—
Stearic acid	—	—	44.63	—	1.62	—	—	—	—	—
Oleic acid	—	—	44.63	—	—	—	—	—	—	—
Arachidic acid	—	—	48.96	—	—	—	—	—	—	—
Salicyluric acid dimethyl	—	—	—	—	1.40	—	—	—	—	—
Salicyluric acid monomethyl	—	—	—	—	1.45	—	—	—	—	—
4-Hydroxyhippuric acid	—	—	—	—	1.55	—	—	—	—	—
4-Hydroxyphenobarbital trimethyl	—	—	—	—	1.64	—	—	—	—	—
Cinnamylglycine	—	—	—	—	1.68	—	—	—	—	—
1,2-Benzenedicarboxylic acid di-2′ethyl-hexylester	—	—	—	—	2.22	—	—	—	—	—

[a] Measured on chromatogram.
[b] For retention index calculation, internal standards used were alkanes n-C_{12} to n-C_{16}, n-C_{18} to n-C_{20}, n-C_{22} and n-C_{24}.

Packing P1 = 5% OV-17 on Supercoport (80-100 mesh)
P2 = 10% OV-17 on Gas-Chrom® Q (100-120 mesh)

Table GC-36 (continued)
PHENOLS AND ORGANIC ACIDS

P3 = 10% OV-17 on Chromosorb® Q (80-100 mesh)
P4 = OV-1, capillary
P5 = 6% Dexsil® 300 on Chromosorb® W, HP (80-100 mesh)
P6 = GE SE-52, capillary
P7 = OV-101, capillary
P8 = OV-17 (packed)

REFERENCES

1. **Gates, S. C., Dendramis, N., and Sweeley, C. C.,** Automated profiling of organic acids in human urine. I. Description of methods, *Clin. Chem.*, 24, 1674, 1978.
2. **Fitch, W. L., Anderson, P. J., and Smith, D. H.,** Isolation, identification and quantitation of urinary organic acids, *J. Chromatogr.*, 162, 249, 1979.
3. **Williams, V. P., Ching, D. K., and Cederbaum, S. D.,** Adsorption of organic acids from amniotic fluid and urine onto silica gel before analysis by gas chromatography and combined gas chromatography/mass spectrometry, *Clin. Chem.*, 25, 1814, 1979.
4. **Muskiet, F. J. and Groen, A.,** Urinary excretion of conjugated homovanillic acid, 3,4-dihydroxyphenylacetic acid, p-hydroxyphenylacetic acid, and vanillic acid by persons on their usual diet and patients with neuroblastoma, *Clin. Chem.*, 25, 1281, 1979.
5. **Brown, G. K., Stokke, O., and Jellum, E.,** Chromatographic profile of high boiling point organic acids in human urine, *J. Chromatogr.* 145, 177, 1978.
6. **Lewis, S., Kenyon., C. N., Meili, J., and Burlingame, A. L.,** High resolution gas chromatographic/real-time high resolution mass spectrometric identification of organic acids in human urine, *Anal. Chem.*, 51, 1275, 1979.

Table GC-37
ORGANIC ACIDS

Packing	P1		P2		P3		P4		P5		P6	P7
Column												
Length (cm)	183		183		183		183		183		270	270
Diameter (mm)	4		4		4		4		4		6.2 o.d.	6.2 o.d.
Material	G		G		G		G		G		G	G
Gas	N_2		N_2		N_2		N_2		N_2		N_2	N_2
Flow rate (mℓ/min)	40		40		45		41		41		40	40
Temperature (°C)	140		180		145		180		200		100-(6/min)	
Detector	FI		FI		FI		FI		FI		FI	FI
Reference	1		1		1		1		1		2	2
	methyl		methyl		methyl		methyl		methyl		TMS	TMS
Compound	Tr	min	Tr	min	Tr	min	Tr	min	Tr	min	min[a]	min[a]
3-Hydroxy-4-methoxymandelic acid	0.84		1.02		1.65		2.19		3.64		—	—
4-Hydroxy-3-methoxymandelic acid	0.71		0.81		1.32		1.72		2.73		—	—
Anthraquinone	1.00	16	1.00	10.9	1.00	8.5	1.00	9.8	1.00	7.5	—	—
4-Hydroxybenzoic acid	—		—		—		—		—		12.5	—
4-Hydroxy-3-methoxybenzoic acid	—		—		—		—		—		15.0	—
4-Hydroxy-3,5-dimethoxybenzoic acid	—		—		—		—		—		18.0	—
4-Hydroxycinnamic acid	—		—		—		—		—		18.5	—
4-Hydroxy-3-methoxycinnamic acid	—		—		—		—		—		22.0	—
4-Hydroxy-3,5-dimethoxycinnamic acid	—		—		—		—		—		23.5	—
2,3-Dihydroxybenzoic acid	—		—		—		—		—		—	16.8
2,6-Dihydroxybenzoic acid	—		—		—		—		—		—	17.0
2,5-Dihydroxybenzoic acid	—		—		—		—		—		—	17.5
2,4-Dihydroxybenzoic acid	—		—		—		—		—		—	18.5
3,4-Dihydroxybenzoic acid	—		—		—		—		—		—	18.5
3,5-Dihydroxybenzoic acid	—		—		—		—		—		—	18.5

[a] Measured on chromatogram.

Table GC-37 (continued)
ORGANIC ACIDS

Packing
P1 = 3% OV-1 on GasChrom® Q (100-120 mesh)
P2 = 3% OV-17 on GasChrom® Q (100-120 mesh)
P3 = 3% OV-210 on GasChrom® Q (100-120 mesh)
P4 = 3% OV-225 on GasChrom® Q (100-120 mesh)
P5 = 1% HI-EFF-8BP on GasChrom® Q (100-120 mesh)
P6 = 3% UCW-98 on Chromosorb® W, HP (100-120 mesh)
P7 = 3% OV-1 on Chromosorb® W, HP (100-120 mesh)

REFERENCES

1. **Thomas, H., Lax, E. R., and Ockenfels, H.**, Dünnschift- und gaschromatographische Trennung von 4-Hydroxy-3-methoxy- und 3-Hydroxy-4-methoxy-mandelsäure, *J. Chromatogr.*, 84, 187, 1973.
2. **Morita, H.**, Identification of phenolic acids by gas chromatography-mass spectrometry, *J. Chromatogr.*, 71, 149, 1972.

Table GC-38
AROMATIC HYDROXY ACIDS

Packing	P1	P2	P3	P4	P4	P5	P6	P6
Column								
Length (cm)	180	180	210	180	180	100	170	170
Diameter (mm)	3.5	6.2 o.d.	3.2	4	4	3	3	3
Material	G	ss	ss	G	G	G	G	G
Gas	N_2	N_2	Ar-Me	N_2	N_2	He	N_2	N_2
Flow rate (mℓ/min)	25	18	40	45	45	40	40	40
Temperature (°C)	170	171-230	200	115	145	120	210	260
Detector	FI	FI	EC	EC	EC	MS	EC	EC
Reference	1	2	3	4	4	5	6	6
Derivative	methyl	TMS	D1	D2	D2	D3	D4	D4
Compound	min	min	min[a]	min	min	min[a]	min[a]	min[a]
4-Hydroxy-3-methoxyphenylacetic acid	3.6	—	—	14.3	—	7.5	8.0	—
4-Hydroxy-3-methoxymandelic acid	8.0	24	—	7.7	—	7.0	—	8.0
Methyl-14-methylhexadecanoate	—	34.5	—	—	—	—	—	—

4-Hydroxyphenylacetic acid	—	—	—	4.4	—	—	4.0	—
3,4-Dihydroxyphenylacetic acid	—	—	—	6.2	—	—	—	—
2-Hydroxybenzoic acid	—	—	2.5	—	—	—	—	—
4-Hydroxybenzoic acid	—	—	3.3	—	—	—	—	—
5-Hydroxyindoleacetic acid	—	—	—	—	10	—	—	9.5
3-Hydroxy-4-methoxyphenylacetic acid	—	—	—	—	—	9.0	—	—
2-Hydroxyphenylacetic acid	—	—	—	—	—	—	2.5	—
3-Hydroxyphenylacetic acid	—	—	—	—	—	—	3.5	—
4-Hydroxy-3-methoxybenzoic acid	—	—	4.2	—	—	—	5.5	—
4-Hydroxymandelic acid	—	—	—	—	—	—	—	5.3
4-Hydroxyphenyllactic acid	—	—	—	—	—	—	—	6.5
3,4-Dihydroxybenzoic acid	—	—	4.7	—	—	—	—	—

[a] Measured on chromatogram.

Packing P1 = 3% OF-1 on GasChrom® Q (100-120 mesh)
P2 = 4% SE-30 on Chromosorb® G, HP (80-100 mesh)
P3 = 3% UCW-98 on Chromosorb® W, HP (100-120 mesh)
P4 = 3% OV-17 on Gas Chrom® Q (80-100 mesh)
P5 = 2% OV-1 on Chromosorb® W, AW, DMCS
P6 = 5% OV-17 on GasChrom® P (80-100 mesh)

Derivatization D1 = chloromethylsilyl
D2 = Pentafluoropropionyl-2,2,3,3,3-pentafluoro-1-propionylester
D3 = trifluoroacetylhexafluoroisopropionol ester
D4 = *O*-pentafluorobenzoylmethylester

REFERENCES

1. **Roginsky, M. S., Gordon, R. D., and Bennett, M. J.,** A rapid and simple gas-liquid chromatographic procedure for homovanillic and vanillylmandelic acid in urine, *Clin. Chim. Acta,* 56, 261, 1974.
2. **Lun, A., Hacker, R. R., Brown, R. G., and Hurnik, J. F.,** Determination of vanilmandelic acid in pig urine and chicken feces by gas-liquid chromatography, *Anal. Biochem.,* 73, 267, 1976.
3. **Morita, H. and Montgomery, W. G.,** Gas chromatography of chloromethylsilylated phenolic acids, *J. Chromatogr.,* 123, 454, 1976.
4. **Watson, E. and Wilk, S.,** Derivatization and gas chromatographic determination of some biologically important acids in cerebrospinal fluid, *Anal. Biochem.,* 59, 441, 1974.
5. **Takahashi, S., Toshioka, M., Yoshiue, S., and Tamura, Z.,** Mass fragmentographic determination of vanilmandelic acid, homovanillic acid and isohomovanillic acid in human body fluids, *J. Chromatogr.,* 145, 1, 1978.
6. **Makita, M., Yamamoto, S., Ono, K., and Katagi, T.,** Electron-capture gas chromatography of phenolic acids as *O*-pentafluorobenzoyl methyl esters, *J. Chromatogr.,* 166, 294, 1978.

Table GC-39
AROMATIC ACIDS

Packing	P1	P2
Column		
Length (cm)	3000	210
Diameter (mm)	0.5	6.2 o.d.
Material		G
Gas	N_2	
Flow rate (mℓ/min)		27
Temperature (°C)	90-200 (32/min)	150-(4/min)
Detector		EC
Reference	1	2
Compound	TMS min[a]	pentafluorobenzyl min[a]
2-Hydroxyphenylacetic acid	19.2	—
3-Hydroxyphenylacetic acid	21.0	12.0
4-Hydroxyphenylacetic acid	22.5	12.8
3-Hydroxyphenylpropionic acid	26.0	—
4-Hydroxy-3-methoxyphenylacetic acid	31.9	14.7
4-Hydroxymandelic acid	33.7	—
3-Hydroxyphenylhydroacrylic acid	37.0	—
4-Hydroxyphenyllactic acid	44.0	—
4-Hydroxy-3-methoxymandelic acid	46.0	—
3,4-Dihydroxydihydrocinnamic acid	49.0	—
3,4-Dihydroxymandelic acid	51.5	—
Hippuric acid	38.5	—
Phenylacetic acid	—	5.0
Phenylpropionic acid	—	6.2
4-Hydroxyphenylpropionic acid	—	13.9

[a] Measured on chromatogram.

Packing P1 = SE-52 capillary
P2 = 40% OV-17 on GasChrom® Q

REFERENCES

1. **Goodwin, B. L., Ruthven, C. R. J., Fellows, L. E., and Sandler, M.,** A specific method for the recovery of aromatic acids from biological fluids, *Clin. Chim. Acta,* 73, 191, 1976.
2. **Davis, B.,** Crown ether catalyzed derivatization of carboxylic acids and phenols with pentafluorobenzyl bromide for electron capture gas chromatography, *Anal. Chem.,* 49, 832, 1977.

Table GC-40
AROMATIC ACIDS

Packing	P1	P1	P2
Column			
Length (cm)			240
Diameter (mm)			3.2
Material			
Gas	N_2	N_2	N_2
Flow rate (mℓ/min)	30-50	30-50	27
Temperature (°C)	110-180	110-180	75-220 (2/min)
Detector	FI	FI	FI
Reference	1	1	2

Compound	methyl min	methyl min	TMS min[a]
4-Phenylbutyric acid	6.3	4.7	56.7
4-Phenyl-3-butenoic acid	8.4	9.3	—
4-Phenyl-2-butenoic acid	—	7.9	—
trans-Cinnamic acid	6.3	7.0	—
Phenylacetic acid	6.3	7.0	—
3-Phenylpropionic acid	—	3.7	—
Urea	—	—	37
Succinic acid	—	—	42.5
4-Hydroxyphenylacetic acid	—	—	63
Vanilacetic acid	—	—	70.5
Hippuric acid	—	—	74
Vanilglycolic acid	—	—	77
Ascorbic acid	—	—	80.2

[a] Measured on chromatogram.

Packing P1 = 3% SE-30 on GasChrom® Q (100-120 mesh)
P2 = 5% SE-52 on Chromosorb® W, AW,DMCS (100-120 mesh)

REFERENCES

1. **Sariaslani, F. S., Harper, D. B., and Higgins, J. I.,** Microbial degradation of hydrocarbons. Catabolism of 1-phenylalkans, by Nocardia salmonicolor, *Biochem. J.*, 140, 31, 1974.
2. **Wadman, S. K., Ketting, D., and Voute, P. A.,** Gas chromatographic determination of urinary vanilglycolic acid, vanilglycol, vanilacetic acid and vanillactic acid. Chemical parameters for the diagnosis of neurogenic tumours and the evaluation of their treatment, *Clin. Chim. Acta,* 72, 49, 1976.

Table GC-41
DI- AND TRICARBOXYLIC BENZOIC ACIDS

Packing	P1	P2	P3	P4
Column				
Length (cm)	180	360	120	150
Diameter (mm)	4	6	6	3.2 o.d.
Material	G	ss	ss	
Gas	N_2	N_2	N_2	N_2
Flow rate (mℓ/min)	40	45	45	40
Temperature (°C)	90-310 (7.5/min)	200	210	100-250(8/min)
Detector	FI	FI	FI	FI
Reference	1	2	2	3
Compound	methyl	methyl	methyl	methyl
	min[a]	Tr	Tr	Tr
1,2-Benzenedicarboxylic acid	15.0	—	—	—
1,3-Benzenedicarboxylic acid	15.3	—	—	—
3,4,5-Trimethoxyphenylacetic acid	19.5	—	—	—
3,4,5-Trimethoxybenzoic acid	19.7	—	—	—
N-Methylbenzylsulfoamide	21.0	—	—	—
1,2,3-Benzenetricarboxylic acid	21.7	—	—	—
1,2,4-Benzenetricarboxylic acid	21.9	—	—	—
1,3,5-Benzenetricarboxylic acid	22.3	1.00	1.00	—
1,2,3,5-Benzenetetracarboxylic acid	27.3	—	—	—
Dimethoxybenzenetetracarboxylic acid	30.0	—	—	—
Benzenepentacarboxylic acid	31.2	—	—	—
Acetobenzenepentacarboxylic acid	32.2	—	—	—
Benzenehexacarboxylic acid	34.3	—	—	—
Ethyl-1,3,5-benzenetricarboxylate	—	1.95	1.08	—
Ethyl-*cis*-1,3,5-cyclohexanetricarboxylate	—	1.39	0.90	—
Ethyl-*trans*-1,3,5-cyclohexanetricarboxylate	—	1.13	0.56	—
1,3,5-Tris(hydroxymethyl)benzenetriacetate	—	1.58	2.75	—
1,3,5-Tris(hydroxymethyl)cyclohexanetriacetate	—	1.53	1.55	—
cis-1,3,5-Cyclohexanetricarboxylic acid	—	0.73	0.86	—
trans-1,3,5-Cyclohexanetricarboxylic acid	—	0.62	0.56	—
1-Bromomethyl-3,5-hydroxymethylcyclohexanediacetate	—	1.41	1.42	—

1,3-Bromomethyl-5-hydroxymethylcyclohexanediacetate	—	1.32	1.29	—
1,3,5-Tris(bromomethyl)cyclohexane	—	1.22	1.18	—
1,3,5-Tris(bromomethyl)benzene	—	1.35	2.24	—
1,3,5-Tris(hydroxymethyl)benzene TMS	—	1.12	1.79	—
1,3,5-Tris(hydroxymethyl)cyclohexane TMS	—	0.85	1.79	—
Phthalaldehydecarboxylic acid	—	—	—	0.90
Biphenyl-2,2′-dicarboxylic acid	—	—	—	2.26
Quinoxalinedicarboxylic acid	—	—	—	2.27
Quinolinedicarboxylic acid	—	—	—	2.29
Pyridinealdehydetricarboxylic acid	—	—	—	2.56
Biphenyltetracarboxylic acid	—	—	—	3.53
Anthraquinonedicarboxylic acid	—	—	—	3.70
Biphenylhexacarboxylic acid	—	—	—	4.18
1,4-Benzenedicarboxylic acid	—	—	—	1.00
Benzenetetracarboxylic acid	—	—	—	2.75

[a] Measured on chromatogram.

Packing P1 = 3% OV-17 on Chromosorb® W, HMDS (60-80 mesh)
P2 = 3% SE-30 on Chromosorb® G, silanized (60-80 mesh)
P3 = 2% butanediolsuccinate on Chromatoron N, AW, HMDS (0.16-0.20 mm)
P4 = 3% SE-30 on Chromosorb® W.

REFERENCES

1. **Schnitzer, M. and Ortiz de Serra, M. I.**, The chemical degradation of a humic acid, *Can. J. Chem.*, 51, 1554, 1973.
2. **Šafář, M., Kafka, Z., Galik, V., and Landa, S.**, Gas chromatographic analysis of 1,3,5-benzenetricarboxylic and 1,3,5-cyclohexanetricarboxylic acid and their derivatives, *Collect. Czech. Chem. Commun.*, 39, 3390, 1974.
3. **Meineke, I. and Klamberg, H.**, Zum Abbau von polycyclischen aromatischen Kohlenwasserstoffen. II. Gas-chromatographisch-massenspectrometrische Identifizierung von aromatischen Polycarbonsäuren, *Z. Anal. Chem.*, 293, 205, 1978.

Table GC-42
ORGANIC ACIDS

Packing	P1	P2	P3
Column			
Length (cm)	120	150	180
Diameter (mm)	3.2	3.2	2
Material		ss	G
Gas	He	N_2	Ar-Me
Flow rate (ml/min)		30	30
Temperature (°C)	55-195 (6/min)	130	235
Detector	FI	FI	EC
Reference	1	2	3

Compound	methyl min[a]	methyl min	pentafluorobenzyl min
3-Cyclohexane-1-carboxylic acid	3	—	—
Phenylacetic acid	8	—	—
2,4,6-Trimethylbenzoic acid	10	—	—
Linolic acid	19	—	—
11-Oxopalmitic acid	22	—	—
6-Hydroxyhexanoic acid	—	5.5	—
Benzoic acid	—	7.0	—
Cyclohexa-1,4-dienecarboxylic acid	—	5.75	—
Cyclohexa-2,5-dienecarboxylic acid	—	5.0	—
Cyclohex-1-enecarboxylic acid	—	4.0	—
Cyclohex-3-enecarboxylic acid	—	3.0	—
Cyclohexanecarboxylic acid	—	3.0	—
2-Hydroxycyclohexane carboxylic acid	—	24.0	—
Heptanedioic acid	—	26.0	—
Hexanedioic acid	—	16.0	—
2,2-Diphenylglycolic acid	—	—	5.95
2-Cyclohexyl-2-phenylacetic acid	—	—	3.53
2-Cyclohexyl-2-phenylglycolic acid	—	—	4.61
2,2-Diphenylacetic acid	—	—	4.75
3-Hydroxy-2-phenylpropionic acid	—	—	5.05

[a] Measured on chromatogram.

Packing P1 = 7.5% ethyleneglycol adipate and 2% H_3PO_4 on Anakrom ABS (90-100 mesh)
P2 = 10% Carbowax 20M on Chromosorb® W, AW, DSCS (100-120 mesh)
P3 = 3% OV-17 on Chromosorb® W, HP (80-100 mesh)

REFERENCES

1. **Schwartz, D. P. and Bright, R. S.,** A column procedure for the esterification of organic acids with diazomethane at the microgram level, *Anal. Biochem.*, 61, 271, 1974.
2. **Williams, R. J. and Evans, W. C.,** The metabolism of benzoates by Moraxella species through anaerobic nitrate respiration. Evidence for a reductive pathway, *Biochem. J.*, 148, 1, 1975.
3. **Greving, J. E., Jonkman, J. H. G., and De Zeeuw, R. A.,** Determination of carboxylic acids in the picomole range after derivatization with pentafluorobenzyl bromide and electron capture gas chromatography, *J. Chromatogr.*, 148, 389, 1978.

Table GC-43
SUGAR ACIDS

Packing	P1	
Column		
Length (cm)	6000	
Diameter (mm)	0.3	
Material	G	
Gas	He	
Flow rate	18-20 cm/sec[b]	
Temperature (°C)	165-(1/min)	
Detector	FI	
Reference	1	1
Compound	TMS min[a]	TMS min[a]
L-Threonic acid lactone	6.0	14.0
D-Erythronic acid lactone	7.5	13.0
D-Arabonic acid lactone	17.0	19.5
L-Xylonic acid lactone	18.0	—
D-Ribonic acid lactone	20.5	—
D-Lyxonic acid lactone	13.5	19.5
D-Gluconic acid lactone	36.5	37.5
D-Galactonic acid lactone	37.0	—
D-Talonic acid lactone	39.5	—
L-Mannonic acid lactone	43.5	—
D-Glycero-D-guloheptonic acid lactone	57.5	—

[a] Measured on chromatogram.
[b] Split ratio 5:1.

Packing P1 = SE-30 containing Silanox® 101, capillary

REFERENCE

1. **Szafranek, J., Pfaffenberger, C. D., and Hornig, E. C.,** Separation of aldonic, deoxyaldonic, hexuronic and hexaric lactones and acids using thermostable open-tubular glass capillary columns, *J. Chromatogr.*, 88, 149, 1974.

Table GC-44
ORGANIC ACIDS

Packing	P1	P2	P2	P2
Column				
Length (cm)	4500	200	200	200
Diameter (mm)	0.5	4	4	4
Material		G	G	G
Gas	He	N_2	N_2	N_2
Flow rate (mℓ/min)	4	45	45	45
Temperature (°C)	90—160 (1/min)	210	220	200
Detector		FI	FI	FI
Reference	1	2	2	2
Compound	S-(+)-2-butyl min[a]	min[a]	min[a]	min[a]
S-(−)-Methylsuccinic acid	18	—	—	—
R-(+)-Methylsuccinic acid	18.5	—	—	—
Succinic acid	20	—	—	—
5-Methyl-3-phenylisoxazole-4-trimethylsilylester	—	2.0	—	—
5-Methyl-3(2-chlorophenyl)isoxazoly-4-trimethylsilylester	—	2.7	—	—
Hexacosane	—	3.7	—	—
5-Methyl-3-phenylisoxazoly-4*N,N′*-diethylamide	—	8.2	—	—
5-Methyl-3-(2,6-dichlorophenyl)isoxazole-4-trimethylsilylester	—	—	2.6	—
Triacontane	—	—	6.6	—
5-Methyl-3-(2,6-dichlorophenyl)isoxazole-4-*N,N′*-diethylamine	—	—	13.2	—
2,6-Dimethoxybenzene-1-trimethylsilylester	—	—	—	2.4
Tetracosane	—	—	—	3.7
2,6-Dimethoxybenzene-1-*N,N′*-diethylamine	—	—	—	7.1
5-Methyl-3-(2-chlorophenyl)isoxazol-4-*N,N′*-diethylamide	—	13.8	—	—

[a] Measured on chromatogram.

Packing P1 = DEGS, capillary WCOT
P2 = 4% OV-225 on GasChrom® Q (80-100 mesh)

REFERENCES

1. **Zeitman, B. and Lawless, J. G.,** The optical nature of methylsuccinic acid in human urine, *Biochem. Med.*, 13, 111, 1975.
2. **Lauback, R. G., Balitz, D. F., and Mays, D. L.,** Gas Chromatographic determination of carboxylic acid chloride and residual carboxylic acid precursors used in the production of some penicillins, *J. Chromatog. Sci.*, 14, 240, 1976.

Table GC-45
ORGANIC ACIDS

Packing	P1	P2	P3
Column			
Length (cm)	2000	200	180
Diameter (mm)	0.5	3	2
Material	G	G	G
Gas		He	He
Flow rate (mℓ/min)	1	60	30
Temperature (°C)	65—260 (4/min)	140	190
Detector	FI	FI	FI
Reference	1	2	3
Compound	min	min	min
Pyruvic acid methoxine silylester	5.1	—	—
Oxalacetic acid methoxime disilylester	12.6	—	—
2-Oxalacetic acid methoxime disilylester	14.6	—	—
Trimethylsilyl-2-aminobenzoic acid	—	1.2	—
4-Bromobiphenyl	—	2.2	—
Four isomers of 3,4-Dihydro-9,10-dimethoxycarbonyl-4-methyl-1-phenyl-1,4-ethano-1H-2-benzopyran	—	—	27.95, 30.34 34.85, 40.76

Packing P1 = 10 OV-101 capillary
P2 = 3% XE-60 on GasChrom® Q (100-120 mesh)
P3 = Carbowax® 20M (Aue packing)

REFERENCES

1. **Andersson, I., Norkrans, B., and Odham, G.,** Determination of the Krebs cycle related keto acids of microorganisms by capillary gas chromatography on glass columns, *Anal. Biochem.*, 53, 629, 1973.
2. **Douglas, J. F. and Philopoulous, Y.,** Determination of anthranilic acid in plasms, *J. Pharm. Sci.*, 66, 1645, 1977.
3. **Karasek, F. W., Tchir, M. F., and Bowers, W. D.,** Separation and identification of cis-trans isomerie dimethylesters by a gas chromatography-mass spectrometry-calculator system, *J. Chromatogr.*, 151, 321, 1978.

Table GC-46
ORGANIC ACIDS

Packing	P1		P1	P2		P2	P3		P3	P4	P4	P5	P5	P6	P6	P7	P7	P8	P8	P9		P9	P10	P11
Column																								
Length (cm)	180		180	300		300	180		180			180	180	180	180	240	240	330	330	300		300	4500	4500
Diameter (mm)	3.1		3.1	3.1		3.1	3.1		3.1			6.2	6.2	3.1	3.1	3.1	3.1	3.1	3.1	3.1		3.1	0.25	0.25
Material	ss		ss	ss		ss	ss		ss	ss	ss	ss	ss	ss	ss	ss	ss	ss	ss	ss		ss		
Gas	N_2		N_2	N_2		N_2	N_2		N_2	N_2	N_2	N_2	N_2	N_2	N_2	N_2	N_2	N_2	N_2	N_2		N_2		
Flow rate (mℓ/min)	90		90	35		35	40		40	35	35	70	70	20	20	50	50	20	20	17		17		
Temperature (°C)	200		200	200		200	200		200	200	200	200	200	200	200	200	200	200	200	200		200	200	200
Detector	FI		FI	FI		FI	FI		FI	FI	FI	FI	FI	FI	FI	FI	FI	FI	FI	FI		FI	FI	FI
Reference	1		1	1		1	1		1	1	1	1	1	1	1	1	1	1	1	1		1	2	2
	methyl		t-butyl	methyl		t-butyl	methyl		t-butyl	methyl	t-butyl	methyl	t-butyl	methyl	t-butyl	methyl	t-butyl	methyl	t-butyl	methyl		t-butyl	methyl	methyl
Compound	Tr	min	Tr	Tr	min	Tr	Tr	min	Tr	Tr	Tr	Tr	Tr	Tr	Tr	Tr	Tr	Tr	Tr	Tr	min	Tr	Tr	Tr
Pimaric acid	1.00	10.4	1.00	1.00	8.6	1.00	1.00		1.00	1.00	1.00	1.00	1.00	1.00	1.00	1.00	1.00	1.00	1.00	1.00	15	1.00	1.00	1.00
Sandaracopimaric acid	1.06		1.05	1.08		1.09	1.07		1.08	1.09	1.10	1.08	1.09	1.10	1.09	1.12	1.11	1.15	1.14	1.15		1.15	1.09	1.10
Levopimaric acid	1.22		1.24	1.21		1.21	1.33		1.32	1.38	1.37	1.29	1.30	1.34	1.33	1.33	1.33	1.42	1.40	1.38		1.46	—	—
Palustric acid	1.22		1.17	1.17		1.11	1.31		1.31	1.39	1.37	1.27	1.22	1.36	1.33	1.34	1.31	1.36	1.31	1.29		1.27	1.36	1.33
Isopalustric acid	1.20		1.09	1.28		1.16	1.28		1.17	1.39	1.23	1.29	1.19	1.39	1.22	1.46	1.25	1.51	1.32	1.54		1.35	—	—
Abietic acid	1.55		1.41	1.60		1.41	1.74		1.63	2.08	1.91	1.75	1.41	2.01	1.83	2.08	1.86	2.13	1.87	2.14		1.92	1.98	2.06
Dehydroabietic acid	1.34		1.27	1.39		1.34	1.67		1.62	2.01	1.98	1.69	1.65	2.04	1.92	2.25	2.15	2.48	2.30	2.56		2.37	1.58	1.80
Neoabietic acid	1.84		1.82	1.77		1.77	2.17		2.20	2.57	2.55	2.12	2.12	2.40	2.38	2.46	2.39	2.53	2.42	2.54		2.49	2.30	2.39
Isopimaric acid	—		—	—		—	—		—	—	—	—	—	—	—	—	—	—	—	—		—	1.36	1.43

Packing P1 = 9% SE-30 EGiP on Chromosorb® W, AW (80-100 mesh)
P2 = 3% SP-2401 on Chromosorb® W, AW (80-100 mesh)
P3 = 3% OV-17 on Chromosorb® W, AW (80-100 mesh)
P4 = 3% polyphenylether (PPE) on Chromosorb® W, AW
P5 = 5% OV-225 on Chromosorb® W, AW (80-100 mesh)
P6 = 5% Carbowax® 20M on Chromosorb® W, AW (80-100 mesh)
P7 = 10% DEGS on Chromosorb® W, AW (80-100 mesh)
P8 = 10% Silar-10 on Chromosorb® W, AW (80-100 mesh)
P9 = 20% OV-275 on Chromosorb® W, AW (80-100 mesh)
P10 = Apiezon N, capillary
P11 = Versamide-930, capillary

REFERENCES

1. **Zinkel, D. F. and Engler C. C.,** Gas-liquid chromatography of resin acid esters, *J. Chromatogr.*, 136, 245, 1977.
2. **Hasson, J. C. and Kulkarni, V.,** Open-tube column gas chromatography of rosin fluxes, *Anal. Chem.*, 44, 1586, 1972.

Table GC-47
PYRAZINOIC ACID

Packing	P1
Column	
Length (cm)	200
Diameter (mm)	3
Material	G
Gas	N_2
Flow rate (mℓ/min)	50
Temperature (°C)	150
Detector	FI
Reference	1

Compound	Tr
Pyrazinoic acid methylester	1.00
Pyrazinoic acid ethylester	1.20
Pyrazinoic acid propylester	1.90
Pyrazinoic acid isopropylester	1.12

Packing P1 = 5% Versamid-900 + 3% Carbowax® 20M on silanized Chromosorb® W (80-100 mesh)

REFERENCE

1. **Felder, E., Tiepolo, U., and Mengassini, A.**, Method for the esterification of carboxylic acids in gas chromatographic analysis, *J. Chromatogr.*, 82, 291, 1973.

Table GC-48
CARBOHYDRATE-RELATED ACIDS

Packing	P1	P2	P3	P3
Column				
Length (cm)	200	200	200	200
Diameter (mm)	2	2	2	2
Material	ss	ss	ss	ss
Gas	N_2	N_2	N_2	N_2
Flow rate (mℓ/min)	30	30	30	30
Temperature (°C)	160	160	120	160
Detector	FI	FI	FI	FI
Reference	1	1	1	1

Compound	TMS Tr[a]	TMS Tr[b]	TMS Tr[c]	TMS Tr[d]
Glyceronic acid	0.052	0.092	0.086	—
2-*C*-Methylglyceronic acid	0.054	0.082	0.072	—
2-Deoxytetronic acid	0.084	0.146	0.145	—
3-Deoxytetronic acid	0.078	0.136	0.129	—
4-Deoxyerythronic acid	0.058	0.094	0.091	—
4-Deoxythreonic acid	0.061	0.099	0.092	—
Erythronic acid	0.151	0.218	0.212	—

Table GC-48 (continued)
CARBOHYDRATE-RELATED ACIDS

Compound	TMS Tr[a]	TMS Tr[b]	TMS Tr[c]	TMS Tr[d]
Threonic acid	0.156	0.251	0.237	—
3,4-Dideoxypentonic acid	0.114	0.211	0.206	—
3,5-Dideoxy-*erythro*-pentonic acid	0.075	0.125	0.118	—
3,5-Dideoxy-*threo*-pentonic acid	0.080	0.140	0.133	—
3-Deoxy-2-*C*-(hydroxymethyl)tetronic acid	0.224	0.326	0.319	—
2-*C*-Methylerythronic acid	0.184	0.246	0.244	—
2-Deoxy-*erythro*-pentonic acid	0.252	0.379	0.395	—
2-Deoxy-*threo*-pentonic acid	0.251	0.373	0.392	—
3-Deoxy-*erythro*-pentonic acid	0.236	0.380	0.375	—
3-Deoxy-*threo*-pentonic acid	0.252	0.419	0.418	—
Ribonic acid	0.438	0.569	0.565	—
Arabinonic acid	0.463	0.646	0.632	—
Xylonic acid	0.428	0.617	0.570	—
Lyxonic acid	0.457	0.599	0.621	—
2,4-Dideoxy-3-*C*-methylpentonic acid	0.147	0.260	0.266	—
2,6-Dideoxy-*ribo*-hexonic acid	0.286	0.402	0.425	—
3,4-Dideoxyhexonic acid	0.400	0.676	0.720	—
3,6-Dideoxy-*ribo*-hexonic acid	0.266	0.396	0.403	—
3,6-Dideoxy-*arabino*-hexonic acid	0.275	0.425	0.442	—
3-Deoxy-2-*C*-hydroxymethyl-*erythro*-pentonic acid	0.707	0.942	1.003	—
3-Deoxy-2-*C*-hydroxymethyl-*threo*-pentonic acid	0.670	0.896	0.933	—
2-*C*-Methylribonic acid	0.564	0.689	0.742	—
2-*C*-Methylarobinoic acid	0.554	0.674	0.710	—
2-Deoxy-*arabino*-hexonic acid	0.659	0.912	1.007	—
2-Deoxy-*lyxo*-hexonic acid	0.673	0.935	1.035	—
3-Deoxy-*ribo*-hexonic acid	0.673	0.946	1.007	—
3-Deoxy-*arabino*-hexonic acid	0.676	0.971	1.004	—
3-Deoxy-*xylo*-hexonic acid	0.659	0.995	1.026	—
3-Deoxy-*lyxo*-hexonic acid	0.644	0.894	0.973	—
6-Deoxymannonic acid	0.528	0.629	0.687	—
6-Deoxygalactonic acid	0.615	0.809	0.801	—
3-*O*-Methylgulonic acid	0.728	1.038	0.990	—
2-*C*-(Hydroxymethyl)pentonic acid	1.118	1.235	1.333	—
Allonic acid	1.115	1.276	1.410	—
Altronic acid	1.247	1.531	1.568	—
Gluconic acid	1.286	1.569	1.639	—
Mannonic acid	1.113	1.254	1.385	—
Gulonic acid	1.087	1.256	1.309	—
Idonic acid	1.368	1.743	1.812	—
Galactonic acid	1.256	1.561	1.570	—
Talonic acid	1.235	1.507	1.656	—
2,5-Anhydro-3,4-dideoxypentonic acid	0.087	0.325	0.465	—
1,4-Anhydro-3-deoxypentitol-2-carboxylic acid	0.206	0.436	0.456	—
2,5-Anhydrogluconic acid	0.612	1.271	1.928	—
2,5-Anhydromannonic acid	0.477	0.986	1.416	—
2,5-Anhydrotalonic acid	0.590	1.241	1.776	—
Glyceraric acid	0.064	0.152	0.190	—
C-Methylglyceraric acid	0.058	0.101	0.110	—
Dexoytetraric acid (malic acid)	0.103	0.212	0.283	—
O-Methylerythraric acid	0.134	0.313	0.372	—
O-Methylthrearic acid	0.146	0.358	0.478	—
C-(Hydroxymethyl)glyceraric acid	0.152	0.261	0.273	—
Erythraric acid	0.174	0.313	0.354	—
Threaric acid	0.212	0.432	0.471	—

Table GC-48 (continued)
CARBOHYDRATE-RELATED ACIDS

Compound	TMS Tr[a]	TMS Tr[b]	TMS Tr[c]	TMS Tr[d]
C-Ethylglyceraric acid	0.080	0.142	0.136	—
2-Deoxy-3-*C*-methyltetraric acid	0.100	0.185	0.213	—
2,3-Dideoxypentaric acid	0.152	0.333	0.489	—
2,4-Dideoxypentaric acid	0.152	0.331	0.467	—
C-(2-Hydroxymethyl)tetraric acid	0.250	0.444	0.491	—
2-Deoxy-3-*C*-(hydroxymethyl)tetraric acid	0.271	0.466	0.582	—
2-Deoxy-*erythro*-pentaric acid	0.278	0.517	0.655	—
2-Deoxy-*threo*-pentaric acid	0.291	0.573	0.736	—
3-Deoxy-*erythro*-pentaric acid	0.280	0.538	0.673	—
3-Deoxy-*threo*-pentaric acid	0.295	0.593	0.774	—
2-*O*-Methylxylaric acid	0.364	0.805	0.905	—
3-*O*-Methylarabinaric acid	0.357	0.744	0.840	—
Ribaric acid	0.532	0.873	1.037	—
Arabinaric acid	0.522	0.864	0.945	—
Xylaric acid	0.545	0.978	1.034	—
2,4-Dideoxy-3-*C*-methylpentaric acid	0.173	0.356	0.461	—
2,3,4-Trideoxyhexaric acid	0.235	0.539	0.823	—
C-(3-Hydroxypropyl)glyceraric acid	0.348	0.657	0.693	—
3,4-Dideoxyhexaric acid	0.471	0.982	1.401	—
C-(2,3-Dihydroxypropyl)glyceraric acid	0.767	1.278	1.368	—
3-Deoxy-2-*C*-hydroxymethyl-*erythro*-pentaric acid	0.816	1.357	1.940	—
3-Deoxy-2-*C*-hydroxymethyl-*threo*-pentaric acid	0.737	1.179	1.504	—
2-Deoxy-*arabino*-hexaric acid	0.777	1.277	1.565	—
2-Deoxy-*lyxo*-hexaric acid	0.847	1.467	1.751	—
3-Deoxy-*ribo*-hexaric acid	0.732	1.290	1.555	—
3-Deoxy-*arabino*-hexaric acid	0.767	1.425	1.675	—
3-Deoxy-*xylo*-hexaric acid	0.761	1.317	1.639	—
3-Deoxy-*lyxo*-hexaric acid	0.789	1.470	1.713	—
3-*O*-Methylgularic acid	0.909	1.610	1.770	—
C-(Erythro-1,2,3-trihydroxypropyl)glyceraric acid	1.310	1.801	1.762	—
C-(Threo-1,2,3-trihydroxypropyl)glyceraric acid	1.377	1.904	1.822	—
Allaric acid	1.242	1.669	1.904	—
Altraric acid	1.419	2.098	2.195	—
Glucaric acid	1.295	1.814	1.920	—
Mannaric acid	1.074	1.346	1.401	—
Idaric acid	1.602	2.518	2.589	—
Galactaric acid	1.500	2.272	2.347	—
Shikimic acid	0.482	0.990	1.050	—
Quinic acid	0.668	0.906	0.852	—
(Hydroxyethanoyl)hydroxyethanoic acid	0.063	0.187	0.237	—
2-Hydroxypropanol-2-hydroxypropanoic acid	0.061	0.153	0.183	—
3-Hydroxypropanol-3-hydroxypropanoic acid	0.142	0.432	0.709	—
2-Hydroxybutanoyl-2-hydroxybutanoic acid	0.113	0.281	0.326	—
Ethanedioic acid (oxalic acid)	0.018	0.046	0.077	—
Propanedioic (malonic acid)	0.025	0.064	0.092	—
Methylpropanedioic acid	0.029	0.066	0.083	—
Butanedioic acid (glutaric acid)	0.042	0.109	0.145	—
Dimethylpropanedioic acid	0.027	0.056	0.071	—
Ethylpropanedioic acid	0.039	0.088	0.112	—
Methylbutanedioic acid	0.045	0.106	0.143	—
Pentanedioic acid (glutaric acid)	0.064	0.167	0.242	—
2,2-Dimethylbutanedioic acid	0.047	0.103	0.127	—
meso-2,3-Dimethylbutanedioic acid	0.057	0.121	0.167	—

Table GC-48 (continued)
CARBOHYDRATE-RELATED ACIDS

Compound	TMS Tr[a]	TMS Tr[b]	TMS Tr[c]	TMS Tr[d]
2-Methylpentanedioic acid	0.070	0.173	0.239	—
3-Methylpentanedioic acid	0.072	0.180	0.257	—
Hexanedioic acid (adipic acid)	0.101	0.272	0.419	—
3,3-Dimethylpentanedioic acid	0.078	0.177	0.227	—
Heptanedioic acid	0.158	0.441	0.677	—
Octanedioic acid	0.245	0.708	1.075	—
Nonanedioic acid	0.384	1.117	1.718	—
cis-Butanedioic acid (maleic acid)	0.041	0.121	0.138	—
trans-Butanedioic acid (fumaric acid)	0.049	0.108	0.196	—
cis-Methylbutanedioic acid	0.050	0.148	0.165	—
Methylenebutanedioic acid	0.048	0.132	0.159	—
Propane-1,2,3-tricarboxylic acid	0.299	0.823	1.569	—
trans-Propane-1,2,3-tricarboxylic acid	0.323	0.915	1.656	—
1-Hydroxypropane-1,2,3-tricarboxylic acid	0.497	1.162	1.880	—
2-Hydroxypropane-1,2,3-tricarboxylic acid (citric acid)	0.488	1.060	1.461	—
Phosphoric acid	0.037	0.088	0.197	—
2-Deoxytetronic acid 1,4-lactone	0.030	0.116	—	0.57
3-Deoxytetronic acid 1,4-lactone	0.024	0.100	—	0.31
Erythronic acid 1,4-lactone	0.078	0.271	—	0.92
Threonic acid 1,4-lactone	0.063	0.173	—	0.56
3,5-Deoxy-*erythro*-pentonic acid 1,4-lactone	0.024	0.088	—	0.32
3,5-Deoxy-*threo*-pentonic acid 1,4-lactone	0.026	0.104	—	0.36
3-Deoxy-2-*C*-(hydroxymethyl)tetronic acid 1,4-lactone	0.065	0.184	—	0.49
2-C-Methylerythronic acid 1,4-lactone	0.076	0.202	—	0.68
2-Deoxy-*erythro*-pentonic acid 1,4-lactone	0.110	0.33	—	1.48
2-Deoxy-*threo*-pentonic acid 1,4-lactone	0.124	0.44	—	1.81
3-Deoxy-*erythro*-pentonic acid 1,4-lactone	0.100	0.34	—	0.98
3-Deoxy-*threo*-pentonic acid 1,4-lactone	0.108	0.38	—	1.02
Ribonic acid 1,4-lactone	0.252	0.64	—	2.06
Arabinonic acid 1,4-lactone	0.202	0.49	—	1.13
Xylonic acid 1,4-lactone	0.218	0.54	—	1.58
Lyxonic acid 1,4-lactone	0.321	0.83	—	2.47
2,6-Dideoxy-*ribo*-hexaonic acid 1,4-lactone	0.125	0.37	—	1.55
3,6-Dideoxy-*arabino*-hexonic acid 1,4-lactone	0.115	0.36	—	0.93
3-Deoxy-2-*C*-hydroxymethyl-*erythro*-pentonic acid 1,4-lactone	0.257	0.60	—	1.29
3-Deoxy-2-*C*-hydroxymethyl-*threo*-pentonic acid 1,4-lactone	0.272	0.64	—	1.45
2-*C*-Methylribonic acid 1,4-lactone	0.211	0.45	—	1.24
2-*C*-Methylarabinonic acid 1,4-lactone	0.208	0.42	—	1.01
2-Deoxy-*arabino*-hexonic acid 1,4-lactone	0.42	1.28	—	3.64
2-Deoxy-*lyxo*-hexonic acid 1,4-lactone	0.39	1.00	—	4.17
3-Deoxy-*ribo*-hexonic acid 1,4-lactone	0.38	0.92	—	2.40
α-Deoxy-*arabino*-hexonic acid 1,4-lactone	0.37	1.01	—	2.24
3-Deoxy-*xylo*-hexonic acid 1,4-lactone	0.40	1.13	—	2.26
3-Deoxy-*lyxo*-hexonic acid 1,4-lactone	0.41	1.09	—	2.99
6-Deoxymannoic acid 1,4-lactone	0.38	0.94	—	2.57
6-Deoxygalactonic acid 1,4-lactone	0.25	0.56	—	1.31
Allonic acid 1,4-lactone	0.81	1.70	—	4.80
Altronic acid 1,4-lactone	0.62	1.19	—	2.28
Gluconic acid 1,4-lactone	0.71	1.48	—	3.19
Mannoic acid 1,4-lactone	1.01	2.21	—	4.64
Gulonic acid 1,4-lactone	0.78	1.72	—	3.58

Table GC-48 (continued)
CARBOHYDRATE-RELATED ACIDS

Compound	TMS Tr[a]	TMS Tr[b]	TMS Tr[c]	TMS Tr[d]
Idonic acid 1,4-lactone	0.79	1.63	—	4.31
Galactonic acid 1,4-lactone	0.69	1.37	—	2.82
Talonic acid 1,4-lactone	0.80	1.70	—	4.87
Xylonic acid1,5-lactone	0.26	0.67	—	1.79
Gluconic acid 1,5-lactone	0.67	1.49	—	2.99
Mannoic acid 1,5-lactone	0.65	1.35	—	3.27
erythro-Hex-2-enono-1,4-lactone	0.91	2.02	—	4.25
threo-Hex-2-enono-1,4-lactone (L-form = L-ascorbic acid)	0.88	2.01	—	4.72

[a] Retention time of glucitol (Tr = 1.000) is 12 min.
[b] Retention time of glucitol (Tr = 1.000) is 7 min.
[c] Retention time of gluciton (Tr = 1.000) is 15 min.
[d] Retention time of gluciton (Tr = 1.00) is 3 min.

Packing P1 = 0.5% OV-1 on Chromosorb® G (100-120 mesh)
P2 = 0.5% OV-17 on Chromosorb® G (100-120 mesh)
P3 = 3% DC QF-1 on GasChrom® Q (100-120 mesh)

From Petersson, G., *J. Chromatogr. Sci.*, 15, 245, 1977. With permission.

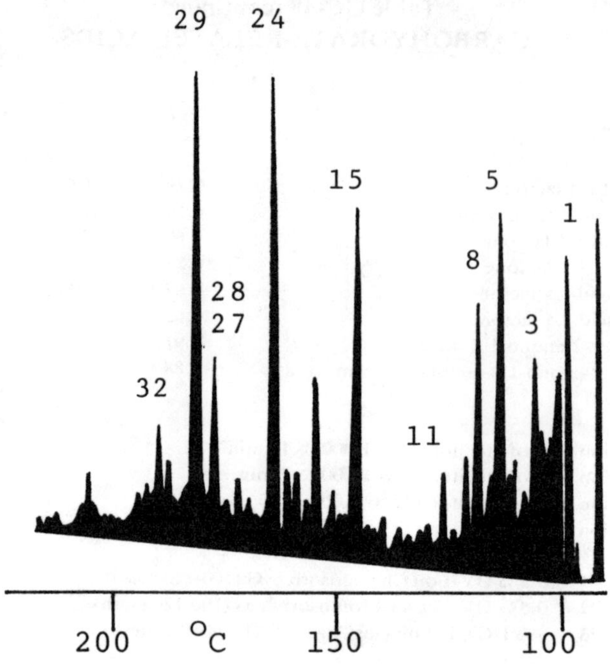

FIGURE GC-F1. Urinary organic acids as TMS derivatives. Packing = 3% OV-1 on Chromosorb® W, HP (100-120 mesh). Column = 200 cm long, 6 mm i.d., glass column. Temperature = 90 − 280 ° C (5/min). Detector = MS. Compound: 1 = silylation reagent; 2 = Lactic acid; 3 = 2-Hydroxy-isobutanoic acid; 4 = 2-Hydroxybutanoic acid; 5 = 3-Hydroxy-isobutanoic acid; 6 = 3-Hydroxybutanoic acid; 7 = 3-Hydroxy-2-methylbutanoic acid; 8 = 3-Hydroxy-isovaleric acid; 9 = 2-Ethylhydracrylic acid; 10 = 2-Methylacetoacetic acid; 11 = Ethylmalonic acid; 12 = Succinic acid; 13 = Methylsuccinic acid; 14 = Mesaconic acid; 15 = Capric acid, (internal standard); 16 = Adipic acid; 17 = 3-Methyladipic acid; 18 = 5-Hydroxymethyl-2-furoic acid; 19 = *N*-Triglycine; 20 = 2-Hydroxyphenylacetic acid; 21 = Hexanedioic acid; 22 = 3-Hydroxyphenylacetic acid; 23 = Heptanedioic acid; 24 = 4-Hydroxyphenylacetic acid; 25 = Octenedioic acid; 26 = Suberic acid (octanedioic acid); 27 = 4-Hydroxy-3-methoxybenzoic acid; 28 = 4-Hydroxy-3-methoxyphenylacetic acid; 29 = Hippuric acid; 30 = 4-Hydroxymandelic acid; 31 = 2-Oxoglutaric acid; 32 = 4-Hydroxy-3-methoxymandelic acid; 33 = 4-Hydroxyphenylacetic acid; 34 = 3-Indoleacetic acid; 35 = 3-Vanillylhydracryric acid; 36 = Palmitic acid; 37 = Ferulic acid. (From Mamer, O.A. and Tjoa, S.S., *Clin. Chim. Acta.*, 55, 199, 1974. With permission.)

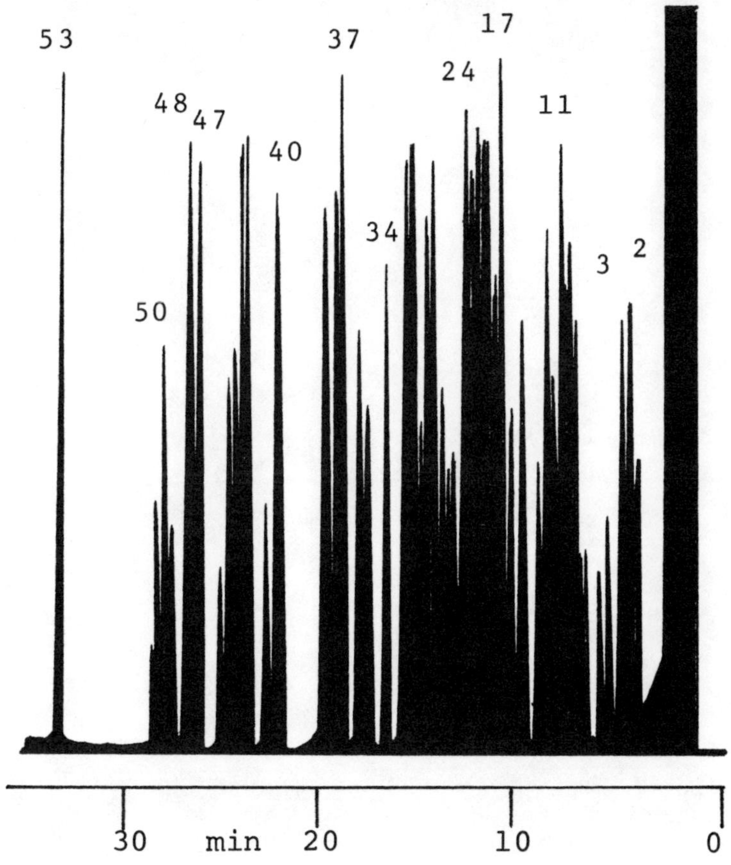

FIGURE GC-F2. Gas chromatogram of TMS and methoxime-TMS derivatives of organic acids.
Packing = 3% Dexsil® 300 on Veraport 30 (100-120 mesh)
Column = 250 cm long, 3.1 mm o.d., ss.
Temperature = 80 to 250 °C (4/min)
Detector = FI
Compound 1 = 2,3-Dihydroxy-3-methylbutyric acid; 2 = D- + L-lactic acid; 3 = glycolic acid; 4 = Oxaloacetic acid; 5 = Glyoxylic acid; 6 = 2-ketoglutaric acid; 7 = Pyromucic acid; 8 = Heptanoic acid (internal standard); 9 = Pyruvic acid; 10 = Sorbic acid; 11 = Dimethylmalonic acid; 12 = Malonic acid; 13 = 2-hydroxy-isocaproic acid; 14 = Oxalic acid; 15 = 2-hydroxyhexanoic acid; 16 = 2-ketoisocaproic acid; 17 = Succinic acid; 18 = Nicotinic acid + glyceric acid; 19 = Methyl-succinic acid; 20 = Itaconic acid; 21 = 2,3-dihydroxy-2-methylbutyric acid (*threo*) + citaconitic acid; 22 = 2,3-dihydroxybutyl-2-methylbutyric acid (*erythro*); 23 = Fumaric acid; 24 = 2,3-dimethylsuccinic acid; 25 = Glutaric acid; 26 = Mesaconic acid; 27 = Mesoxalic acid; 28 = 3-methylglutaric acid; 29 = 2-hydroxyoctanoic acid; 30 = Citramalic acid; 31 = Glutaconic acid; 32 = 3.3-dimethylsuccinic acid; 33 = Malic acid; 34 = Adipic acid; 35 = Furan-3,4-dicarbonic acid; 36 = 2-hydroxyglutaric acid; 37 = Pimelic acid; 38 = 1,2-cyclohexanedicarbonic acid; 39 = Tartaric acid; 40 = Tricarballylic acid; 41 = *cis*- + *trans*-aconitic acid; 42 = Azelaic acid; 43 = Citric acid; 44 = Isocitric acid; 45 = 2-hydroxylauric acid; 46 = 4-methylphthalic acid; 47 = Sebacic acid; 48 = Gluconic acid + α-gluconic acid + α-galacturonic acid; 49 = Ascorbic acid; 50 = Mucic acid; 51 = β-galacturonic acid; 52 = β-glucuronic acid; 53 = Stearic acid (internal standard). (From Drawert, F. and Zeupold, A., *Chromatographia*, 9, 397, 1976. With permission.)

Liquid Chromatography Tables

Section I.II.

LIQUID CHROMATOGRAPHY TABLES

Table LC-1
Alkylphenols

Packing	P1	P2	P3	P4	P5	P5	P5	P5
Column								
Length (cm)	40	40	50	30	25	25	25	25
Diameter (mm)	2.1	2.1	2.1	5	2.1	2.1	2.1	2.1
Material	ss	ss	ss	ss	ss	ss	ss	ss
Eluent	E1	E1	E2	E3	E4	E4	E4	E5
Flow rate (mℓ/min)	0.6	0.6	0.62	0.15	0.56	0.54	0.61	0.53
Temperature (°C)	20±0.1	20±0.1	rt	rt	46	53	63	42
Detection	D1	D1	D1	D1	D2	D2	D2	D2
Reference	1	1	1	1	2	2	2	2
Compound	k′	k′	k′	k′	min	min	min	min
Phenol	17.0	17.0	12.7	0.28	19.6	16.5	10.6	24.2
2-Ethylphenol	6.4	7.0	6.0	0.74	9.2	8.0	5.2	10.8
3-Ethylphenol	10.7	12.2	9.8	0.67	18.9	15.8	10.0	23.5
4-Ethylphenol	11.7	13.9	10.9	0.66	20.8	17.3	10.9	25.7
2-Methylphenol	7.6	8.2	7.2	0.44	11.7	10.0	6.6	14.2
3-Methylphenol	12.8	13.7	11.3	0.44	20.8	17.3	10.9	25.7
4-Methylphenol	13.0	14.1	11.4	0.42	21.8	18.4	10.9	27.1
2-*tert*.-Butylphenol	4.9	5.7	4.0	1.87	—	—	—	—
4-*tert*.-Butylphenol	10.7	12.7	9.4	1.21	—	—	—	—
4-*sec*.-Butylphenol	9.9	13.0	9.1	1.36	—	—	—	—
2-Propylphenol	5.5	6.1	5.5	1.14	—	—	—	—
4-Propylphenol	10.8	12.3	10.0	1.01	—	—	—	—
2-Isopropylphenol	5.7	6.3	5.5	1.07	—	—	—	—
4-*tert*.-Pentylphenol	9.7	12.0	8.8	1.65	—	—	—	—
2,3-Dimethylphenol	7.0	7.5	6.5	0.69	12.2	10.5	6.8	14.9
2,4-Dimethylphenol	6.2	6.6	6.5	0.69	13.3	11.5	7.3	16.0
2,5-Dimethylphenol	5.6	6.0	6.3	0.69	11.7	10.0	6.6	14.2
2,6-Dimethylphenol	1.9	2.0	2.6	0.69	5.7	5.0	3.6	6.4
3,4-Dimethylphenol	11.7	12.4	10.3	0.57	24.3	19.8	12.6	30.0
3,5-Dimethylphenol	10.1	11.1	10.1	0.64	21.8	18.4	10.9	27.1
2,5-Di-*tert*.-butylphenol	2.3	2.8	1.8	6.04	—	—	—	—
2,6-Di-*tert*.-butylphenol	—	—	—	7.40	—	—	—	—
3,5-Di-*tert*.-butylphenol	6.4	7.5	6.0	3.73	—	—	—	—
2,6-Di-*sec*.-butylphenol	0.40	0.59	0.43	5.11	—	—	—	—
2,6-Diisopropylphenol	0.67	0.83	0.78	2.58	—	—	—	—
2-*tert*.-Butyl-5-methylphenol	3.4	3.9	—	2.50	—	—	—	—
2-*tert*.-Butyl-4-methylphenol	3.8	4.4	3.2	2.28	—	—	—	—
2,3,5-Trimethylphenol	5.2	5.9	5.0	0.95	—	—	—	—
2,3,6-Trimethylphenol	—	—	—	—	—	—	—	—
2,4,5-Trimethylphenol	4.9	5.7	5.3	1.00	—	—	—	—
2,4,6-Trimethylphenol	1.5	1.9	2.2	0.96	—	—	—	—
3,4,5-Trimethylphenol	—	12.0	—	0.79	—	—	—	—
2,3,5,6-Tetramethylphenol	1.4	1.6	1.6	1.26	—	—	—	—
2,6-Di-*tert*.-butyl-4-methylphenol	—	—	—	9.41	—	—	—	—
6-Hydroxymethyl-2,3,5-trimethylphenol	—	—	—	—	—	—	—	—
4-Hydroxymethyl-2,3,5-trimethylphenol	—	—	—	—	—	—	—	—
4,6-Dihydroxymethyl-2,3,5-trimethylphenol	—	—	—	—	—	—	—	—
4-Hydroxymethyl-2,3,6-trimethylphenol	—	—	—	—	—	—	—	—

Table LC-1 (continued)
ALKYLPHENOLS

Compound	k′	k′	k′	k′	min	min	min	min
2-Ethyl-6-hydroxymethylphenol	—	—	—	—	—	—	—	—
2-Ethyl-4-hydroxymethylphenol	—	—	—	—	—	—	—	—
2-Ethyl-4,6-diHydroxymethylphenol	—	—	—	—	—	—	—	—
6-Hydroxymethyl-2,3-dimethylphenol	—	—	—	—	—	—	—	—
4-Hydroxymethyl-2,3-dimethylphenol	—	—	—	—	—	—	—	—
4,6-Dihydroxymethyl-2,3-dimethylphenol	—	—	—	—	—	—	—	—

[a] Measured on chromatogram.

Packing	P1 =	Porasil® C (37-73 μm) coated with 1,2,3-tris(cyanoethoxy)propane
	P2 =	Porasil® C coated with 1,2,3,4,5,6-hexakis(2′-cyanoethoxy)hexane
	P3 =	Cyano Sil-XI (13±5 μm)
	P4 =	μBondapak C18 (10±2 μm)
	P5 =	Zorbax Sil (5 μm)
	P6 =	LiChrosorb® Si 60 (30 μm)
	P7 =	LiChrosorb® Si 60 (5 μm)
Eluent	E1 =	isooctane
	E2 =	isooctane + 0.5% isopropanol
	E3 =	ethanol-water (60:40)
	E4 =	cyclohexane-methylenedichloride (3:1)
	E5 =	cyclohexane-methylenedichloride (4:1)
	E6 =	cyclohexane-methylenedichloride (6:1)
	E7 =	cyclohexane-methylenedichloride (7.5:1)
	E8 =	cyclohexane-methylenedichloride (1:1)
	E9 =	cyclohexane-methylenedichloride (2:1)
	E10 =	methanol-dioxane-dichloromethane (8:8:250)
Detection	D1 =	UV
	D2 =	UV 254 nm

REFERENCES

1. **Callmer, K., Edholm, L. E. and Smith, B. E. F.**, Study on retention behaviour of alkylphenols in straight- and reversed-phase liquid chromatography and application to the analysis of complex phenolic mixtures in conjugation with gas chromatography, *J. Chromatogr.*, 136, 45, 1977.
2. **Husain, S., Kunzelmann, F. and Schilaknecht, H.**, Separation of isomeric alkylphenols by high-performance liquid chromatrgraphis and gas-liquid chromotographic techniques, *J. Chromatogr.*, 137, 53, 1977.

Table LC-2
PHENOLS AND AROMATIC CARBOXYLIC ACIDS

Packing	P1	P1	P1	P1	P1	P1	P1	P1	P1	P1
Column										
Length (cm)	35	35	35	35	35	35	35	35	35	35
Diameter (mm)	1.9	1.9	1.9	1.9	1.9	1.9	1.9	1.9	1.9	1.9
Material										
Eluent	E1	E2	E3	E4	E5	E6	E7	E8	E9	E10
Flow rate (cm/min)	18	18	18	18	18	18	18	18	18	18
Temperature (°C)	30	30	30	30	30	30	30	30	30	30
Detection	D1[a]	D1[a]	D1[a]	D1[a]	D1[a]	D1[a]	D1[a]	D1[a]	D1[a]	D1[a]
Reference	1	1	1	1	1	1	1	1	1	1
Compound	Dv	Dv	Dv	Dv	Dv	Dv	Dv	Dv	Dv	Dv
Phenol	26	3.34	1.09	2.70	1.1	1.09	3.50	1.2	—	—
1,3-Dihydroxybenzene	88	10.2	7.01	33	7.3	7.01	—	6.1	—	—
1,2-Dihydroxybenzene	46	8.02	5.32	24.7	5.5	5.32	—	4.8	—	—
1,4-Dihydroxybenzene	38	5.70	4.37	19.9	4.4	4.37	—	3.7	—	—
2-Chlorophenol	19	7.78	1.85	5.27	2.1	1.85	—	2.2	—	—
2,5-Dichlorophenol	300	18.5	2.61	8.30	4.3	2.61	—	—	—	—
2,6-Dichlorophenol	300	8.90	1.73	3.60	9.0	1.73	—	—	—	—
R′CH(OH)CH(OH)CH_2OH	1.58	0.59	1.79	14	1.97	1.79	0.59	—	—	—
R′CH(OH)CH(OH)R′	14.5	1.27	2.02	27	2.13	2.02	1.27	—	—	—
1,3,5-Trihydroxybenzene	—	—	—	—	54	47	33	—	—	—
R′CH_2OH	—	—	—	—	0.92	0.86	1.14	—	—	—
R′$CH_2CH_2CH_2$OH	—	—	—	—	0.48	0.46	1.02	—	—	—
4-Hydroxy-3-methoxybenzaldehyde	—	—	—	—	2.05	1.02	2.45	—	—	—
R′CH(OH)CH(CH_2OH)OR″	—	—	—	—	0.48	0.41	0.75	—	—	—
R′CH(OH)CH_2OR″	—	—	—	—	0.58	0.50	2.18	—	—	—
R‴CH(OH)CH(CH_2OH)OR″	—	—	—	—	0.10	0.08	0.27	—	—	—
R′CH(OH)CH(CH_2OH)R′	—	—	—	—	1.60	1.46	1.08	—	—	—
R′COCH_2R′	—	—	—	—	3.45	2.55	8.54	—	—	—
(R′ = 4-hydroxy-3-methoxyphenyl, R″ = 2-methoxyphenyl, R‴ = 3,4-dimethoxyphenyl)										
3-Chlorophenol	—	—	—	—	1.8	1.60	—	1.9	—	—
4-Chlorophenol	—	—	—	—	1.6	1.50	—	1.7	—	—

Table LC-2 (continued)
PHENOLS AND AROMATIC CARBOXYLIC ACIDS

Compound	Dv	Dv	Dv	Dv	Dv	Dv	Dv	Dv	Dv	Dv
2,3-Dichlorophenol	—	—	—	—	4.2	3.15	—	—	—	—
2,4-Dichlorophenol	—	—	—	—	3.8	2.75	—	—	—	—
2,4,6-Trichlorophenol	—	—	—	—	110	11.5	—	—	—	—
2,4,5-Trichlorophenol	—	—	—	—	16.4	5.04	—	—	—	—
3,4,5-Trichlorophenol	—	—	—	—	5.6	3.35	—	—	—	—
2,3,4,5-Tetrachlorophenol	—	—	—	—	110	15.5	—	—	—	—
2-Nitrophenol	—	—	—	—	8.7	2.25	—	9.7	—	—
3-Nitrophenol	—	—	—	—	3.0	2.75	—	3.3	—	—
4-Nitrophenol	—	—	—	—	17.3	5.55	—	19.8	—	—
Benzoic acid	—	—	—	—	—	33	—	—	1.9	1.2
2-Hydroxybenzoic acid	—	—	—	—	—	200	—	—	83	37
3-Hydroxybenzoic acid	—	—	—	—	—	170	—	—	7.0	5.9
4-Hydroxybenzoic acid	—	—	—	—	—	58	—	—	5.0	4.3
4-Hydroxy-3-methoxybenzoic acid	—	—	—	—	—	41	—	—	2.4	2.3
2-Chlorobenzoic acid	—	—	—	—	—	200	—	—	10.8	6.5
3-Chlorobenzoic acid	—	—	—	—	—	145	—	—	5.9	2.4
4-Chlorobenzoic acid	—	—	—	—	—	97	—	—	4.0	1.6

[a] Volume distribution coefficient.

Packing P1 = Dowex® 1-X10 (7-10 μm)
Eluent E1 = water
E2 = 40% ethahol
E3 = 80% ethanol
E4 = 98% ethanol
E5 = 0.01 *M* acetic acid/80% ethanol
E6 = 0.1 *M* acetic acid/80% ethanol
E7 = 0.1 *M* acetic acid/40% ethanol
E8 = 0.01 *M* acetic acid/73% ethanol
E9 = 5 *M* acetic acid/40% ethanol
E10 = 5 *M* acetic acid/93% ethanol
Detection D1 = UV 280 nm

From Jahangir, L. M. and Samuelson, O., *Anal. Chim. Acta*, 92, 329, 1977. With permission.

Table LC-3
HYDROXY- AND AMINOPHENOLS AND ORGANIC ACIDS

Packing	P1	P1	P1	P1	P1	P1	P1	P1	P1	P1	P1	P1	P1	P1	P1	P1	P1	P1	P1	P1	P1	P1
Column																						
Length (cm)	18.5	18.5	18.5	18.5	18.5	18.5	18.5	18.5	18.5	18.5	18.5	18.5	18.5	18.5	18.5	18.5	18.5	18.5	18.5	52.5	18.5	52.5
Diameter (mm)	8	8	8	8	8	8	8	8	8	8	8	8	8	8	8	8	8	8	8	8	8	8
Material	G	G	G	G	G	G	G	G	G	G	G	G	G	G	G	G	G	G	G	G	G	G
Eluent	E1	E2	E3	E4	E5	E6	E7	E8	E9	E10	E11	E12	E13	E14	E15	E16	E17	E18	E19	E19	E20	E21
Flow rate (mℓ/min)	0.5-0.7																					
Temperature (°C)																						
Detection	D1	D1	D1	D1	D1	D1	D1	D1	D1	D1												
Reference	1	1	1	1	1	1	1	1	1	1	1	1	1	1	1	1	1	1	1	1	1	1
Compound	Dv	Dv	Dv	Dv	Dv	Dv	Dv	Dv	Dv	Dv	Dv	Dv	Dv	Dv	Dv	Dv	Dv	Dv	Dv	Dv	Dv	Dv
1,2-Dihydroxybenzene	4.13	1.10	1.00	1.22	—	—	—	7.61	0.58	—												
1,3-Dihydroxybenzene	13.2	12.1	12.0	21.5	31.5	12.8	13.3	13.4	6.06	—	0.63	—	0.90	—	1.10	—	1.01	—	—	1.19	—	—
2-Methoxyphenol	4.76	4.56	4.12	8.96	10.0	—	4.99	4.60	1.98	—	3.86	4.34	4.99	5.20	8.64	9.93	8.68	9.29	—	8.88	11.2	19.2
3,4-Dihydroxymethylbenzene	—	0.79	—	1.01	—	—	—	8.31	0.47	—	1.44	1.65	1.28	1.60	2.90	3.05	2.41	3.37	—	—	2.84	8.00
1,3,5-Trihydroxybenzoic acid	—	21.4	—	48.3	52.6	—	21.4	—	12.7	—	0.36	—	0.31	—	—	—	0.79	—	—	0.56	—	—
4-Aminophenol	—	0.17	—	0.27	0.84	—	0.90	0.18	0.17	—	5.20	—	11.4	12.1	18.8	—	11.2	13.9	—	—	21.1	—
4-Methoxyphenol	—	7.89	—	18.9	27.1	—	8.00	7.81	4.45	—	0.05	0.26	0.06	0.30	0.07	0.33	0.21	0.58	0.22	—	1.87	0.58
4-Amino-2-hydroxybenzoic acid	—	18.7	—	—	—	—	—	—	9.29	—	2.94	—	2.41	2.41	6.60	7.35	6.06	7.89	—	4.62	6.28	12.1
2-Hydroxybenzaldehyde	5.52	5.31	4.23	9.83	7.78	4.34	3.27	6.50	2.30	—	5.09	—	4.77	—	16.2	—	—	—	—	—	—	—
2-Hydroxybenzoic acid	—	15.3	—	—	—	24.5	30.8	—	5.63	10.8	1.55	1.65	1.01	0.79	3.80	2.62	3.70	3.16	—	2.95	2.51	7.35

Packing: P1 = Dowex® 1-X8, Cl^- form

Eluent:
E1 = 0.025 M $CuCl_2$-45% methanol, pH 1.3
E2 = 0.05 M $CuCl_2$-45% methanol, pH 1.3
E3 = 0.10 M $CuCl_2$-45% methanol, pH 1.3
E4 = 0.05 M $CuCl_2$-30% methanol, pH 1.3
E5 = 0.05 M $CuCl_2$-30% methanol, pH 3.5
E6 = 0.05 M $CuCl_2$-45% methanol, pH 2.3
E7 = 0.05 M $CuCl_2$-45% methanol, pH 3.5
E8 = 0.10 M KCl-45% methanol, pH 1.3
E9 = 0.05 M $CuCl_2$-60% methanol, pH 1.3
E10 = 0.05 M $CuCl_2$-60% methanol, pH 3.5
E11 = 0.05 M $CuCl_2$-45% acetone, pH 1.3
E12 = 0.05 M $CuCl_2$-45% acetone, pH 3.5
E13 = 0.05 M $CuCl_2$-45% propanol, pH 1.3
E14 = 0.05 M $CuCL_2$-45% Propanol, pH 3.5
E15 = 0.05 M $CuCl_2$-45% ethanol, pH 1.3
E16 = 0.05 M $CuCl_2$-45% ethanol, pH 3.5
E17 = 0.05 M $CuCl_2$-30% acetone, pH 1.3
E18 = 0.05 M $CuCl_2$-30% acetone, pH 3.5
E19 = 0.05 M $CuCl_2$-30% propanol, pH 1.3
E20 = 0.05 M $CuCl_2$-30% propanol, pH 3.5
E21 = 0.05 M $CuCl_2$-30% ethanol, pH 3.5

Detection: D1 = Phenolic compounds in $CuCl_2$-organic solvent produced yellowish green color as a result of reaction with mercuric acetate-sodiumnitrite solution.
Salycylaldehyde and 2-hydroxybenzoic acid are analyzed using ferric chloride solution.

From Lee, K. S., Lee, D. W., and Chung, Y. S., *Anal. Chem.*, 45, 396, 1973. With permission.

Table LC-4
BISPHENOLS

Packing	P1
Column	
Length (cm)	120
Diameter (mm)	8
Material	
Eluent	E1
Flow rate (ml/min)	0.67
Temperature (°C)	
Detection	
Reference	1

Compound	Ve
3,3'-Di-*tert*.-butyl-2,2'-dihydroxy-5,5'-dimethylbiphenyl	68.2
2,2'-Dihydroxy-3,3',5,5'-tetramethylbiphenyl	75.2
5,5'-Di-*tert*.-butyl-2,2'-dihydroxy-3,3'-dimethylbiphenyl	67.7
3,3',5,5'-Tetra-*tert*.-butyl-2,2'-dihydroxybiphenyl	64.6
4,4'-Dihydroxybiphenyl	72.4
4,4'-Dihydroxy-3,3'-dimethylbiphenyl	72.1
4,4'-Dihydroxy-3,3',5,5'-tetramethylbiphenyl	71.2
5,5'-Di-*tert*.-butyl-4,4'-dihydroxy-3,3'-dimethylbiphenyl	64.6
3,3',5,5'-Tetra-*tert*.-butyl-4,4'-dihydroxybiphenyl	62.4
1-(3'-*tert*.-Butyl-2'-hydroxy-5'-methylbenzyl)-3-*tert*.-butyl-2-hydroxy-5-methylbenzene	66.7
1-(2'-Hydroxy-3',5'-dimethylbenzyl)-2-hydroxy-3,5-dimethylbenzene	73.7
1-(5'-*tert*.-Butyl-2'—hydroxy-3'-methylbenzyl)-5-*tert*.-butyl-2-hydroxy-3-methylbenzene	66.3
1-(3',5'-Di-*tert*.-butyl-2'-hydroxybenzyl)-3,5-di-*tert*.-butyl-2-hydroxybenzene	61.8
1-(4'-Hydroxybenzyl)-4-hydroxybenzene	72.7
1-(4'-Hydroxy-3',5'-dimethylbenzyl)-4-hydroxy3,5-dimethylbenzene	69.6
1-(5'-*tert*.-Butyl-4'-hydroxy-3'-methylbenzyl)-5-*tert*.-butyl-4-hydroxy-3-methylbenzene	64.2
1(3',5'-Di-*tert*.-butyl-4'-hydroxybenzyl)-3,5-di *tert*.-butyl-4-hydroxybenzene	62.0
1,1-Di-(3'-*tert*.-butyl-2'-hydroxy-5'-methylphenyl)ethane	66.8
1,1-Di-(5'-*tert*.-butyl-2'-hydroxy-3'-methylphenyl)ethane	65.6
1,1-Di-(3',5'-di-*tert*-butyl-2'-hydroxyphenyl)ethane	62.0
1,1-Di-(5'-*tert*.-butyl-4'-hydroxy-3'-methylphenyl)ethane	65.9
1,1-Di-(3',5'-di-*tert*.-butyl-4'-hydroxyphenyl)ethane	62.0
1,1-Di-(3'-*tert*.-butyl-2'-hydroxy-5'-methylphenyl)butane	65.5
1,1-Di-(4'-hydroxyphenyl)butane	69.8
1,1-Di-(5'-*tert*.-butyl-4'-hydroxy-3'-methylphenyl)butane	62.5
Phenyl-1,1-di-(3'-*tert*.-butyl-2'-hydroxy-5'-methylphenyl)methane	65.9
Phenyl-1,1-di-(5'-*tert*.-butyl-4'-hydroxy-3'-methylphenyl)methane	62.8
2,2-Di-(3'-*tert*.-butyl-2'-hydroxy-5'-methylphenyl)propane	70.0
2,2-Di(2'-hydroxy-3',5'-dimethylphenyl)propane	74.7
2,2-Di-(5'-*tert*.-butyl-2'-hydroxy-3'-methylphenyl)propane	67.1
2,2-Di-(3',5'-di-*tert*.-butyl-2'-hydroxyphenyl)propane	62.5
2,2-Di-(4'-hydroxyphenyl)propane	69.0
2,2-Di-(4'-hydroxy-3',5'-dimethylphenyl)propane	68.0
2,2-Di-(5'-*tert*.-butyl-4'-hydroxy-3'-methylphenyl)propane	66.0
2,2-Di-(3',5'-di-*tert*.-butyl-4'-hydroxyphenyl)propane	61.3
2,2-Di-(3'-*tert*.-butyl-2'-hydroxy-5'-methylphenyl)butane	69.4
2,2-Di-(4'-hydroxyphenyl)butane	69.4
2,2-Di-(4'-hydroxy-3'-methylphenyl)butane	68.2
2,2-Di-(5'-*tert*.-butyl-4'-hydroxy-3'-methylphenyl)butane	63.2
1,1-Di-(4'-hydroxy-3'-methylphenyl)cyclohexane	68.7
1,1-Di-(5'-*tert*.-butyl-'-hydroxy-3'-methylphenyl)cyclohexane	63.3
Di-(3-*tert*.-butyl-2-hydroxy-5-methylphenyl)sulfide	67.9
Di-(2-hydroxy-3,5-dimethylphenyl)sulfide	74.0
Di-(5-*tert*.-butyl-2-hydroxy-3-methylphenyl)sulfide	66.2
Di-(3,5-di-*tert*.-butyl-2-hydroxyphenyl)sulfide	62.4
Di-(4-hydroxy-3,5-dimethylphenyl)sulfide	69.8

Table LC-4 (continued)
BISPHENOLS

Compound	Ve
Di-(5-*tert*.-butyl-4-hydroxy-3-methylphenyl)sulfide	64.3
Di-(3,5-di-*tert*.-butyl-4-hydroxyphenyl)sulfide	61.8
Di-(3-*tert*.-butyl-2-hydroxy-5-methylphenyl)disulfide	67.7
Di-(5-*tert*.-butyl-4-hydroxy-3-methylphenyl)disulfide	64.2
Di-(5-*tert*.-butyl-4-hydroxy-3-methylphenyl)trisulfide	63.4

Packing P1 = styrene-divinylbenzene copolymer S-GEL-832
Eluent E1 = tetrahydrofuran

REFERENCE

1. **Cŏupek, J., Pororný, S., Jiráčková, L., and Pospišil, J.,** Gel chromatographic behaviour of bisphenols, *J. Chromatogr.*, 75, 87, 1973.

Table LC-5
ALKYL AND BROMINATED ALKYL PHENOLS

Packing	P1	P2
Column		
Length (cm)	30	50
Diameter (mm)	4	2.6
Material	ss	ss
Eluent	E1	E1
Flow rate (mℓ/min)	0.67	
Temperature (°C)		
Detection	D1	D1
Reference	1	1

Compound	k′	k′
2,6-Dimethylphenol	0.6	2.0
2,3,5,6-Tetramethylphenol	.01	—
2-Methyl-6-propylphenol	1.12	1.1
2-*tert*.-Butyl-6-methylphenol	1.52	0.7
2,6-Diisopropylphenol	1.71	0.6
2,6-Di-*sec*.-butylphenol	2.87	0.3
2,6-Di-*tert*.-butylphenol	4.08	—
2,4-Dimethylphenol	0.61	4.3
2,4,6-Trimethylphenol	0.78	—
2-*tert*.-Butyl-4-methylphenol	1.52	2.4
4-*tert*.-Hexyl-2-methylphenol	2.08	3.7
4-Methylphenol	0.44	8.8
4-Ethylphenol	0.60	7.9
4-Propylphenol	0.82	7.3
4-*tert*.-Butylphenol	0.92	7.5
4-*sec*.-Butylphenol	1.00	6.8
4-*tert*.-Pentylphenol	1.23	6.4
4-Cyclohexylphenol	1.47	7.5
4-Octylphenol	2.53	5.5
4-Bromo-2,6-dimethylphenol	1.07	3.5
4-Bromo-2,3,5,6-tetramethylphenol	1.75	—
4-Bromo-2-methyl-6-propylphenol	1.92	2.1
4-Bromo-2-*tert*.-butyl-6-methylphenol	2.51	1.4
4-Bromo-2,6-diisopropylphenol	2.67	1.2
4-Bromo-2,6-di-*sec*.-butylphenol	4.09	0.7

Table LC-5 (continued)
ALKYL AND BROMINATED ALKYL PHENOLS

Compound	k′	k′
4-Bromo-2,6-di-*tert*.-butylphenol	5.70	—
6-Bromo-2,4-dimethylphenol	1.02	0.15
6-Bromo-2,4,5-trimethylphenol	1.43	—
6-Bromo-2-*tert*.-butyl-4-methylphenol	2.94	0.05
6-Bromo-4-*tert*.-hexyl-2-methylphenol	3.15	0.10
2-Bromo-4-methylphenol	0.74	1.6
2-Bromo-4-ethylphenol	0.89	1.4
2-Bromo-4-propylphenol	1.21	1.2
2-Bromo-4-*tert*.-butylphenol	1.30	1.1
2-Bromo-4-*sec*.-butylphenol	1.42	1.1
2-Bromo-4-*tert*.-pentylphenol	1.68	1.1
2-Bromo-4-cyclohexylphenol	2.27	1.2
2-Bromo-4-octylphenol	3.27	1.1
2,6-Dibromo-4-methylphenol	1.05	0.6
2,6-Dibromo-4-ethylphenol	1.38	0.5
2,6-Dibromo-4-propylphenol	1.92	0.4
2,6-Dibromo-4-*tert*.-butylphenol	2.02	0.35
2,6-Dibromo-4-*sec*.-butylphenol	2.20	0.4
2,6-Dibromo-4-*tert*.-pentylphenol	2.55	0.35
2,6-Dibromo-4-cyclohexylphenol	3.78	0.4
2,6-Dibromo-octylphenol	4.83	0.35

Packing P1 = μBondapak C18
P2 = Cyano Sil-X-1
Eluent E1 = ethanol-water (60:40)
E2 = 0.5% 2-propanol in isooctane
Detection D1 = UV 280 nm

REFERENCE

1. **Edholm, L. -E. and Callmer, K.**, Liquid chromatographic study of brominated alkylphenols and investigation of product formation in the coulometric bromination of some alkyl-substituted phenols, *J. Chromatogr.*, 137, 363, 1977.

Table LC-6
PHENOLS

Packing	P1		P1		P2		P2		P3	P4	P3	P5	P5	P5	P5
Column															
Length (cm)	200		200		100		100		61	100	61	30	30	30	30
Diameter (mm)									4.83	2.1	4.83	19	19	19	19
Material															
Eluent	E1		E2		E3		E4		E5	E6	E7	E8	E9	E10	E11
Flow rate (mℓ/min)	0.5		0.65		0.7		0.65		1.05	1.05		1	1	1	1
Temperature	46		28		28		28								
Detection	D1		D1		D1		D1		D2	D2	D2	D3	D3	D3	D3
Reference	1		1		1		1		2	2	2	3	3	3	3
Compound	Tr	min	Tr	min	Tr	min	Tr	min	min[a]	min[a]	min[a]	mℓ	mℓ	mℓ	mℓ
Phenol	1.00	8.7	1.00	6.7	1.00	7.0	1.00	5.0	9.0	5.5	24	7.6	—	—	—
Benzene	2.83		2.16		—		—		—	—	—	—	—	—	—
Bromobenzene	>10.00		2.32		—		—		—	—	—	—	—	—	—
Pyridine	2.53		1.65		—		—		—	—	—	—	—	—	—
Aniline	1.28		1.00		0.36		0.50		—	—	—	—	—	—	—
2-Methoxyphenol	1.55		1.12		0.65		0.78		—	—	35.0	—	—	—	—
2-Methylphenol	1.86		1.42		1.57		1.80		—	—	—	—	—	—	—
3-Methylphenol	1.72		1.35		—		—		—	—	—	—	—	—	—
4-Methylphenol	1.86		1.35		1.50		1.70		—	—	—	—	—	—	—
2-Bromophenol	1.77		1.50		>10.00		5.10		—	—	—	—	—	—	—
3-Bromophenol	2.90		2.20		>10.00		12.40		—	—	—	—	—	—	—
4-Bromophenol	3.42		2.24		>10.00		13.50		—	—	—	—	—	—	—
3-Chlorophenol	2.30		1.79		>10.00		—		—	—	—	—	—	—	—
4-Chlorophenol	2.42		1.87		>10.00		7.00		14.0	7	55.0	16.5	—	—	—
1,2-Dihydroxybenzene	0.74		0.79		10.00		0.8	—	—	—	—	—	—	—	—
1,3-Dihydroxybenzene	0.67		0.79		1.80		1.1		—	—	—	—	—	—	—
1,4-Dihydroxybenzene	0.63		0.79		0.36		0.5		—	—	—	—	—	—	—
1,4-Benzoquinone	0.63		0.79		0.36		0.4		—	—	—	—	—	—	—
1,2,4-Trihydroxybenzene	0.63		0.79		>10.00		0.68		—	—	—	—	—	—	—
1,3,5-Trihydroxybenzene	0.63		0.79		>10.00		1.20		—	—	—	—	—	—	—
2-Aminophenol	0.94		0.79		0.60		0.7		—	—	—	—	—	—	—

Table LC-6 (continued)
PHENOLS

Compound	Tr	min	Tr	min	Tr	min	Tr	min	min[a]	min[a]	min[a]	mℓ	mℓ	mℓ	mℓ
4-Aminophenol	0.78		0.79		0.36		0.4		—	—	—	—	—	—	—
Biphenyl	>10.00		>8.00		—		—		—	—	—	—	—	—	—
2-Hydroxybiphenyl	>10.00		9.5		—		—		—	—	—	—	—	—	—
2,2′-Dihydroxybiphenyl	1.65		1.09		—		—		—	—	—	—	—	—	—
4,4′-Dihydroxybiphenyl	6.00		1.80		—		—		—	—	—	—	—	—	—
Indole	3.10		2.02		—		—		—	—	—	—	—	—	—
2-Hydroxynaphthalene	>8.00		3.42		—		—		—	—	—	—	—	—	—
1-Hydroxynaphthalene	>8.00		3.90		—		—		23.0	14.0	—	—	—	—	—
5-Bromo-2,4′-dihydroxybiphenyl	>10.00		—		—	—	—	—	—	—	—	—	—	—	—
2-Bromo-5,4′-dihydroxybiphenyl	>10.00		—		—		—		—	—	—	—	—	—	—
Phenolic terphenyls	>10.00		—		—		—		—	—	—	—	—	—	—
4-Hydroxy-3,5-dimethoxybenzoic acid	—		—		—		—		—	—	12.0	—	—	—	—
o-Vanillin	—		—		—		—		—	—	49.0	—	—	—	—
4-Allyl-2-methoxyphenol	—		—		—		—		—	—	74.0	—	—	—	—
2-Methoxy-4-propenylphenol	—		—		—		—		—	—	74.0	—	—	—	—
2-Chlorophenol	—		—		—		—		—	—	47.0	3.7	—	—	—
2,4-Dichlorophenol	—		—		—		—		—	—	79.0	—	—	25.4	—
2,4,6-Trichlorophenol	—		—		—		—		—	—	92.0	—	—	—	—
3,5-Dichlorophenol	—		—		—		—		—	—	—	—	—	34.5	—
2,4-Dinitrophenol	—		—		—		—		—	—	—	—	3.8	—	--
4-Nitrophenol	—		—		—		—		—	—	—	—	8.1	1.9	—
3-Nitrophenol	—		—		—		—		—	—	—	—	13.3	12.1	—
2,6-Dichlorophenol	—		—		—		—		—	—	—	—	—	6.9	—
2,5-Dichlorophenol	—		—		—		—	—	—	—	—	—	—	19.2	—
2-Isopropylphenol	—		—		—		—		—	—	—	—	—	—	1.3
2,6-Diisopropylphenol	—		—		—		—		—	—	—	—	—	—	2.6
2,6-Di-*tert*.-butylphenol	—		—		—		—		—	—	—	—	—	—	6.8
2-*tert*.-Butyl-6-methylphenol	—		—		—		—		—	—	—	—	—	—	2.5

[a] Measured on chromatogram.

Packing P1 = Corasil C18

	P2 = Zipax AX
	P3 = Bondapak C18 Corasil
	P4 = Permaphase ODS
	P5 = XAD-2 (45-65 μm)
Eluent	E1 = water
	E2 = 25% methanol
	E3 = 0.02 *M* boric acid
	E4 = 0.01 *M* Naborate + 0.01 *M* $NaNO_3$
	E5 = 25% acetonitrile
	E6 = 5% acetonitrile
	E7 = water-34% acetonitrile (linear gradient)
	E8 = 0.1 *M* NaOH into 30 mℓ of 0.1 *M* phosphate buffer(pH 10.1)/10% ethanol
	E9 = 0.2 *M* NaOH into 30 mℓ of 0.1 *M* phosphate buffer (pH 7.8)/10% ethanol
	E10 = 0.1 *M* NaOH into 40 mℓ of 0.05 *M* carbonate buffer (pH 8.95)/10% ethanol
	E11 = 66% acetonitrile
Detection	D1 = UV 254 nm & RI
	D2 = UV 254 nm and cerium sulfate reaction detector (Ex 260 nm, Em 350 nm)

REFERENCES

1. **Bhatia, K.**, Determination of trace phenol in aqueous solution by aqueous liquid chromatography, *Anal. Chem.*, 45, 1344, 1973.
2. **Wolkoff, A. W. and Larose, R. H.**, A highly selective technique for the liquid chromatographic analysis of phenols and other environmental pollutants, *J. Chromatogr.*, 99, 731, 1974.
3. **Pietrzyk, D. J. and Chi-Hong Chu,** Separation of organic acids an Amberlite XAD copolymer by reversed phase high-pressure liquid chromatography, *Anal. Chem.*, 49, 860, 1977.

Table LC-7
DERIVATIVES OF 2-(4-HYDROXYPHENYL)-2-PHENYLPROPANE

Packing	P1	P1
Column		
Length (cm)	47.7	47.6
Diameter (cm)	2.7	2.7
Material		
Eluent	E1	E2
Flow rate (mℓ/min)	1	1
Temperature (°C)		
Detection	D1	D1
Reference	1	1
Compound	Kd	Kd
2-(4′-Hydroxyphenyl)-2-phenylpropane	1.39	0.57
2-(4′-Hydroxy-3′-nitrophenyl)-2-phenylpropane	1.41	0.56
2-(4′-Hydroxy-3′,5′-dinitrophenyl)-2-phenylpropane	1.83	0.55
2-(4′-Hydroxy-3′,5′-dinitrophenyl)-2-(4″-nitrophenyl)propane	1.98	0.53
2-(4′-Ethoxy-3′-nitrophenyl)-2-phenypropane	1.40	0.54
2-(4′-Methoxy-3′,5′-dinitrophenyl)-2-phenylpropane	1.19	0.04
2-(4′-Methoxy-3′,5′-dinitrophenyl)-2-(4″-nitrophenyl)propane	1.16	0.01
2-(4′-Chloro-3′,5′-dinitrophenyl)-2-phenylpropane	1.27	0.03
2-(4′-Chloro-3′,5′-dinitrophenyl)-2-(4″-nitrophenyl)propane	1.19	0.03
2-(3′-Amino-4′-hydroxy-5′-nitrophenyl)-2-phenylpropane	1.83	0.55
2-(4′-Amino-3′,5′-dinitrophenyl)-2-phenylpropane	1.58	0.55
2-(4′-Amino-3′,5′-dinitrophenyl)-2-(4″-nitrophenyl)propane	1.74	0.49

Packing P1 = Sephadex LH-20
Eluent E1 = methanol
E2 = N,N-dimethylformamide
Detection D1 = Abbé refractometer

From Shopova, B. I., Mladenov, I. T., and Kurtev, K. S., *J. Chromatogr.*, 132, 99, 1977. With permission.

Table LC-8
PHENOLS

Packing	P1	P2	P3	P4	P5	P5	P6
Column							
Length (cm)	50	100	50	16.5	42	42	91
Diameter (cm)	1.8	2.3	2.6	25	5.0	5.0	2
Material (mm)				G			
Eluent	E1	E2	E3	E4	E5	E6	E7
Flow rate (mℓ/min)	0.9	0.5	1.4	0.5			1
Temperature (°C)					22	22	
Detection	D1	D2	D2	D2	D2	D2	D2
Reference	1	2	2	3	4	4	5
Compound	min[a]	min[a]	min[a]	mℓ	mℓ[a]	mℓ[a]	min[a]
2-Hydroxybiphenyl	4.2	—	—	—	—	—	5.2
2-*tert.*-Butyl-4,6-dimethylphenol	—	24	9	—	—	—	—
Butylated hydroxymethoxybenzene	—	—	—	43	—	—	—
2-*tert.*-Butyl-4-methoxyphenol	—	—	—	378	—	—	—
3-*tert.*-Butyl-4-methoxyphenol	—	—	—	478	—	—	—
1,2-Dihydroxybenzene	—	—	—	—	28	11	—
1,4-Dihydroxybenzene	—	—	—	—	32	30.5	—
1,3-Dihydroxybenzene	—	—	—	—	41	40	—

[a] Measured on chromatogram.

Packing
P1 = Bondapak C18, Corasil
P2 = Durapak-OPN, Porasil C
P3 = Cyano-SilX
P4 = Sephadex LH-20
P5 = Merckogel PGM2000
P6 = Corasil II

Eluent
E1 = 40% ethanol-water
E2 = 25% water-saturated isooctane
E3 = 0.03% isopropanol-isooctane
E4 = chloroform-cyclohexane (1:1)
E5 = 0.05 *M* phosphate buffer, pH 6.0
E6 = 0.05 *M* phosphate buffer-2/3M borate buffer, pH 6.0
E7 = heptane-dichloromethane-acetic acid (84:15:0.5)

Detection
D1 = LC-10 amperometric detector (Ag/AgCl)
D2 = UV 280 nm

REFERENCES

1. **Ott, D. E.**, Evaluation of amperometric detector for liquid chromatographic determination of 2-phenylphenol residues in orange rind, *J. Assoc. Off. Anal. Chem.*, 61, 1465, 1978.
2. **Cunningham, A. F. and Hillman D. E.**, Direct determination of 2,4-dimethyl-6-*tert.*-butylphenol in aviation turbine fuel by high-performance liquid chromatography, *J. Chromatogr.*, 148, 528, 1978.
3. **Kauffman, P.**, The separation of butylated hydroxyanisol isomers on Sephadex LH-20, *J. Chromatogr.*, 132, 356, 1977.
4. **Seki, T.**, Chromatographic separation of dihydroxybenzenes on a column of Merckogel PGM 2000, *J. Chromatogr.*, 115, 262, 1975.
5. **Reeder, S. K.**, High-speed liquid chromatographic determination of o-phenylphenol residues in citrus products, *J. Assoc. Off. Anal. Chem.*, 59, 162, 1976.

Table LC-9
PHENOLS

Packing	P1	P2	P3
Column			
Length (cm)	15	120	30
Diameter (mm)	6.5 o.d.	2.3	6.2
Material			
Eluent	E1	E2	E3
Flow rate (mℓ/min)	2.5	1	1
Temperature (°C)			
Detection	D1	D1	D2
Reference	1	2	3
Compound	min	min[a]	min
Camphor	2.76	—	—
4-Chlorophenol	10.64	—	—
Phenol	12.61	—	5.5
1-Hydroxynaphthalene	—	12	—
2-Hydroxynaphthalene	—	17	—
2-Naphthylmethylcarbamate	—	24	—
1-Naphthylmethylcarbamate	—	26	—
2,4′-Isopropylidenediphenol	—	—	14.7
3-Trifluoromethylphenol	—	—	6.1
4,4′-Isopropyridinediphenol	—	—	10.2
2,4-*bis*(4-Hydroxy-α,α-dimethylbenzyl)phenol	—	—	19.2
4,4′-Hydroxyphenyl-2,2,4-trimethylchroman	—	—	21.8

[a] Measured on chromatogram.

Packing	P1 = silica (5 μm)	
	P2 = Corasil II	
	P3 = μBondapak C18/Porasil	
Eluent	E1 = heptane-chloroform (60:40)	
	E2 = heptane-chloroform (4:1)	
	E3 = 45-75% acetonitrile/water, gradient	
Detection	D1 = UV 254 nm	
	D2 = UV 280 nm	

REFERENCES

1. **Musto, J. D., Sane, J. N., and Warner, V. D.,** High-pressure liquid chromatographic determination of camhor and parachlorophenol in camphorated parachlorophenol, *J. Pharm. Sci.*, 67, 266, 1978.
2. **Argauer, R. J. and Warthen, Jr., J. D.,** Separation of 1- and 2-naphthols and determination of 2-naphthylmethylcarbamate in carbaryl formulations by high-pressure liquid chromatography with confirmation by spectrofluorometry, *Anal. Chem.*, 47, 2472, 1975.
3. **Szap, P., Kesse, I., and Klapp, J.,** The analysis of bisphenol A by high performance liquid chromatography, *J. Liquid Chromatogr.*, 1, 89, 1978.

Table LC-10
NATURALLY OCCURRING PHENOLIC COMPOUNDS

Packing	P1	P2
Column		
Length (cm)	30	—
Diameter (mm)	4	16
Material	ss	G
Eluent	E1	E2
Flow rate (mℓ/min)	1.0	—
Temperature (°C)		
Detection	D1	D2
Reference	1	2
Compound	min	mℓ
Neochlorogenic acid	8.6	—
Chlorogenic acid (3-*O*-caffeoylquinic acid)	11.4	—
4-*O*-Caffeoylquinic acid	13.9	—
Esculin	14.9	—
Scopoletin (6-methoxy-7-hydroxycoumarin)	24.2	—
Rutin (quercetin-3-rhamnosidogluside)	33.6	—
Quercitrin	42.2	—
Quercetin (3,3′,4′,5,7-pentahydroxyflavone)	67.1	48.5
Scopolin (scopoletin-7-glucoside)	17.7	—
3,4-Dihydroxycinnamic acid	15.5	—
Detran blue	—	14
Tannic acid	—	14
bisFlavanol A	—	19.5
(−)-Epigallocatechin gallate	—	23.5
(−)-Epicatechin gallate	—	24.8
(+)-Catechin	—	29.
1,2,3-Trihydroxybenzene	—	34
Theaflavin digallate	—	21.5
Theaflavin monogallate	—	24.5
Theaflavin	—	27
Epitheaflavic acid	—	23.5
3′,4′-Dihydroxy-6,7-benzotropolone	—	50

Packing P1 = μBondapak C18 (10 μm)
P2 = Sephadex LH-20

Eluent E1 = 16% methanol/0.1 *N* KH_2FO_4 - 40% methanol/0.1N KH_2PO_4, 15 min gradient
E2 = 70% aqueous DMF

Detection D1 = UV 350 nm
D2 = $Fe^{3+}/Fe(CN)_6^{3-}$ solution after TLC

REFERENCES

1. **Court, W. A.**, High-performance reversed-phase liquid chromatography of naturally occurring phenolic compounds, *J. Chromatogr.*, 130, 287, 1977.
2. **Wilkins, C. K.**, Chromatography of tea polyphenols on Sephadex columns as a method of estimation of molecular size, *J. Chromatogr.*, 87, 250, 1973.

Table LC-11
ALKYL AND AMINOPHENOLS

Packing	P1	P1	P2	P3	P4
Column					
Length (cm)	100	100	50	50	100
Diameter (mm)	2.1	2.1	3	3	2.1
Material					
Eluent	E1	E2	E3	E4	E5
Flow rate (mℓ/min)	0.7	0.5	0.48	0.50	0.64
Temperature (°C)	50	50	60	70	25
Detection	D1	D1	D1	D1	D1
Reference	1	1	1	1	2
Compound	mℓ	mℓ	min[a]	min[a]	min[a]
Phenol	1.51	1.50	10.9	13.0	—
2-Methylphenol	1.97	1.77	16.0	17.0	—
3-Methylphenol	1.86	1.71	14.2	16.0	—
4-Methylphenol	1.85	1.72	—	16.0	—
2-Ethylphenol	3.16	2.59	22.0	23.5	—
3-Ethylphenol	0.35	2.27	19.0	20.0	—
4-Ethylphenol	0.35	2.34	—	21.5	—
2-*n*-Propylphenol	0.98	4.89	—	34.5	—
4-*n*-Propylphenol	6.32	4.34	28.0	29.8	—
2-Isopropylphenol	6.16	4.36	—	—	—
3-Isopropylphenol	4.70	3.33	—	—	—
4-Isopropylphenol	5.00	3.58	—	—	—
2-*sec.*-Butylphenol	3.57	11.0	—	45.5	—
2-*tert.*-Butylphenol	22.3	16.0	48.7	56.0	—
3-*tert.*-Butylphenol	8.27	5.32	—	—	—
4-*tert.*-Butylphenol	9.01	5.65	31.2	—	—
4-Aminophenol	—	—	—	—	4.0
2-Amino-4-methylphenol	—	—	—	—	6.0
2-Amino-5-methylphenol	—	—	—	—	10.0
3-Amino-6-methylphenol	—	—	—	—	26.0
4-Aminomethylbenzene	—	—	—	—	49.0

[a] Measured on chromatogram.

Packing
P1 = Permaphase ODS
P2 = Hitachigel 3011 (polystyrene gel)
P3 = Hitachigel 3010-0 (alcoholic polystyrene gel)
P4 = Zipax SCX

Eluent
E1 = methanol-water (2:98)
E2 = methanol-water (1:9)
E3 = acetonitrile-methanol-water (1:6:3)
E4 = acetonitrile-methanol-water (1:5:4)
E5 = 0.1 $M H_3PO_4$-KH_2PO_4 (pH 3)

Detection
D1 = UV 254 nm

REFERENCES

1. **Nomura, N.,** Automatic liquid chromatography of organic compounds. VI. Selection of the composition of eluent to separate alkylphenols. U.S.-Japan seminar on liquid chromatography, Boulder, Co., June, 1978.
2. **Sakurai, H. and Kito, M.,** Determination of aminocresol isomers by high-speed liquid chromatography, *Talanta*, 23, 842, 1976.

Table LC-12
ALKYL PHENOLS

Packing	P1	P2	P3
Column			
Length (cm)	120 × 5	30 × 5	50
Diameter (mm)	8	6.3	10
Material		ss	G
Eluent	E1	E2	E3
Flow rate (mℓ/min)	0.5	1.5	
Temperature (°C)			
Detection	D1	D2	D3
Reference	1	2	3
Compound	c × 2.7mℓ	mℓ	mℓ
Phenol	83.0	—	15
2-Methylphenol	82.8	—	—
2-Ethylphenol	80.7	—	—
2-Isopropylphenol	78.6	—	—
2-*tert.*-Butylphenol	76.6	—	—
2-*tert.*-Octylphenol	73	—	—
2-Hydroxybiphenyl	79.5	—	—
3-Methylphenol	81.8	—	—
3-*tert.*-Butylphenol	76.8	—	—
4-Methylphenol	81.7	—	—
4-Ethylphenol	80.0	—	—
4-Isopropylphenol	77.4	—	—
4-*sec.*-Butylphenol	76.5	—	—
4-*tert.*-Butylphenol	76.2	—	—
4-*tert.*-Amylphenol	75.2	—	—
4-*tert.*-Octylphenol	72.0	—	—
4-Cyclohexylphenol	77.1	—	—
4-Hydroxybiphenyl	78.2	—	—
2,3-Dimethylphenol	81.5	—	—
2,4-Dimethylphenol	81.1	—	—
4-*tert.*-Butyl-2-methylphenol	77.0	—	—
2-Methyl-4-*tert.*-octylphenol	71.3	—	—
2-*tert.*-Butyl-4-methylphenol	75.9	—	—
4-Methyl-2-*tert.*-octylphenol	71.9	—	—
2,4-Di-*tert*-butylphenol	70.8	—	—
4-*tert.*-Butyl-2-*tert.*-octylphenol	68.2	—	—
2,4-Di-*tert.*-octylphenol	65.5	—	—
2,5-Dimethylphenol	81.3	—	—
2-*tert.*-Butyl-5-methylphenol	75.5	—	—
2,6-Dimethylphenol	81.7	—	—
2-*tert.*-Butyl-6-methylphenol	77.5	—	—
2-Methyl-6-*tert.*-octylphenol	73.1	—	—
2,6-Di-*tert.*-butylphenol	75.0	—	—
2,6-Di-*tert.*-octylphenol	67.9	—	—
2,4,6-Trimethylphenol	80.8	—	—
2-*tert.*-Butyl-4,6-dimethylphenol	76.1	—	—
4-*tert.*-Butyl-2,6-dimethylphenol	75.1	—	—
2,6-Dimethyl-4-*tert.*-octylphenol	70.9	—	—
2,4-Di-*tert.*-butyl-6-methylphenol	71.7	—	—
2,6-Di-*tert.*-butyl-4-methylphenol	73.6	—	—
2,4,6-Tri-*tert.*-butylphenol	70.0	—	—
1-Hydroxynaphthalene	81.5	—	—
2-Hydroxynaphthalene	81.0	—	—
1,2-Dihydroxybenzene	78.0	—	—
1,2-Dihydroxy-3-methylbenzene	77.3	—	—
3-Ethyl-1,2-dihydroxybenzene	75.8	—	—

Table LC-12 (continued)
ALKYL PHENOLS

Compound	c × 2.7mℓ	mℓ	mℓ
1,2-Dihydroxy-3-isopropylbenzene	74.5	—	—
3-*sec.*-Butyl-1,2-dihydroxybenzene	73.8	—	—
3-*tert.*-Butyl-1,2-dihydroxybenzene	73.2	—	—
1,2-Dihydroxy-3-*tert.*-octylbenzene	69.5	—	—
1,2-Dihydroxy-4-methylbenzene	77.2	—	—
4-Ethyl-1,2-dihydroxybenzene	75.3	—	—
1,2-Dihydroxy-4-isopropylbenzene	74.3	—	—
4-*n*-Butyl-1,2-dihydroxybenzene	72.3	—	—
4-*sec.*-Butyl-1,2-dihydroxybenzene	72.3	—	—
4-*tert.*-Butyl-1,2-dihydroxybenzene	72.3	—	—
4-*tert.*-Amyl-1,2-dihydroxybenzene	71.9	—	—
1,2-Dihydroxy-4-*tert.*-octylbenzene	69.1	—	—
4-*tert.*-Dodecyl-1,2-dihydroxybenzene	66.1	—	—
1,2-Dihydroxy-3,5-dimethylbenzene	76.2	—	—
5-*tert.*-Butyl-1,2-dihydroxy-3-methylbenzene	77.2	—	—
3,4-Dihydroxy-5-methyl-*tert.*-octylbenzene	68.8	—	—
3-*tert.*-Butyl-4,5-dihydroxy-methylbenzene	72.3	—	—
3,5-Di-*tert.*-butyl-1,2-dihydroxybenzene	69.2	—	—
1,2-Dihydroxy-3,5-di-*tert.*-octylbenzene	64.2	—	—
1,2-Dihydroxy-3,6-dimethylbenzene	77.6	—	—
4-*tert.*-Butyl-2,3-dihydroxy-methylbenzene	73.0	—	—
3,6-Di-*tert.*-butyl-1,2-dihydroxybenzene	70.3	—	—
1,4-Dihydroxybenzene	77.2	—	—
1,4-Dihydroxy-2-methylbenzene	76.4	42	—
1,4-Dihydroxy-2-isopropylbenzene	73.4	—	—
2-*tert.*-Butyl-1,4-dihydroxybenzene	72.1	—	—
1,4-Dihydroxy-2,5-dimethylbenzene	75.7	—	—
4-*tert.*-Butyl-2,5-dihydroxy-methylbenzene	71.2	—	—
2,5-Di-*tert.*-butyl-1,4-dihydroxybenzene	67.8	—	—
1,4-Dihydroxy-2,6-dimethylbenzene	75.5	—	—
3-*tert.*-Butyl-2,5-dihydroxy-methylbenzene	72.1	—	—
2,6-Di-*tert.*-butyl-1,4-dihydroxybenzene	69.7	—	—
1,3-Dihydroxybenzene	77.0	—	—
1,2,3-Trihydroxybenzene	74.3	—	—
1,3,5-Trihydroxybenzene	—	42	—
2,4,6-Trihydroxymethylbenzene	—	42	—
1,3-Dihydroxybenzene	—	42	—
4-Ethyl-1,3-dihydroxybenzene	—	42	—
Catechin	—	38	—
2,2′-Dihydroxybiphenyl	—	—	11

Packing	P1 = S-gel-832 styrene-divinylbenzene copolymer (Institute of Macromolecular Chemistry, Czechoslovak Academy of Science)
	P2 = μStyragel® (10^3Å, 500Å, 100Å)
	P3 = hydrated Bio-gel® P-2
Eluent	E1 = tetrahydrofuran
	E2 = tetrahydrofuran or chloroform
	E3 = water
Detection	D1 = RI
	D2 = RI and UV 254 nm
	D3 = FP 398 nm

REFERENCES

1. **Čoupek, J., Pokorný, S., and Pospíšil, J.,** Gel chromatographic behaviour of mononuclear aromatic hydrocarbons and phenols, *J. Chromatogr.*, 95, 103, 1974.
2. **Hemingway, R. W. and McGraw, G. W.,** Formaldehyde condensation products of model phenols for conifier bark tannins, *J. Liquid Chromatogr.*, 1, 163, 1978.

Table LC-12 (continued)
ALKYL PHENOLS

3. **Danner, D. J., Brignac, P. J., Jr., Arceneaux, D., and Patel, V.,** The oxidation of phenol and its reaction product by horseradish peroxidase and hydrogen peroxide, *Arch. Biochem. Biophys.,* 156, 759, 1973.

Table LC-13
PHENOLIC COMPOUNDS

Packing	P1	P1
Column		
Length (cm)	25 × 2	25 × 2
Diameter (mm)	25	25
Material	G	G
Eluent	E1	E2
Flow rate (mℓ/min)		
Temperature (°C)		
Detection	D1	D1
Reference	1	1
Compound	mℓ	mℓ
Aroclor 1254	160-190	160-190
Aroclor 1016	150-186	150-188
Hexachlorobenzene	168-199	168-198
Naphthalene	168-195	170-195
Hexachlorobutadiene	144-168	148-169
p,p'-DDT	144-171	152-178
2-Chlorophenol	162-178	170-198
Pentachloromethoxybenzene	156-174	172-196
2,4,6-Tribromomethoxybenzene	158-176	172-198
2,4-Dibromomethoxybenzene	152-169	168-202
4-Bromomethoxybenzene	156-174	174-200
2-Bromophenol	156-181	174-208
2,4-Dibromophenol	162-183	182-211
2,4-Dichlorophenol	170-195	190-218
2,4,6-Trichlorophenol	172-197	192-244
Pentachlorophenol	166-196	186-216
3,4-Dichloroaniline	170-195	200-255
Dipentylamine	154-180	184-214
1-Naphthylamine	166-192	206-237
3-Chloroaniline	172-200	216-246
2,4,6-Tribromophenol	147-161	194-214
4-Bromoindole	167-181	224-270
1,2,4-Trichlorobenzene	174-200	234-266
4-Chlorophenol	180-210	240-284
4-Bromophenol	154-176	220-244
Pyrene	176-200	252-294
Phenanthrene	168-195	250-284

Packing P1 = Bio-Rad® SX-2 (100-200 mesh)
Eluent E1 = dichloromethane
E2 = cyclohexane-dichloromethane (50-50)
Detection D1 = UV 254 nm

Table LC-14
PHENOLS AND ORGANIC ACIDS

Packing	P1	P1	P2	P2	P3
Column					
Length (cm)	100	100	100	100	79
Diameter (mm)	2.2	2.2	2.2	2.2	4.3
Material	G	G	G	G	G
Eluent	E1	E2	E1	E2	E1
Flow rate (mℓ/min)	1.2	1.2	1.2	1.2	0.5
Temperature (°C)	30	30	30	30	29
Detection	D1	D1	D1	D1	D2
Reference	1	1	1	1	2
Compound	mℓ	mℓ	mℓ	mℓ	Dv
Benzene	0.50	0.48	0.09	0.09	1.62
Naphthalene	8.96	8.29	2.87	2.55	18.1
Phenol	0.84	0.78	0.16	0.17	4.96
1,2-Dihydroxybenzene	0.77	0.71	0.13	0.14	7.60
1,3-Dihydroxybenzene	1.50	1.35	0.30	0.27	21.5
1,4-Dihydroxybenzene	0.93	0.88	0.17	0.14	13.5
1,3,5-Trihydroxybenzene	2.53	2.36	0.50	0.50	>50
2-Methylphenol	1.53	1.42	0.37	0.32	6.83
3-Methylphenol	1.64	1.49	0.38	0.35	7.06
4-Methylphenol	1.57	1.42	0.39	0.35	6.66
2,4-Dimethylphenol	2.93	2.73	0.87	0.74	9.16
3,4-Dimethylphenol	2.96	2.86	0.82	0.72	9.74
2,6-Dimethylphenol	2.01	1.86	0.52	0.44	6.68
2,4,5-Trimethylphenol	5.92	5.61	1.78	1.56	13.1
2-Ethylphenol	2.58	2.67	0.78	1.56	8.30
3-Ethylphenol	2.88	2.67	0.81	0.74	8.42
4-Ethylphenol	2.72	2.54	0.77	0.70	7.78
2-Methoxyphenol	0.61	0.58	0.09	0.09	4.26
3-Methoxyphenol	1.43	1.30	0.32	0.27	9.11
4-Methoxyphenol	0.95	0.89	0.19	0.16	6.47
2,6-Dimethoxyphenol	0.48	0.45	0.07	0.06	3.85
2-Methoxy-4-propylphenol	3.53	3.24	1.06	0.91	7.19
4-Allyl-2-methoxyphenol	2.48	2.36	0.75	0.64	7.21
Methoxybenzene	0.67	0.61	0.14	0.13	2.40
1,2-Dimethoxybenzene	0.47	0.44	0.08	0.07	2.50
1,3-Dimethoxybenzene	1.06	0.93	0.26	0.24	3.85
1,4-Dimethoxybenzene	0.86	0.84	0.18	0.16	3.24
1,2,3-Trimethoxybenzene	0.42	0.44	0.05	0.05	2.05
1,3,5-Trimethoxybenzene	2.00	1.75	0.49	0.43	7.21
Benzoic acid	2.00	2.40	0.60	0.90	6.36
2-Hydroxybenzoic acid	640	640	18.0	>40	37.2
3-Hydroxybenzoic acid	5.0	5.6	1.5	2.1	34.0
4-Hydroxybenzoic acid	5.4	4.8	1.3	1.4	32.9
2,5-Dihydroxybenzoic acid	650	650	18	>50	650
4-Hydroxy-3-methoxybenzoic acid	4.1	3.8	0.9	1.1	29.0
Benzaldehyde	0.44	0.42	0.05	0.02	1.55
2-Hydroxybenzaldehyde	0.79	0.75	0.15	0.14	2.69
3-Hydroxybenzaldehyde	1.35	1.22	0.26	0.23	6.04
4-Hydroxybenzaldehyde	1.78	1.54	0.36	0.29	6.54
2,4-Dihydroxybenzaldehyde	3.37	3.12	0.85	0.70	14.7
4-Hydroxy-3-methoxybenzaldehyde	1.38	1.22	0.27	0.23	6.51
1,3-Dihydroxynaphthalene	43.8	39.8	13.5	12.2	650
2,7-Dihydroxynaphthalene	41.0	32.8	9.66	8.8	650
2,6-Di-*tert*.-butylphenol	650	650	650	650	—
2,5-Dihydroxy-methylbenzene	—	—	—	—	31.7
Benzquinone	0.10	0.08	.00	0.00	0.60

Table LC-14 (continued)
PHENOLS AND ORGANIC ACIDS

Compound	mℓ	mℓ	mℓ	mℓ	Dv
Benzylalcohol	0.31	0.30	0.02	0.03	1.36
2-Phenylethanol	0.33	0.36	0.02	0.03	1.39
3-Phenyl-1-propanol	0.63	0.64	0.10	0.10	1.88
3,4,5-Trihydroxybenzoic acid	—	—	—	—	650
3,5-Di-*tert*.-butyl-4-hydroxybenzoic acid	—	—	—	—	650
3,4-Dihydroxybenzaldehyde	—	—	—	—	16.5
3-Hydroxy-4-methoxybenzaldehyde	—	—	—	—	6.32
4-Hydroxy-3-methoxyphenylacetaldehyde	—	—	—	—	6.53
4-Methoxybenzaldehyde	—	—	—	—	2.21
3,3′,4′,5,7-Pentahydroxyflavone-3-L-rhamnoside	—	—	—	—	650
5,5-Dimethyl-1,3-cyclohexanedione	—	—	—	—	1.51
Nitrobenzene	—	—	—	—	1.79
3-Nitrophenol	—	—	—	—	8.20
4-Nitrophenol	—	—	—	—	9.07
2-Nitrobenzaldehyde	—	—	—	—	1.60
4-Nitrobenzaldehyde	—	—	—	—	1.25
2,4,6-Trinitrophenol	—	—	—	—	650
3-Nitrobenzoic acid	—	—	—	—	650
Formic acid	—	—	—	—	0.76
Acetic acid	—	—	—	—	0.68
Acetone	—	—	—	—	0.48

Packing P1 = Perisorb® PA-6
P2 = Pellidon
P3 = Polyclar® (polyvinylpyrrolidone, 60-75 μm)
Eluent E1 = 0.001 *M* HCl
E2 = 0.1 *M* acetic acid
Detection D1 = UV 254 nm
D2 = UV 280 nm

REFERENCES

1. **Olsson, L. and Samuelson, O.**, Chromatography of aromatic acids, aldehydes and phenols on pellicular polyamide in aqueous solution, *J. Chromatogr.*, 106, 139, 1975.
2. **Olsson, L. and Samuelson, O.**, Chromatography of aromatic acids, aldehydes and phenols on cross-linked polyvinylpyrrolidone, *J. Chromatogr.*, 93, 189, 1974.

Table LC-15
ACIDIC COMPOUNDS

Packing	P1	P1	P1	P1	P1	P1	P1
Column							
Length (cm)	18.5	18.5	18.5	18.5	18.5	18.5	18.5
Diameter (mm)	3	3	3	3	3	3	3
Material	G	G	G	G	G	G	G
Eluent	E1	E1	E1	E2	E3	E4	E5
Flow rate (mm/sec)	0.7	0.7	0.7	0.7	0.7	0.7	0.7
Temperature (°C)	25	45	65	25	25	25	25
Detection	D1	D1	D1	D1	D1	D1	D1
Reference	1	1	1	1	1	1	1
Compound	k′	k′	k′	k′	k′	k′	k′
Phenol	1.03	0.89	0.72	1.07	1.07	1.11	0.34
3-Methylphenol	3.03	2.50	1.92	3.05	3.06	3.16	0.94

Table LC-15 (continued)
ACIDIC COMPOUNDS

Compound	k′	k′	k′	k′	k′	k′	k′
4-Methylphenol	3.13	2.61	1.99	3.21	3.26	3.33	0.93
2-Methylphenol	3.35	2.82	2.13	3.35	3.37	3.51	1.28
4-Nitrophenol	3.49	2.60	1.85	3.41	3.48	4.29	0.31
2-Nitrophenol	5.16	4.08	2.95	5.17	5.12	5.52	0.15
2,6-Dimethylphenol	6.05	5.63	4.41	6.57	6.62	6.93	1.82
2,5-Dimethylphenol	8.33	7.06	5.10	8.53	8.59	9.03	3.25
2-Aminobenzoic acid	—	—	—	0.11	0.14	0.02	0
2,4-Dihydroxybenzoic acid	0.49	0.37	0.27	0.58	0.80	0.83	0
Benzoic acid	0.85	0.72	0.57	0.92	0.85	0.77	0
4-Nitrobenzoic acid	1.09	0.98	0.73	1.17	2.07	2.74	0
2-Hydroxybenzoic acid	1.73	1.43	1.12	3.12	5.62	5.48	0
4-Methylbenzoic acid	2.43	2.02	1.50	2.47	2.53	2.04	0
2-Amino-4-nitrobenzoic acid	2.89	1.99	1.29	—	—	—	—
Cinnamic acid	3.03	2.39	1.71	3.12	3.37	3.20	0
Benzenesulfonic acid	0.42	0.62	0.82	0.43	0.46	0.42	0
4-Nitrobenzenesulfonic acid	1.34	1.69	1.97	—	—	—	—
4-Methylbenzenesulfonic acid	1.37	1.95	2.40	1.18	1.45	1.30	0
2-Hydroxy-6-naphthalenesulfonic aid	2.82	2.90	2.92	2.96	2.97	2.97	0
1-Hydroxy-2-naphthalenesulfonic acid	3.30	3.85	4.00	3.70	3.68	3.65	0
1-Hydroxy-4-naphthalenesulfonic acid	3.37	3.81	3.95	3.61	3.68	3.63	0
1-Hydroxy-5-naphthalenesulfonic acid	4.10	4.80	4.97	4.55	4.67	4.71	0
1-Hydroxy-3-naphthalenesulfonic acid	10.17	9.54	7.33	10.53	10.65	10.22	0
1-Hydroxy-3,6-naphthalenedisulfonic acid	0.45	0.69	0.98	0.42	0.46	0.40	0
4,5-Dihydroxy-2,7-naphthalenedisulfonic acid	0.53	0.75	1.05	—	—	—	—
2-Hydroxy-3,6-naphthalenedisulfonic acid	2.00	3.85	6.72	1.98	2.29	1.92	0
1-Hydroxy-3,8-naphthalenedisulfonic acid	15.14	22.61	28.37	14.51	15.23	14.75	0

Packing P1 = Spherosil XOC 005, coated with tri-*n*-octylamine
Eluent E1 = 0.05 $M HClO_4$, pH 1.5
E2 = 0.05 $M HClO_4$, pH 2.5
E3 = 0.05 $M HClO_4$. pH 4.1
E4 = 0.05 $M HClO_4$, pH 6.0
E5 = 0.05 $M HClO_4$, pH 7.5
(pH was adjusted with NaOH soln.)
Detection D1 = UV 263 nm

From Kraak, J. C. and Huber, J. F. K., *J. Chromatogr.*, 102, 333, 1974. With permission.

Table LC-16
PHTHALIC ACIDS

Packing	P1	P2	P3	P4
Column				
Length (cm)	30	100	30	30.3
Diameter (mm)		2.1	4	2.6
Material		ss		
Eluent	E1	E2	E3	E4
Flow rate (mℓ/min)	3	0.95	2	2.4
Temperature (°C)		rt		
Reference	1	2	3	4
Compound	min[a]	mℓ[a]	min[a]	Tr
1,2-Benzenedicarboxylic acid didodecylester	6.3	—	—	—
1,2-Benzenedicarboxylic acid diethylhexyl ester	6.3	—	20.0	0.821

Table LC-16 (continued)
PHTHALIC ACIDS

Compound	min[a]	mℓ[a]	min[a]	Tr
1,2-Benzenedicarboxylic acid didecyl ester	7.3	—	—	—
1,2-Benzenedicarboxylic acid dioctyl ester	8.1	—	—	0.812
1,2-Benzenedicarboxylic acid dipentyl ester	12.4	—	17.5	—
1,2-Benzenedicarboxylic acid dibutyl ester	15.0	5.3	16.5	—
1,2-Benzenedicarboxylic acid dipropyl ester	18.1	6.2	—	—
1,2-Benzenedicarboxylic acid diethyl ester	25.7	8.4	—	—
1,2-Benzenedicarboxylic acid dimethyl ester	28.1	11	—	1.020
1,2-Benzenedicarboxylic acid dinonyl ester	—	3.0	21.0	0.798
1,2-Benzenedicarboxylic acid diisodecyl ester	—	3.4	—	—
1,2-Benzenedicarboxylic acid diheptyl ester	—	3.5	—	—
1,4-Benzenedicarboxylic acid diethyl ester	—	3.9	—	—
1,4-Benzenedicarboxylic acid dimethyl ester	—	5.1	—	—
1,2-Benzenedicarboxylic acid diisobutyl ester	—	5.1	—	—
1,2-Benzenedicarboxylic acid diisopropyl ester	—	6.9	—	—
1,2-Benzenedicarboxylic acid dihexyl ester	—	19.0	—	—
Tristearin	—	—	—	0.557
Triolein	—	—	—	0.577
Sudan black	—	—	—	1.000
Phosphatidyl ethanolamine	—	—	—	0.403
Trilaurin	—	—	—	0.641
Cholesterol	—	—	—	0.873

[a] Measured on chromatogram.

Packing P1 = Porasil (10 μm)
P2 = Corasil II
P3 = μBondapak C18
P4 = Biobeads S-X2
Eluent E1 = chloroform-isooctane (50:50)
E2 = ethylacetate-*n*-hexane (1:50)
E3 = methanol-water, gradient
E4 = benzene
Detection D1 = UV 254nm
D2 = RI

REFERENCES

1. **Persiani, C. and Cukor, P.,** Liquid chromatographic method for the determination of phthalate esters, *J. Chromatogr.,* 109, 413, 1975.
2. **Funasaka, W., Hanai, T., and Fujimura, K.,** High-speed liquid chromatographic separations of phthlic esters, carbohydrates, TCA organic acids and organic mercury compounds, *J. Chromatogr. Sci.,* 12, 517, 1974.
3. **Otsuki, A.,** Reversed-phase adsorption of phthalate esters from aqueous solutions and their gradient elution using a high-performance liquid chromatograph, *J. Chromatogr.,* 133, 402, 1977.
4. **Baker, R. W. R.,** Gel filtration of phthalate esters, *J. Chromatogr.,* 154, 3, 1978.

Table LC-17
ORGANIC ACIDS

Packing	P1	P1	P2	P3	P4	P5	P6	P7	P7	P7	P8	P9
Column												
Length (cm)	100	100	25	50	25	30	20	23	22	22	122	15
Diameter (mm)	2.1	2.1	2.5	2.1	2.3	9	9	12.7	12.7	2.8	8	
Material	ss	ss	G	ss	ss				G	G		
Eluent	E1	E2	E3	E4	E5	E6	E7	E7	E8	E7	E9	E10
Flow rate (mℓ/min)	1.2	1.1	frg[a]	1	1.5	1	1.1	0.8	1	1	—	—
Temperature (°C)	rt	rt	rt	rt	rt	19			rt	rt	—	—
Detection	D1	D1	D2	D2	D2	D3	D1	D1	D1	D1	D4	
Reference	1	1	2	2	2	3	4	4	5	5	6	7
Compound	mℓ[b]	mℓ[b]	min[b]	min[b]	min[b]	mℓ	min[b]	min[b]	mℓ[b]	mℓ[b]	mℓ	mℓ
Fumaric acid	2.4	—	3.5	0.8	—	2.56	8.0	14.0	—	22.0	—	—
Succinic acid	4.8	—	4.0	1.1	3.0	9.0	—	—	—	24.0	—	26.0
Glycolic acid	8.0	—	5.2	1.6	4.7	—	—	—	—	—	8.74	16.0
Citric acid	25	—	16.3	7.0	10.7	—	—	—	—	14.0	—	—
Oxalic acid	—	38	—	—	—	—	—	—	—	—	—	60.0
Lactic acid	—	7.4	—	—	—	9.0	—	—	—	—	—	—
2-Oxoglutaric acid	—	12.9	12.1	2.8	4.3	18.0	—	—	—	10.0	—	—
trans-Aconitic acid	—	15.9	—	—	—	—	—	—	—	—	—	—
1-Malic acid	—	52.8	9.0	3.5	6.4	—	4.0	7.0	—	18.0	23.85	—
Acetic acid	—	—	1.7	—	—	2.32	—	17.0	—	—	—	—
Pyruvic acid	—	—	8.	—	—	3.62	—	—	—	—	—	—
cis-Aconitic acid	—	—	23.0	—	—	—	—	—	—	9.0	—	—
Glutaric acid	—	—	—	—	—	4.12	—	—	—	—	—	—
3-Hydroxybutyric acid	—	—	—	—	—	7.76	—	—	—	—	—	—
Citraconic acid	—	—	—	—	—	—	—	9.0	—	—	—	—
Acrylic acid	—	—	—	—	—	—	—	20.0	—	—	—	—
Trichloroacetic acid	—	—	—	—	—	—	—	—	8	—	—	—
Dichloroacetic acid	—	—	—	—	—	—	—	—	12	—	—	—
Chloroacetic acid	—	—	—	—	—	—	—	—	27	—	—	—
Oxaloacetic acid	—	—	—	—	—	—	—	—	—	11	—	—

Isocitric acid	—	—	—	—	—	—	—	—	—	16	—	—
Malonic acid	—	—	—	—	—	—	—	—	—	—	—	38

[a] Flow gradient 1.5 – 3.0 mℓ/min
[b] Measured on Chromatogram.

Packing P1 = Corasil II
P2 = TSKLS140 (8 μm porous polyacryl gel)
P3 = Corasil C18
P4 = Yanapak SIL (5 μm, dimethylsilanized silica)
P5 = silica (25-28 μm)
P6 = Dowex® 50W-X4 (100 μm)
P7 = Aminex 50W-Xr
P8 = IRCG400 formate (100-200 mesh)
P9 = Chelex 100 (100-200 mesh), $Ni(NH_4)_6^{2+}$ form

Eluent E1 = isopropanol-*n*-hexane (3:100)
E2 = isopropanol-*n*-hexane (1:50)
E3 = *n*-hexane-tetrahydrofuran-*tert*.-butanol (100:5:5)
E4 = *n*-hexane *tetrahydrofuran-tert*.-butanol (100:3:2)
E5 = *n*-hexane-tetrahydrofuran-*tert*.-butanol (100:2:2)
E6 = *tert*-amyl alcohol-chloroform-0.1 *N* H_2SO_4 (6:50:50)
E7 = 0.001 *N* HCl
E8 = 0.01 *N* HCl
E9 = 0.2 *M* formaic acid
E10 = 10^{-5} m NH_4OH

Detection D1 = UV 210nm
E2 = UV 220nm
D3 = *o*-Nitrophenol (432nm)
D4 = RI

REFERENCES

1. **Funasaka, W., Hanai, T., and Fujimura, K.**, High-speed liquid chromatographic separation of phthalic esters, carbohydrates, TCA organic acids and organic mercury compounds, *J. Chromatog. Sci.*, 12, 517, 1974.
2. **Hyakutake, H. and Hanai, T.**, Further studies on practical high-speed liquid chromatographic separations of tricarboxylic organic acids and carbohydrates, *J. Chromatogr.*, 108, 385, 1975.

Table LC-17 (continued)
ORGANIC ACIDS

3. **Stahl, K. W., Schäffer, G., and Lamprecht, W.,** Design of a high efficiency liquid chromatograph using specific detection and its evaluation for analysis of tricarboxylic acid cycle intermediates and related compounds on a non-equivalent scale, *J. Chromatogr. Sci.*, 10, 95, 1972.
4. **Richards, H.,** Separation of mono- and dicarboxylic acids by liquid chromatography, *J. Chromatogr.*, 115, 259, 1975.
5. **Turkelson, V. T. and Rochards, M.,** Separation of the citric acid cycle acids by liquid chromatography, *Anal. Chem.*, 50, 1420, 1978.
6. **Linder, D. E.,** The separation and determination of glycolic and malic acids by ion-exchange chromatography, *Anal. Chim. Acta*, 75, 355, 1975.
7. **Bedetti, R., Carunchio, V., and Marino, A.,** An application of the ligand-exchange method to the separation of some aliphatic carboxylic acids, *J. Chromatogr.*, 95, 127, 1974.

Table LC-18
ORGANIC ACIDS

Packing	P1	P2	P3	P3	P3	P4	P5
Column							
Length (cm)	600	45	25	25	25	25	30
Diameter (mm)	4.6	10	4	4	4	4	4
Material							
Eluent	E1	E2	E3	E3	E3	E4	E5
Flow rate (mℓ/min)	2	1	3.6	4.85	3.22	1	0.5
Temperature (°C)	—	—	40	40	40	30	—
Detection	D1	D2	D3	D4	D3	D5	D6
Reference	1	2	3	3	3	4	5
Compound	min	mℓ[a]	min[a]	min[a]	min[a]	mℓ	mℓ
Mesaconic acid	46	—	—	—	—	—	—
Pyruvic acid	94	—	—	—	—	—	—
Fumaric acid	108	32.5	—	18.0	—	—	—
Glutaric acid	148	—	10.8	—	—	—	—
Lactic acid	170	—	—	—	—	3.95	—
3-Hydroxybutyric acid	170	—	—	—	—	—	—
Succinic acid	214	57.5	6.5	7.5	—	—	—
Malonic acid	270	—	4.0	—	4.5	—	—
trans-Aconitic acid	300	—	—	—	—	—	—
Pyrrolidonecarboxylic acid	320	—	—	—	—	—	—
Oxalic acid	360	—	—	—	—	—	—
Citramalic acid	410	—	—	—	—	—	—
cis-Aconitic acid	424	—	—	—	—	—	—
Citric acid	506	—	—	—	—	—	5.5
Isocitric acid	520	—	—	—	—	—	—
Glyceric acid	538.0	—	—	—	—	—	—
Tartaric acid	560.0	—	—	—	—	—	—
2-Oxoglutaric acid	—	95.0	—	—	—	—	—
Malic acid	—	165.0	—	3.0	—	—	3.52
Adipic acid	—	—	18.0	—	—	—	—
Methylmalonic acid	—	—	—	—	9.0	—	—
Glyoxylic acid	—	—	—	—	—	2.54	—
Formic acid	—	—	—	—	—	3.02	—
Acetic acid	—	—	—	—	—	4.27	—
Propionic acid	—	—	—	—	—	8.86	—
Quinic acid	—	—	—	—	—	—	3.02

[a] Measured on chromatogram.

Packing
P1 = Bio-Sil A(20-44 μm)
P2 = DEAE Sephadex A25
P3 = Corasil II C9
P4 = Nucleosil 10, C18
P5 = μBondapak C18

Eluent
E1 = 4.5% *tert.*-amyl alcohol-chloroform-cyclohexane (50:50)
E2 = ethyl ether-formic acid-water (10:2:1)
E3 = methanol-water (32:68)
E4 = pH 3.53 water (H_2SO_4)
E5 = 2% potassium phosphate buffer (pH 2.4)

Detection
D1 = UV 254 nm, 313 nm and 2-nitrophenol reaction detector (435 nm)
D2 = ribose-aniline
D3 = UV 254 nm analyzed as phenacylester
D4 = UV 254 nm analyzed as 4-nitrobenzylester
D5 = UV 210 nm
D6 = RI

Table LC-18 (continued)
ORGANIC ACIDS

REFERENCES

1. **Morot-Gaudy, J. F., Fiala, V., Huet, J. C., and Jolivet, F.,** Improvements in the automatic determination of organic acids from plant tissues by liquid-liquid chromatography on a silica gel column, *J. Chromatogr.*, 117, 279, 1976.
2. **Bruinsma, B. L. and Le Tourneau, D.,** Separation of Krebs cycle acids with beaded diethylaminoethyl cellulose, *J. Assoc. Off. Anal. Chem.*, 61, 751, 1978.
3. **Grushka, E., Durst, H. D., and Kikta, Jr., E. J.,** Liquid chromatographic separation and detection of nanogram quantities of biologically important dicarboxylic acids, *J. Chromatogr.*, 112, 673, 1975.
4. **Distler, W.,** Trennung von Milch-, Glyoxyl-, Ameisen-, Essig- und Propionsäure mit Hilfe der Umkehrphasen-Hochdrükflüssigkeitschchromatographie, *J. Chromatogr.*,152, 250, 1978.
5. **Coppola, E. D., Conrad, E. C., and Cotter, R.,** High pressure liquid chromatographic determination of major organic acids in cranberry juice, *J. Assoc. Off. Anal. Chem.*, 61, 1490, 1978.

Table LC-19
BENZOIC AND PHENOXYACETIC ACIDS

Packing	P1	P1	P1	P1
Column				
Length (cm)	45	45	45	45
Diameter (mm)	2.36	2.36	2.36	2.36
Material				
Eluent	E1	E2	E3	E4
Flow rate (mℓ/min)	0.5	0.5	0.5	0.5
Temperature (°C)				
Detection	D1	D1	D1	D1
Reference	1	1	1	1
Compound	mℓ	mℓ	mℓ	mℓ
Benzoic acid	7.8	4.3	0.8	0.3
4-Aminobenzoic acid	1.8	1.0	0.5	0.2
3-Aminobenzoic acid	1.8	1.0	0.3	0.2
2-Aminobenzoic acid	5.3	3.2	1.0	0.3
4-Nitrobenzoic acid	13.4	2.9	0.5	0.4
3-Nitrobenzoic acid	11.0	2.7	0.5	0.4
4-Hydroxybenzoic acid	1.7	0.9	0.4	0.1
3-Hydroxybenzoic acid	1.9	0.9	0.3	0.2
4-Chlorobenzoic acid	—	9.8	1.3	0.5
4-Bromobenzoic acid	—	13.2	1.5	0.6
4-Chlorophenoxyacetic acid	11.2	2.2	0.7	0.6
2-Chlorophenoxyacetic acid	9.0	1.9	0.6	0.5
2,4-Dichlorophenoxyacetic acid	23.3	3.8	1.1	0.9
2,4,5-Trichlorophenoxyacetic acid	42.8	6.0	1.8	1.6

Packing P1 = XAD-2 (45-65μm)
Eluent E1 = 0.1 *M* carbonate buffer/50% ethanol, pH 3.9
E2 = 0.1 *M* carbonate buffer/50% ethanol, pH 4.9
E3 = 0.1 *M* phosphate buffer/50% ethanol, pH 6.6
E4 = 0.1 *M* phosphate buffer/50% ethanol, pH 9.4
Detection D1 = UV 254 nm

Table LC-20
AROMATIC ACIDS

Packing	P1	P2	P2	P2	P3	P3	P3	P4	P5	P5	P5
Column											
Length (cm)	42	20	15	20	30	30	30	50	75	75	75
Diameter (mm)	9	3.2	3.2	3.2	4	4	3.9	3	10	10	10
Material								G	G	G	G
Eluent	E1	E2	E3	E3	E4	E5	E6	E7	E8	E9	E10
Flow rate (mℓ/min)	3	0.58	0.33	0.80	1	2	1.5	1.3	1	1	1
Temperature (°C)		25	25	25	rt			30	50	50	50
Detection	D1	D2	D2	D2	D2	D2	D2	D3	D4	D4	D4
Reference	1	2	2	2	3	4	5	6	7	7	7
Compound	min[a]	min[a]	min[a]	min[a]	mℓ[a]	min	mℓ[a]	mℓ	mℓ	mℓ	mℓ
Cinnamic acid	18	—	—	—	—	—	—	—	124	167	301
4-Hydroxy-3-methoxycinnamic acid	79	—	—	—	—	—	—	—	202	284	563
4-Hydroxy-3-methoxybenzoic acid	96	—	—	—	—	—	—	55	192	242	430
4-Hydroxycinnamic acid	168	—	—	—	—	—	—	—	242	350	808
2,5-Dihydroxybenzoic acid	190	—	—	—	—	—	—	—	—	—	—
3,4-Dihydroxycinnamic acid	268	—	—	—	—	—	—	—	499	652	1496
3,4-Dihydroxybenzoic acid	302	—	—	—	—	—	—	—	516	618	1152
3,4,5-Trihydroxybenzoic acid	415	—	—	—	—	—	—	—	—	—	—
Chlorogenic acid	454	—	—	—	—	—	—	—	—	—	—
4-Hydroxy-5-methoxymandelic acid	—	3	—	—	—	—	—	13	—	—	—
4-Hydroxy-3-methoxyphenylacetic acid		6	3.5	—	—	—	—	42	—	—	—
Mandelic acid	—	10	—	—	—	—	—	—	—	—	—
3,4-Dimethoxyphenylacetic acid	—	—	7	—	—	—	—	—	—	—	—
3,4,5-Trimethoxyphenylacetic acid	—	—	9	—	—	—	—	—	—	—	—
4-Methoxyphenylacetic acid	—	—	12	—	—	—	—	—	—	—	—
Nicotinic acid	—	—	—	3.2	—	—	—	—	—	—	—
Isonicotinic acid	—	—	—	3.8	—	—	—	—	—	—	—
5-Fluoro-3-hydroxymethylpyridine HC1	—	—	—	4.5	—	—	—	—	—	—	—
5-Fluoropyridine-3-carboxylic acid	—	—	—	6.2	—	—	—	—	—	—	—
Phenylacetic acid	—	—	—	—	17	—	—	—	—	—	—
2-Hydroxyphenylacetic acid	—	—	—	—	9	—	—	—	—	—	—
3-Hydroxyphenylacetic acid	—	—	—	—	8.3	—	—	—	—	—	—

Table LC-20 (continued)
AROMATIC ACIDS

Compound	min[a]	min[a]	min[a]	min[a]	mℓ[a]	min	mℓ[a]	mℓ	mℓ	mℓ	mℓ
4-Hydroxyphenylacetic acid	—	—	—	—	7.5	—	—	—	—	—	—
1,4-Benzenedicarboxylic acid	—	—	—	—	—	7.3	—	—	—	—	—
4-Aminobenzoic acid	—	—	—	—	—	13.5	—	—	—	—	—
4-Hydroxybenzoic acid	—	—	—	—	—	17.9	3.7	—	242	328	636
2-Aminobenzoic acid	—	—	—	—	—	23.0	—	—	—	—	—
Benzoic acid	—	—	—	—	—	31.4	—	—	—	—	—
Methyl-4-hydroxybenzoate	—	—	—	—	—	—	6.0	—	—	—	—
Propyl-4-hydroxybenzoate	—	—	—	—	—	—	11.4	—	—	—	—
4-Hydroxy-3,5-dimethoxybenzoic acid	—	—	—	—	—	—	—	—	159	185	301

[a] Measured on chromatogram.

Packing
- P1 = silicagel (100 mesh)
- P2 = LiChrosorb ® SI 60, RP2, stationary phase = 1-pentanol
- P3 = μBondapak C18
- P4 = Hitachi 3010 (25 μm porous polystyrene gel)
- P5 = Polyclar® AT (polyvinylpyrrolidone)

Eluent
- E1 = 0.3- 22% *tert.*-amyl alcohol/chloroform-cyclohexane (50:50)
- E2 = 0.1 *M* TBA, 0.04 *M* phosphate buffer pH 7.4
- E3 = 0.03 *M* TBA, 0.04 *M* phosphate buffer, pH 7.4
- E4 = methanol-3% acetic acid (1:3)
- E5 = 30 mℓ 1.0 *M* acetic acid + 10.5 mℓ 1.0 *M* NaOH + 19.5 mℓ 1.0 *M* NaCl/500 mℓ water
- E6 = 0.01 *M* heptanesulfonic acid − acetonitrile (65:35)
- E7 = 0.05 *M* tartrate buffer pH 2.75-4.8 -methanol (4:1)
- E8 = 80% ethanol
- E9 = 50% ethanol
- E10 = 30% ethanol

Detection
- D1 = UV 313, 254nm
- D2 = UV 254nm
- D3 = UV 280nm
- D4 = modified spectrophotometer 235, 265, 320 nm

REFERENCES

1. **Morot-Gaudre, J. F., Nicol, M. Z., and Jolivet, E.,** Analyse quantitative des acids phenoliques par chromatographie de partage sur gel de silice, *J. Chromatogr.*, 100, 206, 1974.
2. **Wahlund, K. -G. and Lund, U.,** Reversed-phase ion-pair partition chromatography of phenylacetic, mandelic and pyririnecarboxylid acid derivatives, *J. Chromatogr.*, 122, 269, 1976.
3. **Warthen, J. D., Jr. and Mandava, N.,** Separation of isomeric (hydroxyphenyl) acetic acids by reversed-phase high performance liquid chromatography, *J. Chromatogr.*, 144, 263, 1977.
4. **Deming, S. N. and Turoff, M. L. H.,** Optimization of reversed-phase liquid chromatographic separation of weak organic acids, *Anal. Chem.*, 50, 546, 1978.
5. **Brown, N. D., Hall, L. L., and Sleeman, H. K.,** Determination of methyl p-hydroxybenzoate and propyl p-hydroxybenzoate in an anticholinergic antidote by ion-pair high-performance liquid chromatography, *J. Chromatogr.*, 166, 316, 1978.
6. **Yoshida, A., Yoshida, M. T., Tanimura, T., and Tamura, Z.,** Determination of vanilmandelic acid and homovanillic acid in urine by high-speed liquid chromatography, *J. Chromatogr.*, 116, 240, 1976.
7. **Alibert, G.,** Séparation et dosage automatisés des acides phénoliques végétaux par chromatographie sur colonne de polyclar et spectrophotométrie dans l'ultra violet, *J. Chromatogr.*, 80, 173, 1973.

Table LC-21
AROMATIC ACIDS

Packing	P1	P2	P3	P4	P3
Column					
Length (cm)	50	100	50	200	50
Diameter (mm)	2.1	2.1	2.1	2.1	2.1
Material	G		G		
Eluent	E1	E2	E3	E4	E5
Flow rate (mℓ/min)	0.3	0.56	0.4	1	0.49
Temperature (°C)	45	—	—	40	rt
Detection	D1	D2	D3	D2	D3
Reference	1	2	3	4	5
Compound	min	min	min	min[a]	min
Chlorogenic acid	60	—	∞	—	15.3
3,4,5-Trihydroxybenzoic acid	2.8	—	35	—	4.4
4-Hydroxymandelic acid	2.3	—	—	—	2.9
2,5-Dihydroxyphenylacetic acid	5.5	—	—	—	30.2
4-Hydroxyphenyllactic acid	2.2	—	—	—	3.5
4-Hydroxy-3-methoxymandelic acid	1.8	—	—	—	2.5
3,5-Dihydroxybenzoic acid	5.0	—	—	—	6.8
3,4-Dihydroxyphenlacetic acid	2.5	—	17	—	3.9
4-Hydroxy-3-methoxyphenyllactic acid	2.8	—	—	—	2.7
4-Hydroxy-3-methoxyphenylpyruvic acid	7.0	—	—	—	6.4
2,5-Dihydroxybenzoic acid	23.1	—	—	—	36.2
3,4-Dihydroxybenzoic acid	3.2	—	20.8	—	4.9
3,4-Dihydroxymandelic acid	1.2	—	13. 2	—	3.2
3,4-Dihydroxycinnamic acid	17.0	—	67	—	16.7
5-Hydroxyindoleacetic acid	8.0	—	—	—	—
3,4-Dihydroxyphenylpropionic acid	4.0	—	24.1	—	23.5
2,3-Dihydroxybenzoic acid	18.0	—	—	—	21.7
2-Hydroxyhippuric acid	12.5	—	—	—	—
3-Hydroxyphenylacetic acid	3.5	—	—	—	4.2
4-Hydroxyphenylacetic acid	19.2	—	—	—	3.5
4-Hydroxybenzoic acid	4.0	—	—	22	9.0
2-Hydroxyphenylacetic acid	9.2	—	—	—	11.3
3-Hydroxybenzoic acid	6.0	—	—	28	9.9
4-Hydroxycinnamic acid	17.5	—	—	—	13.7
3-Hydroxyphenylpropionic acid	6.0	—	—	—	15.8
4-Hydroxyphenylpropionic acid	5.0	—	—	—	10.3
4-Hydroxy-3-methoxyphenylacetic acid	3.2	—	—	—	3.0
4-Hydroxy-3,5-dimethoxybenzoic acid	7.1	—	—	—	2.8
4-Hydroxy-3-methoxycinnamic acid	19.8	—	—	—	8.9
4-Hydroxy-3-methoxybenzoic acid	5.3	—	—	—	3.7
2-Hydroxybenzoic acid	30.0	—	—	82	26.7
Abscisic acid	—	3	—	—	—
Indoleacetic acid	—	8	—	—	—
Indolepropionic acid	—	10	—	—	—
Indolebutanoic acid	—	12	—	—	—
Phenol	—	—	—	4	—
Benzoic acid	—	—	—	14	—
2,4,6-Trihydroxybenzoic acid	—	—	—	—	3.2
2,4-Dihydroxybenzoic acid	—	—	—	—	12.2
4-Hydroxy-3-methoxycinnamic acid	—	—	—	—	5.4
4-Hydroxy-3-methoxyphenylpropionic acid	—	—	—	—	6.7
4-Hydroxy-phenylpyruvic acid	—	—	—	—	8.0
1,2,3-Trihydroxybenzene	—	—	—	—	2.8
1,2,4-Trihydroxybenzene	—	—	—	—	2.9
1,3,5-Trihydroxybenzene	—	—	—	—	3.1
1,2-Dihydroxybenzene	—	—	—	—	2.6

Table LC-21 (continued)
AROMATIC ACIDS

Compound	min	min	min	min[a]	min
D-Catechin	—	—	—	—	31.9
L-Epicatechin	—	—	—	—	28.5
Kaemferol	—	—	—	—	18.5
Quercetin	—	—	—	—	26.4
Rutin	—	—	—	—	28.3

[a] Measured on chromatogram.

Packing	P1 = Vydac anion-exchanger
	P2 = Permaphase AAX
	P3 = Zipax SAX
	P4 = Reeve Angel anion-exchanger
Eluent	E1 = 0.025 *M* acetate buffer (pH 4.7)
	E2 = NaH_2PO_4 solution (0.9 g/*l*)
	E3 = 0.1 *M* acetate buffer (pH 4.7)
	E4 = 7 m*M* NH_4Cl (pH 4.2)
	E5 = 0.1 *M* acetate buffer (pH 4.7)-0.25 *M* citric acid (pH 1.8)-water (1:1:8)
Detection	D1 = electrochemical detector (+ 0.75 V)
	D2 = UV 254 nm
	D3 = amperometric detector (+0.60 V)

REFERENCES

1. **Felice, L. J. and Kissinger, P. T.**, Determination of homovanillic acid in urine by liquid chromatography with electrochemical detection, *Anal. Chem.*, 48, 794, 1976.
2. **Duering, H.**, Analysis of abscisic acid and indole-3-acetic acid form fruits of Vitis vinifera L. by high pressure liquid chromatography, *Experientia*, 33, 1666, 1977.
3. **Felice, L. J., Bruntlett, C. S., and Kissenger, P. T.**, Liquid-chromatography assay for 3,4-dihydroxyphenylacetic acid in urine, *J. Chromatogr.*, 143, 407, 1977.
4. **Kramer, U. and Bhatia, K.**, Determination of trace phenol and isometric hydroxybenzoic acids in aqueous benzoic acid solutions by aqueous liquid chromatography, *J. Chromatogr.*, 89, 348, 1974.
5. **Kenyhercz, T. M. and Kissinger, P. T.**, Determination of selected acidic, neutral and basic natural products in Cacao beans and Processed Cocoa. Liquid chromatography with electrochemical detection, *Lloydia*, 41, 130, 1978.

Table LC-22
AROMATIC ACIDS

Packing	P1	P2	P2	P2	P2	P3	P4	P4	P5	P6	P6
Column											
Length (cm)	60	15	15	15	23	50	50	50	14	14	14
Diameter (mm)	9	10	10	10	12	9	9	9	8	8	8
Material		G	G	G	G	G	G	G	G	G	G
Eluent	E1	E2	E3	E2	E3	E4	E5	E6	E7	E8	E9
Flow rate (mℓ/min)	0.62	0.2	0.4	0.2	0.4	0.21	0.16	0.16	1.17	0.58	0.58
Temperature (°C)	55	40	40	40	40				60		
Detection	D1	D2	D2	D2	D2	D2	D2	D2	D2	D2	D2
Reference	1	2	2	2	2	3	3	3	4	4	4
Compound	mℓ	mℓ	mℓ	mℓ	mℓ	mℓ	mℓ	mℓ	mℓ	mℓ	mℓ
4-Hydroxy-3-methoxymandelic acid	42	—	—	—	—	—	—	—	—	—	—
3-Hydroxy-4-methoxymandelic acid	42	—	—	—	—	—	—	—	—	—	—
4-Hydroxy-3-methoxymandelic acid methylester	84	—	—	—	—	—	—	—	—	—	—
3-Hydroxy-4-methoxymandelic acid methylester	96	—	—	—	—	—	—	—	—	—	—
2-Hydroxybenzoic acid	—	15	25	15	45	35	—	50	280	8	8
3-Hydroxybenzoic acid	—	35	45	40	80	65	—	70	8	70	8*
4-Hydroxybenzoic acid	—	—	—	—	—	85	—	200	8	175	8*
2-Hydroxy-1-naphthoic acid	—	—	—	—	—	—	30	—	—	—	—
1-Hydroxy-2-naphthoic acid	—	—	—	—	—	—	110	—	—	—	—
4-Hydroxy-2-naphthoic acid	—	—	—	—	—	—	30	—	—	—	—

Packing
- P1 = PA 28 (16± 6 μm, cation-exchanger from Beckman)
- P2 = Amberlite® CG 50, Na^+ form (200-400 mesh)
- P3 = Amberlite® CG 120, Ti(IV) form (200-400 mesh)
- P4 = Mitsubishi Diaion SK-1, Fe(III) form (200-400 mesh)
- P5 = Amberlite® CG 50, Fe(III) form
- P6 = Amberlite® CG 50, Hg(II) form

Eluent
- E1 = sodium- citrate-borate, 0.2*N*NaOH
- E2 = 2 *M*NaC1, pH 1.8
- E3 = 4 *M*NaC1, pH 2.8

E4 = pH 4.5 water
E5 = pH 5.5 water
E6 = pH 4.7 water
E7 = ethanol
E8 = chloroform
E9 = chloroform/ethanol * Retention volume by stepwise elution gradient
Detection D1 = UV 280 nm
D2 = UV

REFERENCES

1. **Thomas, H., Ponnath, H., and Müller-Enoch, D.,** Säulenchchromatographische Trennung 3,4-isomerer hydroxymethoxymandelsäuren, *J. Chromatogr.*, 103, 197, 1975.
2. **Funasaka, W., Fujimura, and Kushida, S.,** Salting-out chromatographic separation of hydroxybenzoic acid isomers, *J. Chromatogr.*, 64, 95, 1972.
3. **Fujimura, K., Koyama, T., Tanigawa, T., and Funasaka, W.,** Studies in ligand exchange chromatography. IV. Separation of hydroxybenzoic and hydroxynaphthoic acid isomers, *J. Chromatogr.*, 85, 101, 1973.
4. **Funasaka, W., Hanai, T., Fujimura, K., and Ando, T.,** Non-aqueous solvent chromatography III. Complex chromatography between the metal ion of a cation exchange resin and organic compounds in organic solvents, *J. Chromatogr.*, 78, 424, 1973.

Table LC-23
HYDROXYBENZOIC ACIDS

Packing	P1	P1	P1	P1	P1	P1
Column						
Length (cm)	45.5	45.5	45.5	45.5	45.5	45.5
Diameter (mm)	6	6	6		6	6
Material	G	G	G	G	G	G
Eluent	E1	E2	E3	E4	E5	E6
Flow rate (mℓ/min)	1	1	1	1	1	1
Temperature						
Detection	D1	D1	D1	D1	D1	D1
Reference	1	1	1	1	1	1
Compound	Dv	Dv	Dv	Dv	Dv	Dv
2-Hydroxybenzoic acid	1.17	1.05	0.7	0.31	0.60	4.13
2-Hydroxy-4-methylbenzoic acid	1.72	2.17	1.10	0.39	1.48	7.23
2,4-Dihydroxybenzoic acid	5.54	7.41	3.20	0.86	4.37	14.4
2-Hydroxy-4-nitrobenzoic acid	6.24	3.91	3.51	0.35	—	—
2-Hydroxy-3-methylbenzoic acid	4.80	6.24	2.81	0.40	—	—
2-Hydroxy-3-nitrobenzoic acid	8.19	3.51	4.13	0.31	—	—
5-Chloro-2-hydroxybenzoic acid	8.00	5.07	3.90	0.47	3.67	18.3
2-Hydroxy-5-nitrobenzoic acid	8.50	6.30	4.68	0.70	4.60	27.7
2-Hydroxy-5-sulfobenzoic acid	16.8	4.29	10.9	10.4	4.68	1.56

Packing P1 = Dowex® 1-X8
Eluent E1 = 0.03 $M FeCl_3$
E2 = 0.01 $M FeCl_3$
E3 = 0.03 $M FeCl_3$/ 10% methanol
E4 = 0.03 $M FeCl_3$/70% methanol
E5 = 0.03 $M FeCl_3$ 0.1 $M KCl$/10% methanol
E6 = 0.03 $M FeCl_3$, 2.0 $M KCl$/10% methanol
Detection D1 = UV, na the wavelength

REFERENCE

1. **Lee, K. S. and Lee, D. W.,** Studies on the anion-exchange chromatography of salicyclic acid and its derivatives, *Anal. Chem.,* 46, 1903, 1974.

Table LC-24
ORGANIC ACIDS

Packing	P1	P2	P3	P4	P5
Column					
Length (cm)	50	50	100	25	10
Diameter (mm)		10		4.6	3.2
Material					ss
Eluent	E1	E2	E3	E4	E5
Flow rate (mℓ/min)	1.2	0.1	0.5	2.5	0.72
Temperature (°C)	55		40		25
Detection	D1	D2	D1	D3	D1
Reference	1	2	3	4	5
Compound	min[a]	mℓ[a]	min	min[a]	mℓ
2,4-Dinitrophenylhydrazone-2-oxo-glutaric acid	9.5	—	—	—	—
2,4-Dinitrophenylhydrazone-pyruvic acid	4.5	56	—	—	—
1,2,4-Benzenetricarboxylic acid	2.5	—	—	—	—
2,4-Dinitrophenylhydrazone	—	44	—	—	—
2,4-Dinitrophenylhydrazone-phenylpyruvic acid	—	52	—	—	—
2,4-Dinitrophenylhydrazone-3-indolepyruvic acid	—	60	—	—	—
2,4-Dinitrophenylhydrazone-4-hydroxyphenylpyruvic acid	—	75	—	—	—
2,4-Dinitrophenylhydrazone-3-indolecarboxaldehyde	—	87	—	—	—
3-Benzyl-2-hydroxybenzoquinoxaline	—	—	8	—	—
Chlorothioxanthone	—	—	17.5	—	—
3,4-Dihydroxycinnamic acid tartaric acid-glucose ester	—	—	—	6.0	—
Caffeoyltartaric acid	—	—	—	8.3	—
4-Hydroxycinnamic acid-tartaric acid-glucose ester	—	—	—	10.6	—
4-Coumaroyltartaric acid	—	—	—	13.5	—
Feruloyltartaric acid	—	—	—	20.0	—
3,4-Dihydroxycinnamic acid	—	—	—	24.8	—
6-Hydroxy-2′-methyl-2-naphthaleneacetic acid	—	—	—	—	1.64
6-Methoxy-2-naphthaleneacetic acid	—	—	—	—	3.33
6-Methoxy-2′-methyl-2-naphthaleneacetic acid	—	—	—	—	5.18

[a] Measured on chromatogram.

Packing P1 = Permaphase AAX (30-50μm)
P2 = Sephadex LH 20
P3 = Permaphase ETH
P4 = Zorbax ODS
P5 = LiChrosorb® RP8 (10 μm)

Eluent E1 = 11.1 g KH_2PO_4 + 0.9 g K_2HPO_4/ℓ water
E2 = 95% ethanol
E3 = acetonitrile-water (7:18)
E4 = 6-16% acetonitrile/water (pH 2.6 by phosphoric acid)
E5 = 0.025 *M* tetrapentylammonium, pH 7 phosphate buffer-methanol (60:40)

Detection D1 = UV 254 nm
D2 = 360 nm
D3 = UV 320 nm

REFERENCES

1. **Terada, H., Hayashi, T., Kawai, S., and Ohno, T.,** High-performance liquid chromatographic determination of pyruvic acid and α-ketoglutaric acid in serum, *J. Chromatogr.*, 130, 281, 1977.
2. **Winchester, R. V.,** Separation of 2,4-dinitrophenylhydrazones of some keto acids on Sephadex LH 20, *J. Chromatogr.*, 78, 429, 1973.
3. **Hayashi, T., Sugiura, T., Tereda, H., Kawai, S., and Ohno, T.,** High-speed liquid chromatographic determination of phenylpyruvic acid, *J. Chromatogr.*, 118, 403, 1976.
4. **Ong, B. Y. and Nagel, C. W.,** High-pressure liquid chromatographic analysis of hydroxycinnamic acid-tartaric acid esters and their glucose esters in Vitis vinifera, *J. Chromatogr.*, 157, 345, 1978.
5. **Westerlund, D. and Theodorsen, A.,** Reversed-phase ion-pair chromatography of naphthalene-acetic acid derivatives with water and an organic modifier as the mobile phase, *J. Chromatogr.*, 144, 27, 1977.

Table LC-25
ORGANIC ACIDS

Packing	P1	P2	P2	P2	P2
Column					
Length (cm)	68.5	25	25	25	25
Diameter (mm)	4	2.2	2.2	2.2	2.2
Material					
Eluent	E1	E2	E3	E4	E5
Flow rate (mℓ/min)	3.9	0.5	0.5	0.5	0.5
Temperature (°C)	30				
Detection	D1	D2	D2	D2	D2
Reference	1	2	2	2	2
Compound	mℓ[a]	min[a]	min[a]	min[a]	min[a]
O-(4-*O*-methyl-α-D-glucopyranosyluronic acid)-(1-2)-*O*-β-D-xylopyranosyl-(1-4)-*O*β-D-xylopyranosyl-(1-4)-D-xylonic acid	22	—	—	—	—
O-(4-*O*-methyl-α-D-glucopyranosyluronic acid)-(1-2)-*O*-β-D-xylopyranosyl-(1-4)-D-xylonic acid	46	—	—	—	—
2-*O*-(4-0-methyl-α-D-glucopyranosyluronic acid)-D-xylonic acid	80	—	—	—	—
4-*O*-methylglucaric(3-*O*-methylgularic acid)	248	—	—	—	—
Fenoprop (2-(2,4,5-trichlorophenoxy) propionic acid)	—	5.3	—	—	—
Dichloroprop (2-(2,4-dichlorophenoxy)propionic acid)	—	6.3	—	—	—
Dicamba (3,6-dichloro-2-methoxybenzoic acid)	—	7.6	—	—	—
MCPA (4-chloro-2-methyl-phenoxyacetic acid)	—	9.6	—	—	—
2,4,5-T (2,4,5-trichlorophenoxyacetic acid)	—	12.1	—	—	—
2,4-D (2,4-dichlorophenoxyacetic acid)	—	13.7	—	—	—
Chloramben (3-amino-2,5-dichlorobenzoic acid)	—	21.2	—	—	—
Dinoterb (2-*tert*.-butyl-4,6-dinitrophenol)	—	—	2.3	2.3	2.5
Dinoseb (2-*sec*.-butyl-4,6-dinitrophenol)	—	—	2.7	2.7	3.0
DNPC (2-methyl-4,6 dinitrophenol)	—	—	6.5	7.3	8.4
Dinobuton (2-*sec*.-Butyl-4,6-dinitrophenyl-isopropylcarbonate)	—	—	10.1	13.4	16.9

[a] Measured on chromatogram.

Packing	P1	= Dowex® 1-X8
	P2	= LiChrosorb® SI 60 (5 μm)
Eluent	E1	= 0.3 *M* sodium acetate (pH 5.9)
	E2	= 10% dichloromethane/2,2,4-trimethylpentane with 0.5 *M* acetic aid
	E3	= 0.025 *M* acetic acid/2,2,4-trimethylpentane
	E4	= 0.025 *M* propionic acid/2,2,4-trimethylpentane
	E5	= 0.025 *M* butyric acid/2,2,4-trimethylpentane
Detection	D1	= chromic acid oxidation detector
	D2	= UV 280 nm

REFERENCES

1. **Ericsson, T. and Samuelson, O.,** Anion-exchange chromatography of dicarboxylic hydroxy acids, *J. Chromatogr.*, 134, 337, 1977.
2. **Lawrence, J. F. and Leduc, R.,** High performance liquid chromatography of some acidic and basic organic compounds on silica gel with mobile phases containing organic acids, *Anal. Chem.*, 50, 1161, 1978.

Table LC-26
PHENOLS AND ORGANIC ACIDS

Packing	P1	P2	P3	P4
Column				
Length (cm)	25	15	30	25
Diameter (mm)	3.2	3.2	3.9	3.1
Material	ss	ss	ss	ss
Eluent	E1	E2	E3	E4
Flow rate (mℓ/min)	1-4	1	3	1.25
Temperature (°C)		30		30
Detection	D1	D2	D3	D3
Reference	1	2	3	4
Compound	min	min[a]	min[a]	sec
3-*tert*.-Butyl-4-methoxyphenol	9.6	—	—	—
2-*tert*.-Butyl-4-methoxyphenol	9.6	—	—	—
tert.-Butyl-1,4-dihydroxybenzene	7.1	—	—	—
3,5-Di-*tert*.-butyl-4-hydroxymethylbenzene	13.1	—	—	—
2,6-Di-*tert*.-butyl-4-hydroxymethylbenzene	10.0	—	—	—
2,4,5-Trihydroxybutyrophenone	6.3	—	—	—
3,4,5-Trihydroxybenzoic acid propylester	5.1	—	—	—
3,4,5-Trihydroxybenzoic acid octylester	10.0	—	—	—
3,4,5-Trihydroxybenzoic acid dodecylester	12.7	—	—	—
Nordihydroguaiaretic acid	8.1	—	—	—
α-Tocopherol	—	2.5	4.3	—
β-Tocopherol	—	3.9	—	—
γ-Tocopherol	—	4.3	—	—
Tocol	—	8.3	—	—
Tetrachlorophenol	—	—	—	391
Pentachlorophenol	—	—	—	591

[a] Measured on chromatogram

Packing P1 = LiChrosorb® RP18 (10 μm)
P2 = Silica gel (5 μm)
P3 = μBondapak C18
P4 = Spherisorb ODS
Eluent E1 = 5% acetic acid - 5% acetic acid/acetonitrile (70:30) to 5% acetic acid/acetonitrile, gradient
E2 = *n*-hexane-isopropyl alcohol (99.4 : 0.6)
E3 = methanol-water (97:3)
E4 = water-acetonitrile-acetic acid (65:35:0.1)
Detection D1 = UV 280 nm
D2 = UV 294 nm
D3 = UV 230 nm

REFERENCES

1. **Page, B. D.**, High performance liquid chromatographic determination of nine phenolic antioxidants in oils, lards and shortenings, *J. Assoc. Off. Anal. Chem.*, 62, 1239, 1979.
2. **De Leenheer, A. P., De Bevere, V. O. R. C., and Claeys, A. E.**, Measurement of α-, β–, and γ–Tocopherol in serum by liquid chromatography, *Clin. Chem.*, 25, 425, 1979.
3. **McMurray, C. H. and Blanchflower, W. J.**, Application of a high-performance liquid chromatographic fluoroescence method for the rapid determination of α-tocopherol in the plasma of cattle and pigs and its comparison with direct florescence and high-performance liquid chromatography-ultraviolet detection methods, *J. Chromatogr.*, 178, 525, 1979.
4. **Daniels, C. R. and Swan, E. P.**, Determination of chlorinated phenols in surface-treated lumber by HPLC, *J. Chromatogr. Sci.*, 17, 628, 1979.

Table LC-27
HYDROXYBENZENES AND THEIR SULFATES

Packing	P1	P1	P2	P2
Column				
Length (cm)	30	30	25	25
Diameter (mm)	4.0	4.0	4.6	4.6
Material	ss	ss	ss	ss
Eluent	E1	E2	E3	E4
Flow rate (mℓ/min)	1.5	1.5	1.5	1.5
Temperature (°C)	40	40	40	40
Detection	D1	D1	D1	D1
Reference	1	1	1	1
Compound	k′	k′	k′	k′
Phenol	2.56	1.90	4.22	1.74
Phenyl sulfate	3.31	1.64	19.9	2.11
1,2-Dihydroxybenzene disulfate	3.14	1.13	—	2.10
1,3-Dihydroxybenzene	1.15	0.78	1.24	0.66
1,3-Dihydroxybenzene disulfate	4.25	1.35	—	2.09
1,4-Dihydroxybenzene	0.60	0.39	0.60	0.30
1,4-Dihydroxybenzene disulfate	2.88	0.91	14.2	1.58
1,3,5-Trihydroxybenzene	0.51	0.32	0.51	0.16
1,3,5-Trihydroxybenzene monosulfate	3.13	1.07	2.49	0.55
1,3,5-Trihydroxybenzene disulfate	—	1.74	19.8	1.71
1,3,5-Trihydroxybenzene trisulfate	—	3.59	—	4.00
1,2,3,5-Tetrahydroxybenzene 2,5-disulfate	3.95	1.26	11.7	1.83
Lanosol disulfate	—	ca. 18	—	9.14

Packing P1 = μBondapak C18 (10μm)
P2 = LiChrosorb® RP8 (10μm)
Eluent E1 = water with tetrabutylammonium phosphate-methanol (79:21)
E2 = water with tetrabutylammonium phosphate-methanol (73:27)
E3 = water with tetrabutylammonium phosphate-methanol (74:26)
E4 = water with tetrabutylammonium phosphate-methanol (60:40)
Detector D1 = UV 254 nm

From Ragan, M. A. and Mackinnon, M. D., *J. Chromatogr.*, 178, 505, 1979. With permission.

Table LC-28
ORGANIC ACIDS

Packing	P1	P2	P3	P4
Column				
Length (cm)	100	50	30	25
Diameter (mm)	3	8	7.8	3.2
Material	G	ss	ss	
Eluent	E1	E2	E3	E4
Flow rate (mℓ/min)	0.13	2.3	0.8	0.6
Temperature (°C)	50	80	50	
Detection	D1	D2	D3	D4
Reference	1-6	7	8	9.10
Compound	min	min[a]	k′	min
Glutamic acid	35	—	—	—
Gluconic acid	42	—	0.520	—
Glucuronic acid	48	—	0.547	—
Pyroglutamic acid	55	6.8	—	—

Table LC-28 (continued)
ORGANIC ACIDS

Compound	min	min[a]	k′	min
3-Hydroxybutyric acid	61.8	—	—	—
Lactic acid	62.4	5.5	1.533	—
Glycolic acid	64	5.3	1.240	—
Acetic acid	66	6.3	1.800	—
Levulinic acid	69	—	2.093	—
Tartaric acid	70	4.05	0.547	—
Pyruvic acid	73	—	0.520	—
Formic acid	75.6	5.9	1.507	—
Malic acid	76	4.4	0.813	—
Citric acid	89	3.9	0.467	—
Succinic acid	100	5.0	1.240	—
Isocitric acid	110	—	0.467	—
2-Oxo-glutaric acid	128	—	0.360	—
Methylmalonic acid	138	—	—	—
Fumaric acid	309	—	1.533	—
cis-Aconitic acid	511	—	—	—
Oxalic acid	—	3.2	0.067	3.75
Acrylic acid	—	7.3	—	—
Methacrylic acid	—	9.1	—	—
trans-Crotonic acid	—	9.8	—	—
Glyceric acid	—	—	0.040	—
Maleic acid	—	—	0.280	—
Oxaloacetic acid	—	—	0.333	—
Glyoxylic acid	—	—	0.733	—
Malonic acid	—	—	0.760	—
trans-Aconitic acid	—	—	0.813	—
2-D-Galacturonic acid	—	—	0.867	—
Glutaric acid	—	—	1.587	—
Adipic acid	—	—	2.067	—
L-Pyrrolidine-5-carboxylic acid	—	—	2.120	—
Propionic acid	—	—	2.280	—

[a] Measured on a chromatogram.

Packing P1 = Diaion CA-10S (a basic anion exchange resin)
P2 = Shodex Ionpak
P3 = Aminex HPX-87
P4 = Vydac SCX
Eluent E1 = 0.2 *N*HCl
E2 = 0.1% H_3PO_4
E3 = 0.0045 *N*H_2SO_4
E4 = mixture of 0.0020 *M* sodium acetate-acetic acid (adjusted to pH 2.8 with HCl) and 0.0010 *M* tetrabutylammonium tetrafluoroborate)
Detection D1 = post-column reaction detector using hydroxylamine perchlorate-triethylamine-ferric perchlorate
D2 = RI
D3 = UV 210 nm
D4 = electrochemical detector

REFERENCES

1. **Nakajima, M., Ozawa, Y., Tanimura, T. and Tamura, Z.,** A highly efficient carboxylic acid analyzer and its application, *J. Chromatogr.*, 123, 129, 1976.
2. **Shimazu, Y. and Watanabe, M.,** Qualitative Analyse der organischen der Weine und Moste mit dem Carboxylsäureanalysator, *Die Wein-Wissenschaft*, 31, 45, 1976.

Table LC-28 (continued)
ORGANIC ACIDS

3. **Hirose, Y., Tanaka, H. and Fukuzaki, K.**, Studies on the quality of soy sauce refer to constitution of organic acids. II, *SHOKEN*, 2, 264, 1976.
4. **Shimada, Y.**, Studies on organic acids in vegetables. I. On the composition of organic acids in various vegetables, *CHORIKAGAKU*, 10, 248, 1977.
5. **Shimazu, Y. and Watanabe, M.**, Malo-lactic fermentation in sparkling wine, *J. Ferm. Tech.*, 57, 512, 1979.
6. **Wakasugi, C., Funahashi, K., Uchima, E., and Shikata, I.**, Appearance of formate in blood after ethanol ingestion, *Biochem. Biophys. Res. Commun.* 88, 988, 1979.
7. Shodex application data on Shodex Ionpak. Showa Denko K.K. Tokyo. Technical data 771 (1977).
8. The Liquid Chromatographer, Bio-Rad Lab. Number 2, 1979.
9. **Mayer, W. J. and Greenberg, M. S.**, A comparison of differential pulse and D.C. amperometric detection modes for the liquid chromatographic determination of oxalic acid, *J. Chromatogr. Sci.*, 17, 614, 1979.
10. **Mayer, W. J., McCarthy, J. P., and Greenberg, M. S.**, The determination of oxalic acid in urine by high performance liquid chromatography with electrochemical detection, *J. Chromatogr. Sci.*, 17, 656, 1979.

Table LC-29
KETO ACIDS

Packing	P1	P1
Column		
Length (cm)	30	30
Diameter (mm)	4	4
Material		
Eluent	E1	E2
Flow rate (mℓ/min)	1	1
Temperature (°C)		
Detection	D1	D2
Reference	1	2
Compound	**min[a]**	**min[a]**
Glyoxylic acid	14	—
Pyruvic acid	16	16
2-Oxobutanoic acid	21.5	—
2-Oxavaleric acid	23.5	—
2-Oxohexanoic acid	26	—
2-Oxooctanoic acid	31	—
2-Oxononanoic acid	35.5	—
2-Oxoglutaric acid	—	11.5

[a] Measured on chromatogram.

Packing	P1	= μBondapak C18
Eluent	E1	= 0.05 *M* tetrabutylammonium hydroxide and 1% acetic acid (pH 4.3)-methanol-0.05 *M* tetrabutylammonium hydroxide and 1% acetic acid (pH 4.3) (80:20), gradient
	E2	= 0.35 mole ammonium acetate-methanol (80:20)-methanol, gradient
Detection	D1	= analyzed as 2,4-dinitrophenylhydrazones at 366 nm
	D2	= analyzed as quinoxalones at 254 nm

REFERENCES

1. **Hemming, B. C. and Gubler, C. J.** High-pressure liquid chromatography of α-keto acid 2,4-dinitrophenylhydrozones, *Anal. Biochem.* 92, 31, 1979.
2. **Liao, J. C., Hoffman, N. E., Barboriak, J. J., and Roth, D. A.**, High-performance liquid chromatography of pyruvic and α-ketoglutaric acids and its application to urine samples, *Clin. Chem.*, 23, 802, 1977.

Table LC-30
URINARY COMPOUNDS

Packing	P1	P2	P2	P2	P2	P3	P4	P2	P2	P1
Column										
Length (cm)	30	15	15	15	15	24	25	20	20	30
Diameter (mm)	4	3	3	3	3	6.3	4	3.2	3.2	4
Material	ss	ss	ss	ss	ss	G	ss	ss	ss	ss
Eluent	E1	E2	E3	E4	E4	E5	E6	E7	E8	E9
Flow rate (mℓ/min)	1					0.2	0.3	1.5	1.5	1.5
Temperature (°C)	32	23	23	25	45	55	55	23	23	room
Detection	D1	D2	D2	D2	D2	D2	D2	D2	D2	D3
Reference	1	2	2	2	2	3	3	4	4	5
Compound	min[a]	k′	k′	k′	k′	mℓ[a]	mℓ[a]	min[a]	min[a]	min
4-Hydroxy-3-methoxymandelic acid	3.5	—	—	—	—	—	—	—	—	—
3-Aminobenzoic acid	—	2.25	0.84	1.12	0.75	—	—	—	—	—
6-Hydroxy-1-naphthalenesulfonic acid	—	3.79	1.73	2.25	1.34	—	—	—	—	—
Benzoic acid	—	4.07	1.95	2.29	1.42	—	22.9	13.8	13.3	—
2-Hydroxybenzoic acid	—	6.09	3.22	3.72	2.25	3	21	18	—	7.7
1-Hydroxy-4-naphthalenesulfonic acid	—	6.41	3.64	4.33	2.29	—	—	—	—	—
5-Amino-1,3-benzenedicarboxylic acid	—	10.57	1.63	2.69	1.72	—	—	—	—	—
1,2,3-Benzenetricarboxylic acid	—	22.94	4.25	7.73	4.16	—	—	—	—	—
8-Amino-1-hydroxy-3,6-naphthalenedisulfonic acid	—	23.05	4.88	7.29	3.83	—	—	—	—	
Nicotinic acid	—	—	—	—	—	5	3.3	—	—	—
Xanthine	—	—	—	—	—	6.8	3.2	—	—	—
Caffeine	—	—	—	—	—	6.8	8.6	—	—	—
Trigonellin	—	—	—	—	—	7.9	2.7	—	—	—
Phenacetin	—	—	—	—	—	10.2	40.5	—	—	—
4-Aminobenzoic acid	—	—	—	—	—	12.3	6.0	—	—	—
Salicylamide	—	—	—	—	—	13.2	14.4	—	—	—
Cinnamic acid	—	—	—	—	—	15.8	30	—	—	—
3,4-Dihydroxycinnamic acid	—	—	—	—	—	—	8.6	—	—	—
2-Acetoxybenzoic acid	—	—	—	—	—	—	19.4	11.2	11.5	6.2
4-Hydroxybenzoic acid	—	—	—	—	—	—	—	5	5.7	—
3-Hydroxybenzoic acid	—	—	—	—	—	—	—	6	8	—

Table LC-30 (continued)
URINARY COMPOUNDS

Compound	min[a]	k′	k′	k′	k′	mℓ[a]	mℓ[a]	min[a]	min[a]	min
4-Methylbenzoic acid	—	—	—	—	—	—	—	—	—	12.4
2,5-Dihydroxybenzoic acid	—	—	—	—	—	—	—	—	—	3.6
2-Hydroxyhippuric acid	—	—	—	—	—	—	—	—	—	4.7

[a] Measured on chromatogram.

Packing	P1	= μBondapak C18
	P2	= LiChrosorb® RP8
	P3	= Hamilton cation exchange resin, X4, 10-15 μm, Na^+ form
	P4	= TSK LS 110 P5, macro porous polystyrene gel
Eluent	E1	= phosphate buffer-acetonitrile (97:3); phosphate buffer = 0.01 *M* KH_2PO_4-H_2SO_4 (pH 2.7)
	E2	= 25% *n*-propanol, 0.4% cetrimide, pH 6.0, in water
	E3	= 25% *n*-propanol, 0.4% cetrimide, pH 6.0 containing 0.05 *M* phosphate, in water
	E4	= 25% *n*-propanol, 0.4% cetrimide, 0.036 *M* bromide, pH 6.0, in water
	E5	= 0.1 *M* sodium phosphate (pH 4.5) in 25% ethanol
	E6	= 0.2 *M* phosphoric acid—0.2 *M* Na_2HPO_4 gradient
	E7	= 0.05 *M* sodium phosphate buffer (pH 2.0)-acetonitrile (80:20)
	E8	= 0.032 *M* tetrabutylammonium/acetonitrile — 0.05 *M* sodium phosphate buffer (pH 6.0) (20:80)
	E9	= 20% acetonitrile and 1.0% acetic acid in water
Detection	D1	= post-column reaction detector using alkaline periodate at 365 nm
	D2	= UV 254 nm
	D3	= UV 280 nm and fluorescence detector

REFERENCES

1. **Flood, J. G., Granger, M., and McComb, R. B.,** Urinary 3-methoxy-4-hydroxymandelic acid as measured by liquid chromatography, with on-line post-column reaction, *Clin. Chem.*, 25, 1234, 1979.
2. **Terweij-Grown, C. P., Heemstra, S., and Kraak, J. C.,** Distribution mechanism of ionizable substances in dynamic anion-exchange systems using cationic surfactants in high-performance liquid chromatography, *J. Chromatogr.*, 161, 69, 1978.
3. **Hanai, T., Walton, H. F., Navratil, J. D.,** Liquid chromatography of polar aromatic compounds on cation-exchange resins and porous polymer gels, *J. Chromatogr.*, 155, 261, 1978.
4. **Tilly-Melin, A., Askemark, Y., Wahlund, K. -G., and Schill, G.,** Retention behaviour of carboxylic acids and their quaternary ammonium ion pairs in reversed phase chromatography with acetonitrile as organic modifier in the mobile phase, *Anal. Chem.*, 51, 976, 1979.
5. **Amick, E. N. and Mason, W. D.,** Determination of aspirin, salicylic acid, salicyluric acid, and genstsic acid in human plasma and urine by high pressure liquid chromatography, *Anal. Lett.*, 12, 629, 1979.

Table LC-31
COMPOUNDS IN BIOLOGICAL FLUIDS

Packing	P1	P1	P2	P2	P3	P2	P2
Column							
Length (cm)	25	25	30	30	25	30	30
Diameter (mm)	4.6	4.6	4	4	4.6	4	3.9
Material	ss	ss	ss	ss	ss	ss	ss
Eluent	E1	E2	E3	E4	E5	E6	E7
Flow rate (mℓ/min)	1.2	1.2	2	2.4	2	2	1.5
Temperature (°C)			25			40	20
Detection	D1	D1	D2	D2	D3	D4	D2
Reference	1	1	2	3	4	5	6
Compound	min	min	min[a]	min[a]	min[a]	min[a]	min[a]
cis-4-Hydroxycinnamic acid	13.2	16.0	—	—	—	—	—
cis-4-Hydroxy-3-Methoxycinnamic acid	17.6	—	—	—	—	—	—
trans-4-Hydroxycinnamic acid	21.0	24.5	—	—	—	—	—
trans-4-Hydroxy-3-methoxycinnamic acid	24.8	31.7	—	—	—	—	—
trans-4-Hydroxy-3-methoxycinnamic acid	24.8	31.7	—	—	—	—	—
4-Hydroxybenzoic acid	—	9.8	—	—	—	—	—
4-Hydroxybenzaldehyde	—	11.1	—	—	—	—	—
4-Hdyroxy-3-methoxybenzoic acid	—	12.2	8.5	—	—	—	—
4-Hydroxy-3,5-dimethoxybenzoic acid	—	14.0	—	—	—	—	—
4-Hydroxy-3-methoxybenzaldehyde	—	16.5	—	—	—	—	—
Phenylethylamine	—	—	5	—	—	—	—
β-Phenylpyruvic acid	—	—	5.5	—	—	—	—
Phenylacetic acid	—	—	10	—	—	—	—
3,4-Dihydroxycinnamic acid	—	—	12.5	—	—	—	—
4-Hydroxycinnamic acid	—	—	21.3	—	—	—	—
Hydrocinnamic acid	—	—	32	—	—	—	—
4-Hydroxy-3-methoxycinnamic acid	—	—	34	—	—	—	—
Cinnamic acid	—	—	45.5	—	—	—	—
Nicotinamide-*N*-oxide	—	—	—	2.6	—	—	—
2-Hydroxypyridine-5-carboxylic acid	—	—	—	4.0	—	—	—
Nicotinamide	—	—	—	4.5	—	—	—
Isonicotinic acid	—	—	—	6.5	—	—	—
Nicotinic acid	—	—	—	7.7	—	—	—
Nicotinuric acid	—	—	—	8.3	—	—	—
Orotic acid	—	—	—	—	7.0	—	—
2,2-Dimethyl-3-methylsuccinic acid	—	—	—	—	—	10	—
Ritalinic acid *α*-phenyl-*α*-(2-piperdyl) acetic acid	—	—	—	—	—	14.5	—
Tropic acid	—	—	—	—	—	—	2.3
Atropinesulfate	—	—	—	—	—	—	3.0

[a] Measured on chromatogram.

Packing
- P1 = Spherisorb C18
- P2 = *μ*Bondapak C18
- P3 = Partisil® SAX

Eluent
- E1 = water-acetic acid-*n*-butanol (342:1:14)
- E2 = water-acetic acid-*n*-butanol (347:1:11)
- E3 = 10 mℓ of 2.0 *M* citric acid, 14.53 mℓ of 2.0 *M* NaOH, 4.758 g of NaCl, and 200 mℓ of methanol, and water to make 2.00 mℓ
- E4 = water-methanol (9:1), containing 50 m*M* tetrabutylammonium phosphate
- E5 = 4 m*M* KH_2PO_4, pH 2.8
- E6 = 20 m*M* potassium phosphate buffer, pH 3.8-acetonitrile (93:7)
- E7 = 1-heptanesulfonic acid-acetonitrile (65:35)

Table LC-31 (continued)
COMPOUNDS IN BIOLOGICAL FLUIDS

Detection D1 = UV 275 nm
D2 = UV 254 nm
D3 = UV 280 nm
D4 = UV 192 nm

REFERENCES

1. **Hartley, R. D. and Bushan, H.,** High-performance liquid chromatography of phenolic acids and aldehydes derived from plants of from the decomposition of organic matter in soil, *J. Chromatogr.*, 180, 139, 1979.
2. **Price, W. P., Jr., Edens, R., Hendrix, D. L., and Deming, S. N.,** Optimized reverse-phase high-performance liquid chromatographic separation of cinnamic acids and related compounds, *Anal. Biochem.*, 93, 233, 1979.
3. **Hengen, N., Seiberth, V., and Hengen, M.,** High-performance liquid-chromatographic determination of free nicotinic acid and its metabolite, nicotinuric acid, in plasma and urine, *Clin. Chem.*, 24, 1740, 1978.
4. **Mills, G. C., Schmalstieg, F. C., Newkirk, K. E., and Goldblum, R. M.,** Cytosine and orotic acid in urine of immunodeficient childen, *Clin. Chem.*, 25, 419, 1979.
5. **Soldin, S. J., Hill, B. M., Chan, Y. -P. M., Swanson, J. M., and Hill, J. G.,** A liquid-chromatogrphic analysis for ritalinic acid (α-phenyl-α-(2-piperidyl) acetic acid) in serum, *Clin. Chem.*, 25, 51, 1979.
6. **Brown, N. D. and Sleeman, H. K.,** Ion-pair high-performance liquid chromatographic method for the determination of atoropine sulfate and tropic acid, *J. Chromatogr.*, 150, 225, 1978.

Table LC-32
COMPOUNDS IN BIOLOGICAL FLUIDS

Packing	P1	P2	P3	P4
Column				
Length (cm)	25	25	25	
Diameter (mm)	4	4.6	4.6	4.6
Material	ss	ss	ss	ss
Eluent	E1	E2	E3	E4
Flow rate (mℓ/min)	0.6	2	1	2.25
Temperature (°C)	30-60	room	40	40
Detection	D1	D2	D3	D1
Reference	1	2,3	4	5
Compound	min[a]	min[a]	min[a]	min
Hippuric acid	40	—	5.5	4.4
Uric acid	23	—	—	—
Indoleacetic acid	71.5	—	—	—
Benzoic acid	—	5.0	—	—
Mandelic acid	—	—	6.7	—
Phenyglyoxylic acid	—	—	12	—
Phenylpropionic acid	—	—	—	5.5

[a] Measured on chromatogram.

Packing P1 = Hitachi 301 3N (5-7 μm anion-exchange resin)
P2 = LiChrosorb® SI 60 (5 μm)
P3 = LiChrosorb® RP18 (5 μm)
P4 = Partisil® SAX (10 μm)

Table LC-32 (continued)
COMPOUNDS IN BIOLOGICAL FLUIDS

Eluent	E1 = ammonium chloride-acetonitrile stepwise elution
	E2 = dichloromethane-ethanol-water
	E3 = water-methanol gradient
	E4 = 0.025 M NaH_2PO_4, pH 4.8
Detection	D1 = UV 254 nm
	D2 = UV 240 nm
	D3 = UV 224 or 236 nm

REFERENCES

1. **Miyagi, H., Miura, J., Takata, Y., and Ganno, S.,** High-performance liquid Chromatography for ultraviolet-absorbing constituents in human urine, *Clin. Chem.*, 25, 1617, 1979.
2. **Zschiesche, W., Gossler, K., Schaller, K. H., Triebig, G., Weltle, D., and Angerer, J.,** Biochemische Methoden zur Überwachung Styrol-exponierter Personen. Berufskrankheiten in der keramischen und Glass-Industrie, *Heft*, 28, 25, 1978/1979.
3. **Schoofs, T., Gossler, K., and Schaller, K. H.,** Der Einsatz der Hochdruck-Flüssigkeits-Chromatographie in der arbeitsmedizinischen Toxikologie, *GIT Fachz. Lab.*, 22, 608, 1978.
4. **Gossler, K., Will, W., Schaller, K. H., and Zschiesche, W.,** Quantitative Bestimmung der Metaboliten verschiedener aromatischer Lösungsmittle im Urin mit der Hochdruck-Flüssigkeits-Chromatographie, *Arbeitsmed. Sozialmed. Präventivmed.*, 14, 236, 1979.
5. **Gossler, K., Schaller, K. H., and Zschiesche, W.,** Quantitative Bestimmung von Hippursäure im Urine mit der Hochdruck-Flüssigkeits-Chromatographie, *Clin. Chim. Acta*, 78, 91, 1977.

Thin-Layer Chromatography Tables

Section I. III.

THIN-LAYER CHROMATOGRAPHY TABLES

Table TLC-1
PHENOLS

Layer	L1	L1	L1	L2	L2	L2				
Solvent	S1	S2	S3	S1	S2	S3				
Technique	T1	T1	T1	T1	T1	T1				
Detection										
Reference	1	1	1	1	1	1	D1	D2	D3	D4
Compound	Rf × 100									
Phenol	16	74	100	0	13	92	o	lt-bn	r-bn	lt-y
2-Methylphenol	24	78	100	2	32	95	o	bn	r-bn	lt-bn
3-Methylphenol	16	75	100	0	20	94	o	d-bn	r-bn	lt-bn
4-Methylphenol	15	74	100	0	20	94	vt	lt-bn	lt-bn	lt-bn
2,3-Dimethylphenol	25	80	100	2	39	95	o	d-bn	r-bn	o
2,4-Dimethylphenol	28	83	100	6	28	95	lt-bn	d-bn	y	y
2,5-Dimethylphenol	25	77	100	3	44	95	o	d-bn	bn	bn
2,6-Dimethylphenol	40	76	100	16	52	96	o	bn	r-bn	y
3,4-Dimethylphenol	15	75	100	0	22	92	—	bn	lt-bn	lt-bn
3,5-Dimethylphenol	15	73	100	0	26	90	vt	d-bn	bn	lt-bn
2,3,5-Trimethylphenol	25	80	100	4	46	96	vt	r-bn	bn	lt-bn
2,4,5-Trimethylphenol	19	71	100	0	23	87	vt	d-bn	lt-bn	o
2,4,6-Trimethylphenol	38	79	100	17	51	96	vt	lt-bn	lt-bn	—
3,4,5-Trimethylphenol	18	80	100	3	16	95	—	d-bn	lt-bn	r-bn
2-Ethylphenol	27	77	100	6	32	85	o	y-bn	r-bn	bn
3-Ethylphenol	17	72	100	0	14	80	o	y-bn	y-bn	y-bn
4-Ethylphenol	15	69	100	0	24	84	lt-bn	y-bn	y	y
2-isopropylphenol	31	77	100	7	50	88	o	y-bn	r-bn	y-bn
2-*tert*.-Butylphenol	54	87	100	17	79	96	vt	y-bn	lt-bn	lt-bn
4-*tert*.-Butylphenol	16	75	100	0	26	94	—	lt-bn	lt-bn	lt-bn
4-*tert*.-Pentylphenol	18	76	100	0	26	94	—	lt-bn	lt-bn	lt-bn

Table TLC-1 (continued)
PHENOLS

Compound	Rf × 100									
4-Nonylphenol	24	79	100	0	35	96	vt	lt-bn	gn-y	o
3-Ethyl-5-methylphenol	16	69	100	0	16	85	vt	bn	y	y
2-Methyl-5-isopropylphenol	27	77	100	6	4	85	o	r-bn	bn	bn
3-Methyl-6-isopropylphenol	36	86	100	10	62	97	o	r-bn	r-bn	r-bn
2-Methyl-6-isopropylphenol	51	78	100	31	44	86	o	d-bn	lt-bn	y-gn
4-(1,1,3,3-Tetramethylbutyl)phenol	21	77	100	0	32	95	—	bn	y	y
2-Hydroxybiphenyl	41	76	100	4	24	93	o	y-bn	bn	lt-bn
4-Hydroxybiphenyl	20	76	100	0	6	88	vt	bn	lt-bn	r-bn
2-Benzylphenol	32	72	100	2	36	94	o	y-bn	lt-bn	lt-bn
4-Benzylphenol	17	80	100	0	15	94	o	y-bn	lt-bn	lt-bn
2-Benzyl-4-methylphenol	29	75	100	5	28	86	—	bn	gn-y	o
2-Chlorophenol	42	75	100	0	4	83	1	y-bn	bn-y	bn-y
3-Chlorophenol	20	74	100	0	4	88	o	bn	o	y-bn
4-Chlorophenol	16	69	100	0	4	89	1	y-bn	y	y
2,3-Dichlorophenol	33	63	100	0	0	23	1	y-bn	y	lt-bn
2,4-Dichlorophenol	38	65	100	0	0	24	1	bn	y-bn	y
2,5-Dichlorophenol	41	77	100	0	0	15	1	y-bn	y	y-bn
2,6-Dichlorophenol	56	83	100	0	0	11	1	y-bn	y	r-bn
3,4-Dichlorophenol	15	63	100	0	0	0	o	lt-bn	lt-bn	o
2,3,6-Trichlorophenol	49	76	100	0	0	3	r	r-bn	1	r-bn
2,4,5-Trichlorophenol	32	67	100	0	0	4	r	lt-bn	bn	lt-bn
2,4,6-Trichlorophenol	49	76	100	0	0	3	r	lt-r	—	lt-y
2,3,4,6-Tetrachlorophenol	36	50	100	0	0	3	r	lt-r	—	lt-y
Pentachlorophenol	20	21	100	0	0	3	r	lt-r	—	lt-y
4-Chloro-2-methylphenol	24	74	100	2	13	9	o	bn	y-gn	o
4-Chloro-3-methylphenol	18	69	100	0	6	9	o	y-bn	y-gn	o
2-Chloro-5-methylphenol	41	68	100	0	5	75	o	d-bn	bn	bn
2-Chloro-5-methyl-6-isopropylphenol	38	81	100	7	33	10	o	bn	lt-bn	r-bn
2,2′-Methylene-*bis*-(4-chlorophenol)	3	52	100	0	0	1	o	d-bn	bn	r-bn
2,2′-Methylene-*bis*-(3,4,6-trichlorophenol)	9	14	100	0	0	3	o	lt-r	lt-bn	lt-y
2-Nitrophenol	69	79	100	0	0	97	o	—	lt-bn	—
3-Nitrophenol	7	62	100	0	0	27	o	lt-r	o	o

4-Nitrophenol	4	46	100	0	0	3	lt-y	lt-y	y	—
2,4-Dinitrophenol	7	3	95	0	0	3	y	y	bn	y
2,5-Dinitrophenol	32	18	95	0	0	2	y	y	lt-bn	y
2,6-Dinitrophenol	4	3	96	0	0	2	y	y	y	y
2,4,6-Trinitrophenol	0	0	100	0	0	7	y	y	y	y
2-Methyl-5-nitrophenol	11	78	100	0	3	84	lt-bn	lt-bn	r-bn	lt-bn
3-Methyl-4-nitrophenol	5	54	100	0	0	9	lt-y	lt-bn	y	—
4-Methyl-2-nitrophenol	70	79	100	0	0	2	o	lt-bn	o	y-bn
4-Methyl-3-nitrophenol	8	65	100	0	0	5	lt-y	lt-bn	bn	y-bn
2-Methyl-4,6-dinitrophenol	33	7	100	0	0	4	y	y	y	y
2-Chloro-4-nitrophenol	13	28	100	0	0	4	y	y	y	y
2-Fluorophenol	25	80	100	0	3	7	r-bn	—	lt-bn	lt-bn
3-Fluorophenol	18	73	100	0	6	8	r-bn	lt-bn	y-bn	lt-bn
4-Fluorphenol	15	68	100	0	6	9	r-bn	—	vt	lt-bn
1,2-Dihydroxybenzene	2	49	100	0	0	0	vt	bn	d-bn	d-bn
1,3-Dihydroxybenzene	0	39	100	0	0	77	vt	l	o	o
1,4-Dihydroxybenzene	0	43	100	0	2	90	vt	bn	vt	lt-bn
4-Hexyl-1,3-dihydroxybenzene	0	39	100	0	2	90	bn	vt	o	o
1,2,3-Trihydroxybenzene	0	17	100	0	0	0	vt	r-bn	d-bn	d-bn
1,3,5-Trihydroxybenzene	0	11	100	0	0	28	vt	vt	bn	o
1,2,4-Trihydroxybenzene	0	14	100	0	0	0	vt	bn	d-bn	d-bn
2-Methoxyphenol	30	66	97	0	4	80	r-bn	y-bn	y-bn	y-bn
2-Methoxy-4-methylphenol	26	60	100	0	7	73	o-bn	r-bn	r-y	r-y
4-Allyl-2-methoxyphenol	31	70	100	0	4	87	—	lt-y	H-y	lt-y
2-Methoxy-4-propenylphenol	32	68	100	0	3	81	—	lt-y	lt-y	lt-y
2-Hydroxybenzaldehyde	53	79	97	0	0	2	r-bn	—	y	lt-y
4-Hydroxybenzoic acid	0	22	99	0	0	0	r-bn	—	o	y-bn
4-Hydroxybenzoic acid-propylester	3	64	99	0	0	53	—	lt-bn	lt-bn	lt-bn
4-Hydroxy-3-methoxybenzaldehyde	5	30	100	0	0	7	lt-bn	lt-bn	o	bn
4-Hydroxy-3,5-dimethoxybenzaldehyde	6 & 3	27 & 17	97	0	0	71	lt-bn	lt-bn	lt-bn	lt-bn
4-Methylmercaptophenol	11	63	100	0	8	78	r-bn	bn	bn	r-bn
3-Methyl-4-methylmercaptophenol	12	62	100	0	7	83	bn	bn	bn	r-bn
2′,3′,4′-Trihydroxyacetophenol	0	28	90	0	0	0	bn	r-bn	gn-y	bn
2,4,4′-Trihydroxybenzophenone	0	29	100	0	0	0	lt-bn	l	d-bn	o
2,2′,4,4′-Diphenyltetrol	0	29	95	0	0	63	vt	l	d-bn	r-bn
α,α′-Diethyl-4,4′-stilbenediol	0	56	100	0	0	79	lt-y	lt-bn	lt-bn	bn
3-Ethylamino-4-methylphenol	3	56	100	0	8	86	d-bn	vt	y	y

Table TLC-1 (continued)
PHENOLS

Compound	Rf × 100									
Methoxybenzene	—	—	—	—	53	—	y	y	o	o
2-Methoxymethylbenzene	0	0	100	0	0	100	r-bn	lt-bn	r-bn	bn
3-Methoxymethylbenzene	14	68	100	0	15	81	lt-bn	lt-bn	lt-bn	lt-bn
4-Methoxymethylbenzene	75	0	100	0	0	0	—	lt-bn	—	lt-bn
3-Ethylmethoxybenzene	74	0	100	0	0	100	—	—	—	—
1-Hydroxynaphthalene	29	81	100	0	11	95	r-bn	vt	o	o
2-Hydroxynaphthalene	16	74	100	0	6	94	bn	vt	o	o
1-Hydroxy-2,4-dinitronaphthalene	4	0	100	0	0	3	y	y	y	y
Hippuric acid	0	0	100	0	0	0	lt-bn	r-bn	lt-bn	bn
D,L-Phenylalanine	0	0	0	0	0	0	—	—	—	—
L-Tyrosine	0	0	0	0	0	0	lt-bn	r-bn	lt-bn	lt-bn
L-4-Hydroxyproline	0	0	100	0	0	0	—	o	l	y-bn
Creatinine	0	0	100	0	0	0	—	—	—	lt-y
Indole	59	70	100	44	46	86	r	bn	y	o
2-Methylindole	65	71	100	46	52	85	—	lt-gn	gn-y	y-bn
Xanthine	0	0	0	0	0	0	—	—	—	—
Uric acid	0	0	0	0	0	0	—	—	—	—
4-Propylphenol	17	74	100	0	19	85	—	lt-bn	lt-bn	lt-bn
3-*tert*.-Butylphenol	22	79	100	0	30	93	vt	lt-bn	bn	bn
2-*tert*.-Butyl-6-methylphenol	79	89	100	61	79	95	—	bn	lt-bn	y-bn
2-*tert*.-Butyl-4-methylphenol	56	90	100	18	75	95	—	y-bn	y	bn-y
4-*tert*.-Butyl-2-methylphenol	26	80	100	4	37	95	—	lt-bn	y-bn	bn
6-*tert*.-Butyl-3-methylphenol	59	90	100	21	77	95	l	bn-y	y	y
2-*tert*.-Butyl-4,6-dimethylphenol	76	90	100	61	84	96	—	o	lt-y	—
2,6-Di-*tert*.-butylphenol	97	94	100	89	96	96	y	o	y	y
2,4-Di-*tert*.-butylphenol	59	90	100	20	80	94	—	bn	y-bn	bn
2,6-Di-*tert*.-butyl-4-methylphenol	97	93	100	87	97	97	—	—	—	—
4-*tert*.-Butyl-1,2-dihydroxybenzene	2	49	100	0	0	0	bn	y-gn	lt-bn	lt-bn
4,6-Di-*tert*.-butyl-1,3-dihydroxybenzene	2	49	100	0	0	0	bn	o	y	y
2-*tert*.-Butyl-4-methoxyphenol	32	80	100	10	61	97	vt	lt-bn	bn	vt
3-*tert*.-Butyl-2-hydroxy-5-methylbenzoic acid	0	7	95	0	0	0	—	lt-bn	lt-bn	bn
4-*tert*.-Butyl-2-chlorophenol	47	77	100	3	3	82	—	bn	lt-y	lt-bn

4-Chloro-2-cyclopenthylphenol	39	77	100	4	34	96	vt	lt-bn	y-bn	bn
3-Methyl-6-nitrophenol	69	81	100	0	0	0	d-bn	lt-y	o	lt-bn
2,6-Dimethyl-4-nitrophenol	19	51	100	0	0	26	o	lt-y	y	lt-bn
2-*tert*.-Butyl-4,6-dinitrophenol	73	32	100	0	0	20	y	y	y	y
4-Chloro-2-nitrophenol	73	65	100	0	0	0	lt-y	lt-y	y	lt-bn

Layer L1 = Kieselgel 60F 254
L2 = Aluminium oxide F 254
Solvent S1 = benzene
S2 = diisopropylether
S3 = ethanol
Technique T1 = use one plate with two layers (silica and alumina) and development by two-dimensional way (Rfs were result of each development)
Detection D1 = 4-aminoantipyrine solution
D2 = Genuine blue salt-B-solution
D3 = diazotized sulfanilic acid solution
D4 = diazotized 4-nitroaniline solution

REFERENCE

1. **Dietz, F., Traud, J., Koppe, P., and Rübelt Ch.,** Systeme für die Identifizierung von Phenolen mittels Dünnschicht-Chromatographie, *Chromatographia*, 9, 380, 1976.

Table TLC-2
PHENOLS

Layer	L1	L1	L2	L2	L2
Solvent	S1	S2	S3	S4	S5
Technique	T1	T1	T1	T1	T1
Detection	D1	D1	D1	D1	D1
Reference	1	1	1	1	1
Compound	Rf × 100				
Phenol	66	53	33	42	55
2-Methylphenol	67	50	31	38	—
3-Methylphenol	69	—	45	50	75
4-Methylphenol	—	49	—	—	—
4-Methoxyphenol	64	39	15	25	46
2-Chlorophenol	75	—	25	50	80
2-Nitrophenol	54	32	—	32	—
3-Nitrophenol	63	40	20	—	49
4-Nitrophenol	—	47	—	—	—
2,3-Dimethylphenol	55	51	—	—	—
Octylphenol	79	—	—	—	—
Nonylphenol	76	56	50	59	—
Dodecylphenol	80	—	—	69	80
Hexadecylphenol	80	—	—	70	83
2-*sec*.-Butylphenol	69	51	40	—	—
3-*tert*.-Butyl-2-methylphenol	70	56	47	—	—
4-*tert*.-Butyl-2-methylphenol	73	56	—	—	—
4-*tert*.-Octylphenol	70	57	48	—	—
2-*tert*.-Butyl-4-methylphenol	—	52	—	—	—
2,4-Di-*tert*.-butyl-6-methylphenol	—	48	—	—	—
1-Hydroxynaphthalene	—	—	23	38	59
2-Methyl-3-octylphenol	75	55	45	—	—

Layer	L1	= silicagel CH
	L2	= Kieselgel G
Solvent	S1	= cyclohexane-methylethylketone (7:3)
	S2	= cyclohexane-ethyl acetate (4:1)
	S3	= hexane-ethyl acetate (4:1)
	S4	= benzene-chloroform (1:1)
	S5	= chloroform
Technique	T1	= analyzed as 4-(*N,N*-dimethylamino)benzene 4′-azobenzoate
Detection	D1	= detected less than 0.5 μg of ester

From Churáček, J., Pechová, H., and Mareš, V., *J. Chromatogr.* 67, 97, 1972. With permission.

Table TLC-3
PHENOLS AND BENZOIC ACIDS

Layer	L1	L1	L1	L1	L1	L2	L3	L3	L3	L3	L4	L4	L5	L6	L6	L6	L6	L6	L6	L7	L8	L8	L8	L9	L10	L10	L10	L10	L10	L10	L10
Solvent	S1	S2	S3	S4	S5	S2	S6	S7	S8	S5	S9	S10	S11	S12	S13	S14	S15	S16	S17	S18	S18	S19	S20	S18	S6	S21	S1	S22	S18	S23	S24
Technique	T1	T1	T1	T1	T1	T1	T2	T2	T2	T2	T1	T1	T1	T1	T1	T1	T1	T1	T1	T1	T1	T1	T1	T1	T1	T1	T1	T1	T1	T1	T1
Detection	D1	D1	D1	D1	D1	D1	D1	D1	D1	D1	D2	D2	D2	D2	D2	D2	D2	D2	D2	D1	D1	D1	D1	D1	D1	D1	D1	D1	D1	D1	D1
Reference	1	1	1	1	1	1	1	1	1	1	2	2	2	2	2	2	2	2	2	3	3	3	3	3	3	3	3	3	3	3	3
Compound	Rf × 100																														
Phenol	15	58	22	59	9	71	85	96	85	24	57	92	50	49	53	65	92	90	73	65	61	61	68	45	10	19	39	64	90	30	40
2-Methoxyphenol	11	57	20	57	26	67	84	96	83	73	45	92	45	38	41	58	92	89	56	68	67	69	78	60	5	14	31	54	86	16	31
1,4-Dihydroxybenzene	16	47	7	48	0	61	82	95	78	0	61	94	68	43	nd	nd	nd	nd	nd	45	36	40	45	es	38	44	63	77	91	56	73
1,2-Dihydroxybenzene	9	36	14	29	0	58	82	95	78	0	57	93	63	44	50	70	94	93	es	46	34	39	52	12	20	30	50	68	87	40	57
1,3-Dihydroxybenzene	10	38	18	36	0	58	77	94	75	0	60	94	65	41	49	72	94	94	83	40	27	33	41	14	22	33	53	71	89	45	58
3,5-Dihydroxymethylbenzene	5	39	15	40	0	57	72	94	68	0	48	94	56	33	38	63	94	92	73	42	32	35	43	19	14	20	47	68	88	31	48
1,2,3-Trihydroxybenzene	7	16	es	28	0	45	72	87	69	0	65	89	71	46	71	88	95	95	nd	26	8	19	29	1	33	41	56	68	85	51	65
1,3,5-Trihydroxybenzene	8	20	es	16	0	46	58	89	57	0	70	90	70	39	80	93	95	95	88	19	9	15	18	1	34	44	61	76	88	55	69
2,3-Dihydroxybenzoic acid	0	0	44	6	0	14	5	58	31	0	49	90	50	43	95	95	95	95	nd	1	0	12	30	0	4	5	10	16	es	36	56
3,4,5-Trihydroxybenzoic acid	0	0	43	3	0	7	3	34	20	0	50	90	52	44	95	95	95	95	nd	0	0	6	19	0	4	4	9	14	es	38	57
2-Methylphenol	7	49	14	53	15	60	84	95	83	54	43	92	41	35	35	50	92	82	49	61	60	60	70	57	9	13	30	56	88	12	21
3-Methylphenol	7	49	14	55	11	60	84	95	83	38	43	92	41	35	36	53	92	88	56	61	60	61	71	54	9	14	32	58	89	11	25
2,6-Dimethylphenol	5	51	15	56	27	59	7	90	6	91	2	nd	2	1	1	5	nd	nd	nd	69	68	69	78	67	0	9	19	48	86	0	12
2,3-Dimethylphenol	1	44	9	49	15	53	77	95	76	63	32	90	27	24	24	31	90	72	34	60	60	61	72	55	1	4	15	43	83	6	12
3,4-Dimethylphenol	3	46	13	52	12	56	76	95	75	45	34	90	27	25	25	41	91	78	39	62	61	61	73	56	1	4	19	48	86	8	15
3,5-Dimethylphenol	2	45	13	52	13	57	76	95	75	52	34	90	27	25	25	42	91	78	39	61	60	61	72	56	2	6	26	53	88	6	13
3-Nitrophenol	1	23	20	41	2	45	74	92	69	3	30	89	31	21	58	80	94	90	46	48	34	42	57	4	1	6	24	49	82	9	17
2-Nitrophenol	0	3	22	16	es	52	nd	92	nd	nd	nd	nd	nd	nd	91	92	96	92	65	es	7	59	77	1	0	3	18	34	77	5	14
4-Nitrophenol	0	3	23	26	0	39	72	92	66	1	30	90	29	18	81	84	94	84	44	9	4	35	53	1	0	2	15	30	79	8	17

Table TLC-3 (continued)
PHENOLS AND BENZOIC ACIDS

Compound	Rf × 100																														
2,5-Dinitrophenol	0	0	7	13	10	2	22	72	40	59	22	28	88	21	28	80	80	92	80	45	3	0	8	59	0	0	0	0	1	8	0	6
2,4-Dinitrophenol	0	0	5	10	1	0	2	12	33	35	25	29	88	23	50	72	72	88	72	26	0	0	0	26	0	0	0	0	0	3	0	3
2,6-Dinitrophenol	0	0	4	7	0	0	1	9	35	23	35	38	90	23	79	88	88	94	88	53	0	0	0	32	0	0	0	0	0	2	0	3
2,4,6-Trinitrophenol	0	0	0	3	0	0	0	2	21	0	90	90	96	28	65	67	67	82	66	19	0	0	0	0	0	0	0	0	0	1	0	0
3-Aminophenol	26	52	39	53	0	53	85	93	81	0	0	0	0	71	6	53	68	94	88	71	43	40	60	95	27	32	43	62	69	77	89	90
2-Aminophenol	19	50	31	50	0	56	89	93	83	0	0	0	0	68	3	53	es	es	es	es	46	44	66	96	33	22	34	52	63	75	89	90
4-Aminophenol	38	62	53	66	0	67	91	93	89	0	0	0	0	72	0	es	es	es	es	es	84	64	70	97	41	45	50	66	70	77	95	95
5-Amino-2-hydroxy-benzoic acid	0	0	57	20	0	2	46	26	55	0	0	0	0	55	47	91	91	94	90	65	0	0	8	65	0	0	5	5	6	4	80	89
4-Amino-2-hydroxy-benzoic acid	0	0	50	18	0	2	2	10	43	0	0	0	0	49	53	91	91	94	90	64	0	0	3	37	0	0	3	3	4	3	16	39
2-Amino-3-hydroxy-benzoic acid	0	0	40	16	0	2	5	44	50	0	0	0	0	57	54	92	92	94	92	65	0	0	0	50	0	—	—	—	—	—	—	—
3-Amino-4-hydroxy-benzenesulphonic acid	0	0	56	24	0	1	4	3	48	0	91	85	76	87	96	96	96	96	96	96	0	0	0	0	0	1	1	1	2	2	67	80
4-Amino-2-nitrophenol	6	28	45	23	1	43	70	82	63	3	0	0	0	58	7	74	87	94	88	66	43	40	53	88	2	3	11	27	43	68	64	83
2-Amino-5-nitrophenol	3	20	19	23	0	27	47	81	40	0	0	2	13	53	9	24	65	87	76	49	24	14	25	39	0	1	7	20	41	69	10	21
2-Amino-4-nitrophenol	1	5	27	18	0	30	58	82	51	0	0	0	0	58	10	71	78	89	78	51	10	2	28	55	0	1	5	15	33	65	28	55
2-Amino-4,6-dinitro-phenol	0	0	3	4	0	0	1	9	22	5	0	4	32	22	25	55	56	77	56	18	1	0	1	30	0	0	0	0	1	1	0	4
2-Amino-3,4,6-trichlo-rophenol	0	0	3	7	10	32	30	92	24	56	0	1	44	18	2	54	63	84	63	20	12	0	42	61	0	0	1	4	26	67	0	2
4-Chlorophenol	6	39	11	41	5	52	75	95	75	31	25	30	92	26	19	24	65	94	89	53	56	44	55	66	29	4	7	26	53	83	9	17
3-Chlorophenol	3	37	11	41	6	52	75	95	74	32	25	30	92	25	20	31	72	94	90	66	55	40	52	64	15	2	6	26	53	83	9	17
2-Chlorophenol	4	35	18	38	15	55	nd	95	nd	81	nd	nd	nd	nd	nd	nd	86	96	95	78	55	37	53	65	10	2	6	26	52	82	8	18
4-Bromophenol	2	34	9	40	6	48	70	95	69	32	20	23	91	19	13	21	60	94	88	53	52	40	46	60	21	2	5	21	52	83	6	12
2-Bromophenol	3	30	15	35	19	50	nd	95	nd	86	nd	nd	nd	nd	21	51	84	96	94	74	51	35	43	58	8	2	6	23	53	83	4	14
3,4-Dichlorophenol	0	26	11	33	5	46	59	95	58	33	11	14	91	10	8	24	66	94	82	45	49	30	40	58	6	1	2	10	42	83	3	6
3,5-Dichlorophenol	1	20	4	32	11	46	58	95	es	49	12	16	91	10	8	36	75	94	84	50	49	21	40	58	4	1	2	9	38	82	3	6
2,4-Dichlorophenol	0	14	10	25	14	45	nd	95	42	82	nd	14	91	14	9	58	82	95	90	50	48	19	39	58	3	0	2	9	38	82	3	8
2,3-Dichlorophenol	0	12	11	25	14	49	nd	95	41	82	nd	14	91	14	9	66	84	95	91	56	48	15	38	56	3	0	2	9	81	3	6	
2,5-Dichlorophenol	0	10	11	25	14	47	nd	95	42	82	nd	nd	nd	nnd	11	78	88	96	91	59	48	10	39	58	2	0	1	9	39	82	3	9
2,6-Dichlorophenol	0	1	8	14	25	45	nd	95	45	94	nd	nd	nd	18	12	94	95	96	95	65	es	6	54	70	1	0	1	8	4	77	3	9
2-Hydroxynaphthalene	0	27	2	31	7	37	45	88	41	31	8	14	86	10	4	5	29	86	66	31	39	35	39	54	24	0	1	8	29	72	1	3
1-Hydroxynaphthalene	0	27	2	30	7	38	49	89	44	43	7	13	86	10	4	6	35	92	75	34	38	34	38	51	23	0	1	4	24	69	0	2

1,5-Dihydroxynaphthalene	0	14	2	16	0	26	36	92	27	0	8	14	86	20	3	7	es	es	es	es	16	10	14	21	4	1	3	11	31	70	3	5
2-Hydroxy-1-naphthaldehyde	0	0	0	0	0	8	5	88	3	95	4	7	78	3	4	38	52	nd	nd	nd	es	20	53	77	1	0	0	3	17	64	1	3
7-Amino-2-hydroxynaphthalene	5	31	12	30	0	33	57	84	37	0	0	0	0	34	0	3	26	86	49	28	23	21	35	90	13	2	4	15	42	62	59	81
1-Amino-7-hydroxynaphthalene	2	28	10	25	0	31	59	85	40	0	0	0	0	34	0	5	35	86	61	34	23	20	34	89	11	2	5	18	46	63	62	81
5-Amino-1-hydroxynaphthalene	2	26	7	20	0	27	57	84	40	0	0	0	0	36	0	5	36	86	61	36	24	20	32	89	11	1	3	12	34	58	40	71
4-Hydroxydiphenylamine	0	25	3	29	10	35	57	95	52	15	0	1	3	42	0	0	12	78	51	20	32	30	33	55	26	0	1	4	26	72	4	5
3-Hydroxydiphenylamine	0	23	2	25	8	32	47	95	43	14	0	1	43	21	1	1	15	nd	30	10	27	25	30	47	18	0	1	4	24	71	0	2
4′-Hydroxy-2,4-dinitrodiphenylamine	0	12	1	16	0	24	32	88	18	4	1	7	80	3	0	2	16	68	38	10	36	31	35	63	23	0	0	0	3	38	0	0
4′-Hydroxy-2,4-dinitrodiphenylamine-3′-sulfonic acid	0	0	0	3	0	1	0	3	13	3	90	91	96	20	42	74	86	90	85	30	0	0	0	0	0	0	0	0	0	0	0	0
4-Hydroxyazobenzene	0	13	2	22	27	35	19	92	15	36	1	3	86	1	1	6	es	78	45	13	36	28	36	63	4	0	0	1	22	60	0	0

Note: nd = not determined, es = elongated spot.

Layer	L1	=	Bio-Rad® AG 3-X4A
	L2	=	Bio-Rad® AG 3-X4(Cl^- form)
	L3	=	PEI-cellulose
	L4	=	Dowex® 50-X4 (H^+ form)
	L5	=	Rexyn 102 (H^+ form)
	L6	=	Dowex® 50-X4 (Na^+ form)
	L7	=	AG 1-X4 (Cl^- form)
	L8	=	AG 1-X4 (acetate form)
	L9	=	AG 1-X4 (OH^- form)
	L10	=	BD-cellulose
Solvent	S1	=	50% methanol
	S2	=	95% methanol
	S3	=	0.5 $M NH_3$ in 50% ethanol
	S4	=	0.5 $M NH_3$ in 95% ethanol
	S5	=	acetic acid-cyclohexane (7:93)

Table TLC-3 (continued)
PHENOLS AND BENZOIC ACIDS

S6 = water
S7 = 80% methanol
S8 = 0.1 *M* sodium hydrogen carbonate solution
S9 = 20% ethanol
S10 = ethanol-water (2:1)
S11 = 0.5 *M* HCl in ethanol-water (1:4)
S12 = 0.1 *M* acetate buffer
S13 = 0.1 *M* $NaHCO_3$
S14 = 0.05 *M* Na_2CO_3
S15 = 1 *M* NH_3
S16 = 1 *M* NH_3 + 0.1 *M* CH_3COONa
S17 = 1 *M* NH_3 + 1M CH_3COONa
S18 = methanol
S19 = 1 *M* acetic acid in methanol
S20 = 10 *M* acetic acid in methanol
S21 = water-methanol (4:1)
S22 = water-methanol (3:7)
S23 = 1 *M* acetic acid in water
S24 = 3 *M* acetic acid in water

Technique
T1 = layers were prepared by mixing 3 g or the resin with 9 g microcrystalline cellulose
T1 and T2 = the chromatographic development was carried out at 25 ± 0.5°C and the migration distance.

Detection
D1 = Boute reaction
D2 = after spraying with 1 *M* sodium hydroxide solution, layers are exposed to nitrogen dioxide vapor (Boute reaction)

REFERENCES

1. **Lepri, L., Desideri, P. G., Tanturli, G., and Landini, M.,** Chromatographic behaviour of phenols on thin layers of cation and anion exchangers, 1. Bio-Rad Ag 3-X4A, PEI-CELLULOSE and DEAE-CELLOSE, *J. Chromatogr.*, 108, 169, 1975.
2. **Lepri, L., Desideri, P. G., Landini, M., and Tanturli, G.,** Chromatographic behaviour of phenols on thin layers or cation and anion exchangers. II. Dowex 50-X4 and Rexyn 102, *J. Chromatogr.*, 109, 365, 1975.
3. **Lepri, L., Desideri, P. G., Landini, M., and Tanturli, G.,** Chromatographic behaviour of phenols on thin layers of anion and cation exchangers. III. AG 1-X4 and BD-cellulose, *J. Chromatogr.*, 129, 239, 1976.

Table TLC-4
PHENOLS

Layer	L1	L2		L1	L2
Solvent	S1	S2		S1	S2
Technique	T1	T1		T1	T1
Detection	D1	D1		D1	D1
Reference	1	1		1	1
Compound	Rf × 100			Rf × 100	
Phenol	7	20	3-Fluorophenol	11	26
2-Methylphenol	21	26	2-Chlorophenol	13	—
3-Methylphenol	13	22	3-Chlorophenol	22	28
2-Ethylphenol	45	29	2-Bromophenol	15	25
3-Ethylphenol	29	24	3-Bromophenol	26	27
2-*n*-Propylphenol	62	33	2-Iodophenol	—	30
2-Isopropylphenol	66	34	3-Iodophenol	21	30
2-*sec*-Butylphenol	82	36	2,6-Dichlorophenol	23	35
2-*tert*-Butylphenol	83	44	2,6-Dibromophenol	25	37
3-*tert*-Butylphenol	56	—	2-Bromo-6-chlorophenol	25	36
2-Hydroxybiphenyl	46	30	2,6-Diiodophenol	33	47
3-Hydroxybiphenyl	33	26	2,3,6-Trichlorophenol	47	47
2,5-Dimethylphenol	39	27	2,3,5,6-Tetrachlorophenol	52	—
2,6-Dimethylphenol	48	28	6-Bromo-2-methylphenol	48	47
3,5-Dimethylphenol	49	26	6-Bromo-2,5-dimethylphenol	40	45
3-Methyl-5-isopropylphenol	53	27	2-Chloro-3,5-dimethylphenol	26	—
5-Methyl-2-isopropylphenol	—	45	2,6-Dibromo-3,5-dimethylphenol	43	47
2-*tert*-Butyl-6-methylphenol	92	52	2,6-Diiodo-3,5-dimethylphenol	45	57
2,6-Diethylphenol	84	47	Dichlorophene (2,2′-Methylene-bis-(4-chloropheol))	—	2
2-*tert*-Butyl-6-ethylphenol	—	77	Hexachlorophene (2,2′-Methylene-bis-(3,4,6-tri-chloro-phenol))	—	47
2-Fluorophenol	8	26			

Layer L1 = Lucefol sheets impregnated with 10% formamide in ethanol
L2 = Silufol sheets (cellulose)
(These are precoated sheets from Sklárny Kevalier, Votice, Czechoslovakia.)

Table TLC-4 (continued)
PHENOLS

Solvent	S1	= *n*-heptane-benzene (1:1)
	S2	= benzene-acetone (3:1)
Technique	T1	= analyzed as antipyrilquinoneimine dyes
Detection	D1	= see T1

From Gasparič, J. and Svobodová, D., *J. Chromatogr.*, 153, 153, 1978. With permission.

Table TLC-5
PHENOLS

Layer	L1	L1	L1		L2	L3	L3	L4	L5	
Solvent	S1	S2	S3		S4	S5	S6	S7	S8	
Technique					T2			T1		
Detection	D1	D1	D1	C1	D2	D3	D3	D4	D5	C2
Reference	1	1	1	1	2	3	3	4	5	5
Compound	Rf × 100[a]									
2-Hydroxynaphthalene	47	69	74	cr	—	—	—	—	—	—
1-Hydroxynaphthalene	44	66	71	pk	—	—	—	—	—	—
Phenol	59	64	73	d-bn	—	—	—	70	54	b
1,2,3-Trihydroxybenzene	3	27	46	g	—	—	—	26	—	—
2,4-Dinitrophenol	68	70	76	y	56	—	—	—	—	—
2-Nitrophenol	61	65	73	lt-bn	—	—	—	—	—	—
4-Nitrophenol	30	57	68	y	—	—	—	—	—	—
1,3-Dihydroxybenzene	19	40	48	bn	—	—	—	30	—	—
2-Hydroxybenzoic acid	41	43	55	lt-pk	—	—	—	—	—	—
5-Methyl-2-isopropyl-1-phenol	70	75	86	o	—	—	—	—	—	—
4-Chlorophenol	—	—	—	—	60	—	—	—	—	—
3-Chlorophenol	—	—	—	—	63	—	—	—	—	—
2-Chlorophenol	—	—	—	—	59	—	—	—	—	—
3,4-Dinitrophenol	—	—	—	—	61	—	—	—	—	—
2,4,5-Trichlorophenol	—	—	—	—	79	—	—	—	—	—
Pentachlorophenol	—	—	—	—	83	—	—	—	—	—
2-Octaprenylphenol	—	—	—	—	—	50	70	—	—	—
Ubiquinone 8	—	—	—	—	—	—	20	—	—	—
1,2-Dihydroxybenzene	—	—	—	—	—	—	—	54	—	—
1,3,5-Trihydroxybenzene	—	—	—	—	—	—	—	10	—	—
4-Methylphenol	—	—	—	—	—	—	—	—	56	ft-b
2-Methylphenol	—	—	—	—	—	—	—	—	70	p
4-Ethylphenol	—	—	—	—	—	—	—	—	61	ft-b
2-Ethylphenol	—	—	—	—	—	—	—	—	77	p
4-Isopropylphenol	—	—	—	—	—	—	—	—	58	ft-b
2-Isopropylphenol	—	—	—	—	—	—	—	—	76	p
4-*tert*-Butylphenol	—	—	—	—	—	—	—	—	59	no
2,3-Dimethylphenol	—	—	—	—	—	—	—	—	62	p
2,4-Dimethylphenol	—	—	—	—	—	—	—	—	55	ft-p
2,6-Dimethylphenol	—	—	—	—	—	—	—	—	68	p
2,5-Dimethylphenol	—	—	—	—	—	—	—	—	54	p
2,4,6-Trimethylphenol	—	—	—	—	—	—	—	—	68	p
2,3,4-Trimethylphenol	—	—	—	—	—	—	—	—	68	ft-o
2,4,5-Trimethylphenol	—	—	—	—	—	—	—	—	62	ft-g

[a] Measured on chromatogram.

Layer	L1	=	silica gel G impregnated with anilinium chloride
	L2	=	silica gel 60
	L3	=	silica gel G60
	L4	=	silica gel impregnated 20% polyamide
	L5	=	silica gel FG
Solvent	S1	=	benzene-ethyl acetate (9:1)
	S2	=	benzene-ethyl acetate (3:1)
	S3	=	benzene-ethyl acetate (1:1)
	S4	=	benzene-chloroform (1:1)
	S5	=	light petroleum-diethyl ether (10:4)
	S6	=	benzene
	S7	=	benzene-acetic acid-methanol (45:4:8)
	S8	=	benzene-ethyl acetate (95:5)

Table TLC-5 (continued)
PHENOLS

Technique	T1	= orthogonal temperature gradient
	T2	= analyzed as dansyl derivatives
Detection	D1	= anilium chloride
	D2	= Zeiss PMQ11 Chromatogram Spectrometer after spraying with triethanolamine in isopropanol
	D3	= Gibbs reagent, Telefunken MSRC 61002 (radioactivity ^{14}C)
	D4	= sulfanilic acid in 10% sodium hydroxide soln.
	D5	= Gibbs and Folins reagents

REFERENCES

1. **Chhabra, B. R., Gulati, R. S., Dhillon, R. S., and Kalsi, P. S.,** Modified thin-layer chromatographic method for the detection of phenols, *J. Chromatogr.*, 135, 521, 1977.
2. **Frei-Häusler, M., Frei, R. W., and Hutzinger, O.,** An investigation of fluorigenic labelling of chlorophenols with dansyl chloride, *J. Chromatogr.*, 84, 214, 1973.
3. **Knoell, H. E., Kraft, R., and Knappe, J.,** Dioxygen and temperature dependence of ubiquinone formation in Escherichia coli: studies of cells charged with 2-octaprenyl phenol, *Eur. J. Biochem.*, 90, 107, 1978.
4. **Gocan, S., Konnert, C., and Voik, H.,** Trennung einiger Polyphenole durch Dünnschichtchromatographie mit orthogonalen Temperatur-gradient, *Rev. Roum. Chim.*, 23, 133, 1978.
5. **Kaubisch, N., Daly, J. W., and Jerina, D. M.,** Arene oxides as intermediates in the oxidation metabolism of aromatic compounds. Isomerization of methyl-substituted arene oxide, *Biochemistry*, 11, 3080, 1972.

Table TLC-6
PHENOLS

Layer	L1	L1	L1	L1	L1	L1	L1	L1	L1	L1	L1	L1	L1	L1	L1	L1	L1	L1	L1	L1	L1	L1	L1		
Solvent	S1	S2	S3	S4	S5	S6	S6	S7	S7	S8	S8	S9	S9	S10	S10	S11	S11	S12	S13	S14	S15	S16	S17		
Technique																					—				
Detection	D1	D1	D1	D1	D1	D1	D1	D1	D1	D1	D1	D1	D1	D1	D1	D1	D1	D1	D1	D1	D1	D1	D1	C1	C1
Reference	1	1	1	1	1	1	2	1	2	1	2	1	2	1	2	1	2	2	2	2	2	2	2	1	2
Compound	Rf × 100																								
2-Chloro-4-methylphenol	33	56	23	56	73	63	—	53	—	50	—	52	—	26	—	66	—	—	—	—	—	—	—	y	—
3-Chloro-2-methylphenol	36	56	26	46	66	60	—	50	—	40	—	46	—	26	—	63	—	—	—	—	—	—	—	d-b	—
4-Chloro-2-methylphenol	33	56	26	43	66	60	—	50	—	36	—	43	—	23	—	63	—	—	—	—	—	—	—	bn	—
4-Chloro-3-methylphenol	30	56	26	36	63	60	—	53	—	33	—	43	—	23	—	63	—	—	—	—	—	—	—	bn	—
1,2-Dihydroxybenzene	—	—	—	—	—	—	46	—	20	—	20	—	26	—	10	—	43	56	46	30	53	40	16	—	p
3,4-Dichloro-1,2-dihydroxybenzene	—	—	—	—	—	—	43	—	23	—	17	—	20	—	13	—	40	40	53	30	66	40	16	—	pl-gn
3,5-Dichloro-1,2-dihydroxybenzene	—	—	—	—	—	—	50	—	26	—	20	—	26	—	13	—	46	60	53	30	60	56	23	—	bn
3,6-Dichloro-1,2-dihydroxybenzene	—	—	—	—	—	—	50	—	26	—	23	—	30	—	17	—	46	60	50	36	63	60	23	—	bn
3,4,5-Trichloro-1,2-dihydroxybenzene	—	—	—	—	—	—	43	—	20	—	17	—	20	—	17	—	40	40	53	26	63	46	16	—	bn
Tetrachloro-1,2-dihydroxybenzene	—	—	—	—	—	—	46	—	26	—	17	—	26	—	20	—	43	53	50	36	56	53	23	—	pl-gn
5-Chloro-2,3-dihydroxy-methylbenzene	—	—	—	—	—	—	53	—	30	—	25	—	30	—	26	—	50	60	63	40	63	60	26	—	p

Layer	L2	L2	L2	L2	L2	L2	L2	L2	L2	L2	L2	L2	L2	L2	L2	L2	L2	L2	L2	L2	L2	L2	L2		
Solvent	S1	S2	S3	S4	S5	S6	S6	S7	S7	S8	S8	S9	S9	S10	S10	S11	S11	S12	S13	S14	S15	S16	S17		
Technique																									
Detection	D1	D1	D1	D1	D1	D1	D1	D1	D1	D1	D1	D1	D1	D1	D1	D1	D1	D1	D1	D1	D1	D1	D1	C1	C1
Referece	1	1	1	1	1	1	2	1	2	1	2	1	2	1	2	1	2	2	2	2	2	2	2	1	2
Compound	Rf × 100																								
2-Chloro-4-methylphenol	73	63	36	73	86	76	—	76	—	63	—	60	—	51	—	73	—	—	—	—	—	—	—	y	—
3-Chloro-2-methylphenol	70	63	16	60	83	73	—	70	—	60	—	56	—	43	—	73	—	—	—	—	—	—	—	d-b	—
4-Chloro-2-methylphenol	70	63	16	60	83	73	—	70	—	60	—	56	—	43	—	73	—	—	—	—	—	—	—	bn	—
4-Chloro-3-methylphenol	66	63	13	50	80	73	—	70	—	60	—	56	—	43	—	73	—	—	—	—	—	—	—	y	—
1,2-Dihydroxybenzene	—	—	—	—	—	—	60	—	33	—	33	—	40	—	23	—	60	70	66	36	60	66	20	—	p
3,4-Dichloro-1,2-dihydroxybenzene	—	—	—	—	—	—	—	—	—	—	—	—	—	—	10	—	40	—	—	—	—	—	—	—	pl-gn
3,5-Dichloro-1,2-dihydroxybenzene	—	—	—	—	—	—	36	—	30	—	30	—	36	—	16	—	53	53	46	33	63	—	16	—	bn
3,6-Dichloro-1,2-dihydroxybenzene	—	—	—	—	—	—	33	—	36	—	30	—	36	—	13	—	50	53	36	43	73	—	13	—	bn

Table TLC-6 (continued)
PHENOLS

Compound	Rf × 100																								
3,4,5-Trichloro-1,2-dihydroxybenzene	—	—	—	—	—	—	10	—	13	—	20	—	—	—	6	—	26	—	23	—	—	—	6	—	bn
Tetrachloro-1,2-dihydroxybenzene	—	—	—	—	—	—	6	—	10	—	6	—	10	—	3	—	25	10	6	13	46	23	23	—	pl-gn
5-Chloro -2,3-dihydroxy-methylbenzene	—	—	—	—	—	—	66	—	46	—	43	—	53	—	20	—	60	73	70	46	76	83	23	—	p
Layer	L3	L3	L3	L3	L3	L3	L3	L3	L3	L3	L3	L3	L3	L3	L3	L3	L3	L3	L3	L3	L3	L3	L3		
Solvent	S1	S2	S3	S4	S5	S6	S6	S7	S7	S8	S8	S9	S9	S10	S10	S11	S11	S12	S13	S14	S15	S16	S17		
Technique																									
Detection	D1	D1	D1	D1	D1	D1	D1	D1	D1	D1	D1	D1	D1	D1	D1	D1	D1	D1	D1	D1	D1	D1	D1	C1	C1
Reference	1	1	1	1	1	1	2	1	2	1	2	1	2	1	2	1	2	2	2	2	2	2	2	1	2

Compound	Rf × 100																								
2-Chloro-4-methylphenol	—	—	96	—	—	—	—	—	—	—	—	—	—	—	—	—	—	—	—	—	—	—	—	y	—
3-Chloro-2-methylphenol	93	96	96	96	96	96	—	96	—	96	—	96	—	96	—	96	—	—	—	—	—	—	—	b	—
4-Chloro-2-methylphenol	93	96	—	96	96	—	—	—	—	98	—	96	—	96	—	98	—	—	—	—	—	—	—	bn	—
4-Chloro-3-methylphenol	—	—	—	96	96	—	—	—	—	—	—	—	—	—	—	—	—	—	—	—	—	—	—	y	—
1,2-Dihydroxybenzene	—	—	—	—	—	—	93	—	86	—	96	—	96	—	86	—	96	90	96	53	—	96	90	—	bk
3,4-Dichloro-1,2-dihydroxybenzene	—	—	—	—	—	—	86	—	—	—	—	—	—	—	—	—	—	66	—	—	—	—	—	—	bk
3,5-Dichloro-1,2-dihydroxybenzene	—	—	—	—	—	—	93	—	86	—	96	—	93	—	86	—	—	86	93	80	—	96	90	—	bk
3,6-Dichloro-1,2-dihydroxybenzene	—	—	—	—	—	—	93	—	93	—	96	—	93	—	80	—	90	80	90	93	—	96	83	—	bk
3,4,5-Trichloro-1,2-dihydroxybenzene	—	—	—	—	—	—	86	—	46	—	90	—	73	—	73	—	80	66	66	—	—	—	70	—	bk
Tetrachloro-1,2-dihydroxybenzene	—	—	—	—	—	—	80	—	60	—	73	—	60	—	44	—	66	66	70	—	—	86	56	—	bk
5-Chloro-2,3-dihydroxy-methylbenzene	—	—	—	—	—	—	93	—	90	—	96	—	96	—	90	—	93	90	96	73	—	96	96	—	bk
Layer	L4	L4	L4	L4	L4	L4	L4	L4	L4	L4	L4	L4	L4	L4	L4	L4	L4	L4	L4	L4	L4	L4	L4		
Solvent	S1	S2	S3	S4	S5	S6	S6	S7	S7	S8	S8	S9	S9	S10	S10	S11	S11	S12	S13	S14	S15	S16	S17		
Technique																									
Detection	D1	D1	D1	D1	D1	D1	D1	D1	D1	D1	D1	D1	D1	D1	D1	D1	D1	D1	D1	D1	D1	D1	D1	C1	C1
Reference	1	1	1	1	1	1	2	1	2	1	2	1	2	1	2	1	2	2	2	2	2	2	2	1	2

Compound	Rf × 100																								
2-Chloro-4-methylphenol	50	40	23	56	66	60	—	60	—	43	—	60	—	26	—	70	—	—	—	—	—	—	—	y	—
3-Chloro-2-methylphenol	46	40	23	43	66	60	—	53	—	33	—	53	—	26	—	66	—	—	—	—	—	—	—	d-b	—

4-Chloro-2-methylphenol	43	43	23	40	60	60	—	53	—	33	—	53	—	26	—	66	—	—	—	—	—	—	—	bn	—
4-Chloro-3-methylphenol	43	50	23	33	60	60	—	50	—	30	—	50	—	26	—	63	—	—	—	—	—	—	—	y	—
1,2-Dihydroxybenzene	—	—	—	—	—	—	43	—	23	—	13	—	30	—	13	—	46	50	43	23	73	46	23	—	p
3,4-Dichloro-1,2-dihydroxybenzene	—	—	—	—	—	—	40	—	—	—	13	—		—	20	—	33	—	36	16	—	—	—	—	pl-gn
3,5-Dichloro-1,2-dihydroxybenzene	—	—	—	—	—	—	43	—	23	—	16	—	30	—	13	—	46	50	43	26	—	56	26	—	bn
3,6-Dichloro-1,2-dihydroxybenzene	—	—	—	—	—	—	40	—	33	—	16	—	33	—	20	—	46	46	43	36	70	60	23	—	bn
3,4,5-Trichloro-1,2-dihydroxybenzene	—	—	—	—	—	—	33	—	23	—	13	—	—	—	20	—	26	—	33	13	—	—	—	—	bn
Tetrachloro-1,2-dihydroxybenzene	—	—	—	—	—	—	16	—	16	—	10	—	16	—	10	—	16	33	23	16	23	36	16	—	pl-gn
5-Chloro-2,3-dihydroxy-methylbenzene	—	—	—	—	—	—	53	—	30	—	20	—	40	—	23	—	53	60	53	30	80	63	26	—	p
Layer	L5	L5	L5	L5	L5	L5	L5	L5	L5	L5	L5	L5	L5	L5	L5	L5	L5	L5	L5	L5	L5	L5	L5		
Solvent	S1	S2	S3	S4	S5	S6	S6	S7	S7	S8	S8	S9	S9	S10	S10	S11	S11	S12	S13	S14	S15	S16	S17		
Technique																									
Detection	D1	D1	D1	D1	D1	D1	D1	D1	D1	D1	D1	D1	D1	D1	D1	D1	D1	D1	D1	D1	D1	D1	D1	C1	C1
Reference	1	1	1	1	1	1	2	1	2	1	2	1	2	1	2	1	2	2	2	2	2	2	2	1	2
Compound																									
	Rf × 00																								
2-Chloro-4-methylphenol	46	70	33	66	73	76	—	66	—	—	—	—	—	40	—	66	—	—	—	—	—	—	—	y	—
3-Chloro-2-methylphenol	43	70	33	56	66	73	—	60	—	56	—	—	—	36	—	76	—	—	—	—	—	—	—	y	—
4-Chloro-2-methylphenol	43	63	30	53	56	73	—	63	—	53	—	—	—	36	—	76	—	—	—	—	—	—	—	y	—
4-Chloro-3-methylphenol	40	60	25	46	60	73	—	63	—	53	—	—	—	36	—	76	—	—	—	—	—	—	—	y	—
1,2-Dihydroxybenzene	—	—	—	—	—	—	63	—	30	—	23	—	26	—	20	—	63	66	66	16	50	60	20	—	p
3,4-Dichloro-1,2-dihydroxybenzene	—	—	—	—	—	—	56		30	—	23	—	26	—	—	—	—	53	—	—	—	—	20	—	pl-gn
3,5-Dichloro-1,2-dihydroxybenzene	—	—	—	—	—	—	63	—	36	—	26	—	33	—	23	—	66	70	70	33	66	66	26	—	bn
3,6-Dichloro-1,2-dihydroxybenzene	—	—	—	—	—	—	63	—	43	—	30	—	30	—	23	—	66	70	70	33	66	—	26	—	bn
3,4,5-Trichloro-1,2-dihydroxybenzene	—	—	—	—	—	—	53	—	26	—	26	—	26	—	20	—	—	53	—	33	—	—	23	—	bn
Tetrachloro-1,2-dihydroxybenzene	—	—	—	—	—	—	56	—	33	—	23	—	26	—	20	—	56	60	60	33	66	70	26	—	pl-gn
5-Chloro-2,3-dihydroxy-methylbenzene	—	—	—	—	—	—	70	—	40	—	33	—	60	—	26	—	66	73	76	33	70	73	30	—	p

Layer L1 = silica gel G60
L2 = alumina (150 Saur, Type T)

Table TLC-6 (continued)
PHENOLS

	L3 = Kieselguhr G
	L4 = silica gel G60-alumina (7:3)
	L5 = silica gel G60-Kieselguhr G (3:2)
Solvent	S1 = light petroleum (b.p. 40—60°C)-acetone (80:20)
	S2 = light petroleum-ethyl acetate (70:30)
	S3 = light petroleum-methanol (75:25)
	S4 = dichloromethane
	S5 = dichloromethane-benzene-methanol (60:30:10)
	S6 = benzene-ethanol-acetic acid (85:10:5)
	S7 = benzene-chloroform-acetic acid (60:30:10)
	S8 = *n*-hexane-ethyl acetate-acetic acid (80:15:5)
	S9 = light petroleum-ethyl acetate-acetic acid (80:15:5)
	S10 = light petroleum-acetone-acetic acid (80:15:5)
	S11 = benzene-acetone-acetic acid (80:15:5)
	S12 = chloroform-ethyl acetate-acetic acid (80:15:5)
	S13 = chloroform-diethyl ether-acetic acid (85:10:5)
	S14 = benzene-acetic acid (85:15)
	S15 = chloroform-acetic aid (80:20)
	S16 = light petroleum-dichloromethane-acetic acid (40:40:20)
	S17 = *n*-hexane-acetone-acetic acid (75:20:5)
Detection	D1 = 2% solution of 3,5-Dichloro-*p*-benzoquinonechlorimine in benzene

REFERENCES

1. **Sattar, M. A., Paasivirta, J., Vesterinen, R., and Knuutinen, J.,** Thin-layer chromatography of chlorinated cresols, *J. Chromatogr.*, 136, 379, 1977.
2. **Sattar, M. A., Paasivirta, J., Vesterinen, R., and Knuutinen, J.,** Thin-layer chromatography of chlorinated catechols, *J. Chromatogr.*, 135, 395, 1977.

Table TLC-7
BIPHENYLS

Layer	L1	L1	L1	L1	L1	L1	L1	L1	L1	L1	L1	L1	L2	L2
Solvent	S1	S2	S3	S4	S5	S6	S7	S8	S9	S10	S11	S12	S13	S14
Technique													T1	T1
Detection													D1	D1
Reference	1	1	1	1	1	1	1	1	1	1	1	1	2	2
Compound	Rf × 100													
4-Hydroxybiphenyl	59	66	92	80	29	50	38	32	38	86	67	—	58	62
3-Hydroxybiphenyl	63	70	95	92	38	55	58	35	58	91	75	—	—	—
2-Hydroxybiphenyl	78	77	97	95	63	66	74	64	86	91	86	—	47	62
2′,5′-Dichloro-4-hydroxybiphenyl	76	75	88	90	44	64	54	45	54	92	87	—	—	—
2′,5′-Dichloro-3-hydroxybiphenyl	84	75	93	93	47	62	61	52	53	92	79	—	—	—
2′,5′-Dichloro-2-hydroxybiphenyl	80	78	89	95	67	86	69	70	83	88	90	—	—	—
3-Chloro-4-hydroxybiphenyl	87	79	95	85	55	76	75	69	75	89	85	—	—	—
2-Chloro-4-hydroxybiphenyl	79	75	95	84	48	70	70	63	81	93	82	—	—	—
4′-Chloro-4-hydroxybiphenyl	32	65	93	87	31	54	50	35	47	43	74	—	—	—
3,4′-Dichloro-4-hydroxybiphenyl	80	75	95	80	44	62	65	65	76	93	87	—	—	—
2,4′-Dichloro-4-hydroxybiphenyl	75	75	93	84	41	62	60	39	47	93	86	—	—	—
2,3′,6′-Trichloro-4-hydroxybiphenyl	83	81	91	84	51	84	67	72	81	91	90	—	—	—
3,5-Dichloro-4-hydroxybiphenyl	88	80	98	28	59	85	79	73	83	93	88	—	—	—
3,4′,5-Trichloro-4-hydroxybiphenyl	89	82	90	15	51	88	69	74	85	90	90	—	—	—
4,4′-Dihydroxybiphenyl	29	28	46	50	0	5	5	2	5	87	13	—	20	48
3,3′-Dihydroxybiphenyl	28	33	53	47	7	4	4	2	2	88	20	—	—	—
2,2′-Dihydroxybiphenyl	35	35	55	55	11	10	11	6	5	87	18	—	10	45
3,4-Dihydroxybiphenyl	42	43	69	58	9	20	16	1	16	88	35	—	—	—
2,5-Dihydroxybiphenyl	42	47	78	72	16	25	29	20	33	86	42	—	—	—
3-Chloro-4,4′-dihydroxybiphenyl	57	37	57	47	11	25	12	11	20	89	39	66	—	—
3,3′-Dichloro-4,4′-dihydroxybiphenyl	42	42	68	46	11	32	20	22	35	89	46	45	—	—
3,3′,5-Trichloro-4,4′-dihydroxybiphenyl	58	48	73	7	20	50	33	28	47	87	59	13	—	—
3,3′,5,5′-Tetrachloro-4,4′-dihydroxybiphenyl	60	47	75	0	18	54	64	25	45	90	45	0	—	—
2,6′-Dichloro-4,4′-dihydroxybiphenyl	40	41	64	56	14	27	23	10	14	89	40	78	—	—
4,4′-Dichloro-3-hydroxybiphenyl	85	79	95	87	50	70	71	69	79	93	86	87	—	—
3,5-Dichloro-2-hydroxybiphenyl	92	76	45	94	79	89	88	42	47	93	92	85	—	—
4,4′-Dichloro-3,3′-dihydroxybiphenyl	43	46	63	27	6	25	9	11	20	89	47	21	—	—
3,3′,5,5′-Tetrachloro-2,6′-dihydroxybiphenyl	50	6	6	4	0	49	0	0	0	9	3	43	—	—

Table TLC-7 (continued) BIPHENYLS

Layer															
Solvent															
Technique															
Detection	D2	D3	D4	D5	D6	D7	D8	D9	D10	D11	D12	D13	D14	D15	D16
Reference	1	1	1	1	1	1	1	1	1	1	1	1	1	1	1
Compound	Rf × 100														
4-Hydroxybiphenyl	go-y	lt-g-y	pk	lt-pk	lt-o-bn	lt-y	y	p—bn*	lt-b	lt-pk	o-pk	lt-b	nc	nc	nc
3-Hydroxybiphenyl	pea-gn	g	g-b	pk	lt-o	pk	y	mo—o-y*	w	nc	y	lt-pk	nc	nc	nc
2-Hydroxybiphenyl	y-gn	lt-bn	g-b	lt-mv	lt-o	lt-pk	pk	p—y*	d-r	nc	o-y	nc	nc	nc	nc
2′,5′-Dichloro-4-hydroxybiphenyl	pl-y	lt-y-gn	nc	g	g	lt-g	V-lt-y	nc	lt-b	y	nc	lt-y	—	—	—
2′,5′-Dichloro-3-hydroxybiphenyl	nc	V-lt-be	V-lt-g	lt-pk	nc	lt-pk	y	o	w	lt-o	lt-y	nc	—	—	—
2′,5′-Dichloro-2-hydroxybiphenyl	y	lt-y	mid-b	o	V-lt-pk	lt-y	y-o	o	d-go	lt-o	—	lt-pk	—	—	—
3-Chloro-4-hydroxybiphenyl	o	o	o	r-bn	o	lt-o	nc	lt-pk	nc	pk	o	o	—	—	—
2-Chlor-4-hydroxybiphenyl	y	y	pl-o	mv	V-lt-pk	be	lt-pk	V-lt-mv	lt-mv	V-lt-pk	o	nc	—	—	—
4′-Chloro-4-hydroxybiphenyl	bt-y	bt-y	pk	lt-g	o	pk	pk-o	lt-pk	nc	pk	o	V-lt-o	—	—	—
3,4′-Dichloro-4-hydroxybiphenyl	pl-y	o-y	o-bn	lt-g	pk	V-lt-o	pk-o	lt-pk	nc	pk	bt-o	o	—	—	—
2,4′-Dichloro-4-hydroxybiphenyl	y	V-lt-gn-y	lt-o	nc	nc	be	pk-o	pk	w	pk	bt-y	nc	—	—	—
2,3′,6′-Trichloro-4-hydroxybiphenyl	pl-y	nc	nc	lt-g	lt-g	lt-g	nc	V-lt-mv	V-lt-b-gn	lt-y	nc	nc	—	—	—
3,5-Dichloro-4-hydroxybiphenyl	d-o	o-pk	o-y	lt-g	lt-bn	V-lt-o	nc	w	pk	nc	y-o	o	—	—	—
3,4′,5-Trichloro-4-hydroxybiphenyl	o-rt	o-pk	y	lt-bn	lt-g	lt-y	nc	w	pk	nc	be	y	—	—	—
4,4′-Dihydroxybiphenyl	gn-go	g	nc	lt-g	g-bn	pk	be	nc—y-bn*	pea-gn	g	lt-bn	lt-g-gn	ol—bn	nc	rt-bn
3,3′-Dihydroxybiphenyl	y	g-gn	lt-o	lt-o	lt-y	lt-o-pk	nc	mv—o-bn*	bn	nc	y-o	lt-pk	lt-y	nc	lt-y
2,2′-Dihydroxybiphenyl	d-bn	d-bn	gn	lt-ro	lt-y-o	pk	go	p—o*	bn	lt-pk	y	lt-pk	bn	nc	lt-y
3,4-Dihydroxybiphenyl	bt-o	o	p	g	be	lt-bn	d-pk-bn	cn—bn*	mid-bn	lt-bn-g	bn	g	o-y	y—vt-g	y
2,5-Dihydroxybiphenyl	y	y	y—ol	mid-g-gn	be	lt-g-bn	nc	cn—bn*	lt-b	lt-bn	lt-o	navy-b	y-b-g	g-o-br	b
3-Chloro-4,4′-dihydroxybiphenyl	d-bn	g-be	lt-be	lt-g	be	be	d-be	y	gn-go	lt-mv	pk-be	nc	—	—	—
3,3′-Dichloro4,4′-dihydroxybiphenyl	gn-go	lt-go	be	g	be	lt-be	be	y	lt-go	lt-mv	pk-be	nc	—	—	—
3,3′,5-Trichloro-4,4′-dihydroxybiphenyl	go-y	o-y	lt-y	V-lt-g	lt-g	lt-g	y	y	o-y	nc	lt-y	nc	—	—	—
3,3′,5,5′-Tetrachloro-4,4′-dihydroxybiphenyl	lt-o	bt-o	y	nc	lt-gn	lt-gn	y	mint-gn	o	lt-gn	nc	nc	—	—	—
2,6′-Dichloro-4,4′-dihydroxybiphenyl	nc	be	nc	pk	lt-y	lt-o	o	V-lt-o	lt-mv	salmon-pk	nc	y-o	—	—	—
4,4′-Dichloro-3-hydroxybiphenyl	pl-o	V-pl-y	gn	nc	y	nc	o-y	d-ro	lt-y	nc	y	nc	—	—	—
3,5-Dichloro-2-hydroxybiphenyl	d-o	cherry-r	pl-gn	y	y	y	nc	nc	y	nc	y-gn	o	—	—	—
4,4′-Dichloro-3,3′-Dihydroxybiphenyl	pl-y	pl-y	lt-b-gn	o	o-y	y	y-o	d-ro	V-lt-y	lt-y	y	nc	—	—	—
3,3′,5,5′-Tetrachloro-2,6′-dihydroxybiphenyl	pl-y	w	o-y	lt-g	lt-bn	lt-bn	nc	nc	y	nc	y	nc	—	—	—

Note: V = very.

Layer	L1 = silica gel plate Merck F254
	L2 = silica gel plate Merck
Solvent	S1 = benzene-methanol-98% formic acid (90:9:1)
	S2 = benzene-methanol (90:9)
	S3 = dichloromethane-methanol (95:5)
	S4 = dichloromethane-methanol-conc. NH_3 (95:5:1)
	S5 = carbon tetrachloride-acetone (20:1)
	S6 = carbon tetrachloride-acetone-acetic acid (40:2:1)
	S7 = benzene-ethylacetate (25:1)
	S8 = benzene-chloroform (1:1)
	S9 = chloroform
	S10 = benzene-isopropanol (70:30)
	S11 = chloroform-*n*-hexane-methanol (45:30:5)
	S12 = benzene-isopropanol-conc.-NH_3 (70:30:4)
	S13 = benzene-chloroform (1:1)
	S14 = *n*-hexane-acetone (7:3)
Technique	T1 = analyzed as dansyl derivatives
Detection	D1 = triethanolamine solution, Zeiss PMQ II chromatogram spectrometer at fluorescence
	D2 = conc. nitric acid
	D3 = Millon's reagent
	D4 = 2,6-dibromoquinone-4-chlorimide
	D5 = 9-dicyanomethylene-2,4,7-trinitrofluorene
	D6 = 2,4,5,7-tetranitro-9-fluorene
	D7 = 4,5,7-trinitro-9-fluorene-2-carboxylic acid
	D8 = 2-chloro-4-nitrobenzenediazonium naphthalene-2-sulfonate saturated in chloroform
	D9 = 2-chloro-4-nitrobenzenediazonium naphthalene-2-sulfonate saturated in ethanol
	D10 = cerium (IV) ammonium nitrate
	D11 = chloranil
	D12 = 2,4,7-trinitro-9-fluorene
	D13 = tetracyanoethylene
	D14 = potassium metaperiodate
	D15 = 4-nitrophenylacetonitrile
	D16 = potassium nitrosodisulfonate
	D17 = 2-chloro-4-nitrobenzenediazonium naphthalene-2-sulfonate, 0.1% son. in 50% aquous acetic acid, oversprayed with 0.5 *N* aqueous alcoholic sodium hydroxide soln.

From Heacock, R. A., Hutzinger, O., and Safe, S., *J. Chromatogr.*, 97, 233, 1974. With permission.

Table TLC-8
PHENOLS

Layer	L1	L1	L2	L2		L3	L4	L5	L6	L7	L8	L9	L10	L11	L12	L13
Solvent	S1	S2	S3	S4		S5	S5	S5	S5	S5	S5	S5	S6	S7	S8	S9
Technique													T1			
Detection	D1	D1	D1	D1	C1	D2	D2	D2	D2	D2	D2	D2		D3		
Reference	1	1	1	1	1	2	2	2	2	2	2	2	3	4	5	5
Compound	Rf × 100															
Phenol	63	66	—	—	vt	15	16	t	17	70	1	t	—	70	—	—
2-Methylphenol	92	88	—	—	bn	33	32	t	31	92	1	t	—	—	—	—
3-Methylphenol	84	61	—	—	lt-bn	30	32	t	22	86	1	t	—	—	—	—
4-Methylphenol	79	56	—	—	bk-y	20	21	t	22	81	1	t	—	—	—	—
2-Chlorophenol	56	30	—	—	y-bn	—	—	—	—	—	—	—	—	—	—	—
3-Chlorophenol	47	20	—	—	bn	—	—	—	—	—	—	—	—	—	—	—
4-Chlorophenol	68	27	—	—	vt	—	—	—	—	—	—	—	—	—	—	—
2-Hydroxybenzaldehyde	31	11	—	—	y	—	—	—	—	—	—	—	—	—	—	—
2-Nitrophenol	—	—	96	91	y	—	—	—	—	—	—	—	—	—	—	—
3-Nitrophenol	—	—	80	78	lt-y	—	—	—	—	—	—	—	—	—	—	—
4-Nitrophenol	—	—	65	67	d-y	—	—	—	—	—	—	—	—	—	—	—
1,4-Dihydroxybenzene	—	—	32	38	bk	—	—	—	—	—	—	—	—	—	33	—
1,3-Dihydroxybenzene	—	—	40	45	lt-bn	—	—	—	—	—	—	—	—	33	41	—
1,2-Dihydroxybenzene	—	—	45	50	gn-bk	—	—	6	—	—	—	9	—	—	10	—
2-Aminophenol	—	—	21	30	d-bn	—	—	—	—	—	—	—	—	—	—	—
3-Aminophenol	—	—	25	25	R bn	—	—	—	—	—	—	—	—	—	—	—
4-Aminophenol	—	—	12	19	vt	—	—	—	—	—	—	—	—	—	—	—
1,2,3-Trihydroxybenzene	—	—	5	13	dk-bn	—	—	—	—	—	—	—	—	0	—	0
Pentachlorophenol	—	—	—	—	—	—	—	—	—	—	—	—	64	—	—	—
1,2-Benzoquinone	—	—	—	—	—	—	—	—	—	—	—	—	—	64	—	—
1,4-Benzoquinone	—	—	—	—	—	—	—	—	—	—	—	—	—	89	—	—
1,3,5-Trihydroxybenzene	—	—	—	—	—	—	—	—	—	—	—	—	—	4	—	21
2,4-Dimethylphenol	—	—	—	—	—	—	—	—	—	—	—	—	—	80	—	—
3,4-Dimethylphenol	—	—	—	—	—	—	—	—	—	—	—	—	—	75	—	—
2,6-Dimethylphenol	—	—	—	—	—	—	—	—	—	—	—	—	—	90	—	—
2,3-Dimethylphenol	—	—	—	—	—	—	—	—	—	—	—	—	—	80	—	—
3,5-Dimethylphenol	—	—	—	—	—	—	—	—	—	—	—	—	—	77	—	—
2,5-Dimethylphenol	—	—	—	—	—	—	—	—	—	—	—	—	—	81	—	—
1,2,4-Trihydroxybenzene	—	—	—	—	—	—	—	—	—	—	—	—	—	—	—	26

Layer
- L1 = alumina-calcium hydroxide (40:20)
- L2 = zinc carbonate-silica gel (30:20)
- L3 = Kierselgel G nach Stahl impregnated with formamide-acetone (1:2)
- L4 = Fertigfolien UV254 (Kieselgel on aluminum foil)
- L5 = Fertigfolien UV254 impregnated with formamide-acetone (1:2)
- L6 = Fertigfolien UV254 activated
- L7 = cellulose
- L8 = cellulose activated
- L9 = cellulose impregnated with formamide-acetone (1:2)
- L10 = silica gel G
- L11 = Kieselgel G impregnated with silver nitrate
- L12 = Fertigfolien impregnated with ammonium molybdate

Solvent
- S1 = chloroform
- S2 = benzene-methanol (90:10)
- S3 = benzene-ethyl acetate (90:10)
- S4 = benzene-methylethylketone (90:10)
- S5 = dichloromethane-cyclohexane (55:45)

Table TLC-8 (continued)
PHENOLS

	S6 = benzene-chloroform (1:1)
	S7 = benzene-methanol-acetic acid (8:1:1)
	S8 = toluene-acetone-chloroform (40:35:25)
	S9 = benzene-acetone (70:30)
Technique	T1 = analyzed as dansyl derivatives
Detection	D1 = $K_2S_2O_8$-$AgNO_3$soln. for nitrophenols chromic acid
	D2 = diazotized sulfanilic acid in 5% Na_2CO_3
	D3 = 4-aminophenazon

REFERENCES

1. **Srivastava, S. P. and Dua, V. K.,** Trennung nahe verwandter Phenole durch Dünnschichitchromatographie, *Z. Anal. Chem.,* 275, 29, 1975.
2. **Thielemann, H.,** Dünnschicht-chromatographische Trennung und Identifizierung der isomeren Kresole und des Phenols, *Z. Anal. Chem.,* 265, 271, 1973.
3. **Van Langeveld, H. E. A. M.,** Determination of pentachlorophenol in toy paints, *J. Ass. Oddic. Anal. Chem.,* 58, 19, 1975.
4. **Thielemann, H.,** Dünnschichtchromatographische Trennung und Identifizierung einiger Polyphenole an silbernitratimprägnierten Sorptionsschichten, *Pharmazie,* 32, 244, 1977.
5. **Thieleman, H.,** Dünnschichtchromatographie von Polyphenolen an ammoniummolybdatimprägnierten Fertigfolien, *Pharmazie,* 33, 380, 1978.

Table TLC-9
PHENOLIC COMPOUNDS

Layer	L1	L1	L1	L1	L1				
Solvent	S1	S2	S3	S4	S5				
Technique									
Detection						D1	D2	D3	D4
Reference	1	1	1	1	1	1	1	1	1
Compound	Rf × 100								
Benzophenone	53	48	77	86	80	b	—	—	o
Diphenylmethanol	18	17	64	87	64	b	—	y	—
2-Hydroxybenzophenone	63	60	59	89	53	b	y	y	o
4-Hydroxybenzophenone	0	0	9	35	0	b	y	—	o
2,2′-Dihydroxybenzophenone	44	41	17	61	11	b	y	y	o
4,4′-Dihydroxybenzophenone	0	0	0	8	0	b	y	y	o
2,4-Dihydroxybenzophenone	6	3	0	21	0	b	y	y	o
Diphenylmethane	76	78	85	92	90	—	—	—	—
2-Hydroxydiphenylmethane	33	28	55	71	48	—	—	—	—
4-Hydroxydiphenylmethane	13	12	47	70	36	—	—	—	—
2,2′-Dihydroxydiphenylmethane	0	0	11	52	0	—	—	—	—
4,4′-Dihydroxydiphenylmethane	0	0	26	56	10	—	—	—	—

Layer	L1 = silica gel GF254
Solvent	S1 = benzene
	S2 = benzene-ligroin (9:1)
	S3 = benzene-piperidine (9:1)
	S4 = benzene-ethanol-piperidine (8:1:1)
	S5 = carbon tetrachloride-methyl acetate-piperidine (8:1:1)
Detection:	D1 = UV
	D2 = 0.2% $KMnO_4$ in 5% Na_2CO_3 soln.
	D3 = H_2SO_4
	D4 = 0.5% 2,4-dinitrophenylhydrazine in 10% HCl

From Stocklinski, A. W., Smith, J. L., and Knudson, K. A., *J. Chromatogr.,* 170, 495, 1979. With permission.

Table TLC-10
NATURALLY OCCURRING PHENOLIC COMPOUNDS

Layer	L1	L1	L1	L2	L2	L2	L2	L2	L2	L2	L2	L2	L2	L2	L2	L2
Solvent	S1	S2	S3	S4,5	S5,4	S5	S5,4	S5,4	S6	S5,4	S4,5	S5,4	S5,4	S6	S6	S5,4
Technique																
Detection	D1	D1	D1	D2	D2	D2	D2	D2	D2	D2	D2	D2	D2	D2	D2	D2
Reference	1	2	2	3	3	3	3	3	3	3	3	3	3	3	3	3
Compound	Rf × 100															
2-Methylcardol triene	7.5	—	—	—	—	—	—	—	—	—	—	—	—	—	—	—
2-Methylcardol diene	25	—	—	—	—	—	—	—	—	—	—	—	—	—	—	—
2-Methylcardol monoene	60	—	—	—	—	—	—	—	—	—	—	—	—	—	—	—
Cardanol methylether	—	74	67	—	—	—	—	—	—	—	—	—	—	—	—	—
Cardanol dimethylether	—	72	42	—	—	—	—	—	—	—	—	—	—	—	—	—
2-Methylcardol dimethylether	—	72	50	—	—	—	—	—	—	—	—	—	—	—	—	—
Dimethylanacardate	—	61	17	—	—	—	—	—	—	—	—	—	—	—	—	—
Methylanacardate	—	67	27	—	—	—	—	—	—	—	—	—	—	—	—	—
Cardanolacetate	—	62	32	—	—	—	—	—	—	—	—	—	—	—	—	—
Cardoldiacetate	—	43	13	—	—	—	—	—	—	—	—	—	—	—	—	—
2-Methylcardoldiacetate	—	42	27	—	—	—	—	—	—	—	—	—	—	—	—	—
Cardanol	—	—	—	79	83	95	83	59	79	63	71	72	82	89	67	73
2-Methylcardol	—	—	—	74	76	95	66	50	55	57	61	63	52	66	59	63
Cardol	—	—	—	41	45	66	37	35	26	43	36	58	24	34	40	48
Anacardic acid	—	—	—	0	0	0	0	0	79	0	0	0	—	—	—	0

Layer L1 = silica gel G
L2 = silica gel 60G

Solvent S1 = ethyl acetate-chloroform (1:1)
S2 = chloroform
S3 = chlorofomr-light petroleum (30:70)
S4 = light petroleum (b.p. 40—60°C)-diethylether (7:3) with 2% conc. NH_3
S5 = diethyl ether with 2% conc. NH_3
S6 = light petroleum-diethyl ether (7:3) with 2% formic acid (>98%)

Detection D1 = 0.1% ethanolic dichlorofluorescein
D2 = iodine

REFERENCES

1. **Tyman, J. H. P.,** Long-chain phenols. IX. Compositional studies on the unsaturated phenols of Anacardium occidentale (cashew nut-shell liquid), *J. Chromatogr.*, 138, 97, 1977.
2. **Tyman, J. H. P.,** Long-chain phenols. XII. Compositional studies: the polymeric material in the unsaturated phenols of Anacadium occidentale, *J. Chromatogr.*, 156, 255, 1978.
3. **Tyman, J. H. P.,** Long-chain phenols. XIII. Quantitative analysis of the phenolic composition of natural cashew nut-shell liquid (Anacaridium occidentale) by thin-layer chromatography, densitometry, and ultraviolet spectrophotometry, *J. Chromatogr.*, 166, 159, 1978.

Table TLC-11
PHENOLS AND ORGANIC ACIDS

Layer	L1	L1	L1	L1	L1	L1	L1	L1	L1	L1	L1
Solvent	S1	S2	S3	S4	S5	S6	S7	S8	S9	S10	S11
Technique											
Detection	D1	D1	D1	D1	D1	D1	D1	D1	D1	D1	D1
Reference	1	1	1	1	1	1	1	1	1	1	1
Compound	Rf × 100										
2-Hydroxyacetanilide	—	55	—	—	—	58	—	15	—	—	—
3-Hydroxyacetanilide	—	47	—	—	—	47	—	34	—	—	—
4-Hydroxyacetanilide	—	42	—	—	—	41	—	39	—	—	—
2-Aminophenol	—	55	—	—	—	—	—	46	—	69	—
3-Aminophenol	—	49	—	—	—	—	—	40	—	61	—
4-Aminophenol	—	41	—	—	—	—	—	54	—	66	—
2-Methoxyphenol	65	—	57	—	74	—	—	—	—	—	—
Phenol	48	—	48	—	68	—	—	—	—	—	—
4-Methoxyphenol	42	—	37	—	62	—	—	—	—	—	—
1,4-Dihydroxybenzene	7	—	14	—	39	—	—	—	—	—	—
2-Hydroxybenzoic acid	—	—	—	60	—	—	65	—	—	—	63
3-Hydroxybenzoic acid	—	—	—	38	—	—	47	—	—	—	21
4-Hydroxybenzoic acid	—	—	—	38	—	—	40	—	—	—	34
3,4-Dihydroxybenzoic acid	—	—	—	21	—	—	26	—	—	—	5
2-Hydroxybiphenyl	69	—	—	64	—	—	—	—	52	—	—
4-Hydroxybiphenyl	55	—	—	53	—	—	—	—	32	—	—
2,5-Dihydroxybiphenyl	25	—	—	39	—	—	—	—	28	—	—
4,4′-Dihydroxybiphenyl	15	—	—	27	—	—	—	—	18	—	—
2-Chlorophenol	59	—	—	63	—	—	—	21	—	—	—
3-Chlorophenol	53	—	—	54	—	—	—	34	—	—	—
4-Chlorophenol	48	—	—	52	—	—	—	41	—	—	—
4-Hydroxy-coumarin	7	—	—	—	—	58	—	—	—	12	—
7-Hydroxy-coumarin	18	—	—	—	—	55	—	—	—	30	—
o-Coumarin	3	—	—	—	—	51	—	—	—	4	—
1-Hydroxynaphthalene	50	—	—	56	—	—	—	55	—	—	—
2-Hydroxynaphthalene	45	—	—	52	—	—	—	49	—	—	—
2-Nitrophenol	75	—	—	—	81	—	—	—	—	20	—
3-Nitrophenol	43	—	—	—	60	—	—	—	—	52	—
4-Nitrophenol	35	—	—	—	55	—	—	—	—	16	—
4-Hydroxystilbene	48	—	—	—	60	—	—	50	—	—	—
4,4′-Dihydroxystilbene	16	—	—	—	53	—	—	38	—	—	—
2-Methylphenol	51	—	—	58	—	—	—	—	49	—	—
3-Methylphenol	45	—	—	53	—	—	—	—	38	—	—
4-Methylphenol	45	—	—	50	—	—	—	—	40	—	—

Layer L1 = 250 μm silica gel GF254

Solvent
S1 = benzene-methanol (95:5)
S2 = benzene-methanol (4:1)
S3 = benzene-ethyl acetate (9:1)
S4 = benzene-acetic acid (5:1)
S5 = benzene-acetic acid (5:2)
S6 = benzene-methanol-acetic acid (45:8:4)
S7 = chloroform-ethanol-acetic acid (18:1:1)
S8 = toluene-piperidine (5:2)
S9 = benzene-ethyl acetate-piperidine (6:3:1)
S10 = benzene-isopropanol-conc. NH_4OH (5:4:1)
S11 = isopropanol-benzene-conc. NH_4OH (3:1:1)

Detection D1 = UV 254 nm and diazotized sulfanilic acid

REFERENCE

1. **Smith, R. T., Rosazza, J. P., and Nelson, R. A.**, Thin-layer chromatographic determination of simple phenols in microbial extracts, *J. Chromatogr.*, 95, 247, 1974.

Table TLC-12
PHENOLS AND ORGANICS

Layer	L1	L1	L1	L1	L1	L1	L1	L1	L1	L1	L1	L1	L2	L3	
Solvent	S1	S2	S3	S4	S5	S6	S7	S8	S9	S10	S11	S12	S13	S14	
Technique															
Detection													D1	D2	C
Reference	1	1	1	1	1	1	1	1	1	1	1	1	2	3	3
Compound		Rf × 100													
Acetophenone	—	—	—	—	—	—	24	36	27	—	—	—	—	—	—
Propiophenone	—	—	—	—	—	—	38	49	45	—	—	—	—	—	—
Valerophenone	—	—	—	—	—	—	44	58	51	—	—	—	—	—	—
Decanophenone	—	—	—	—	—	—	51	68	62	—	—	—	—	—	—
3-Aminophenol	2	1	0	—	—	—	—	—	—	—	—	—	—	—	—
4-Aminophenol	0	0	0	—	—	—	—	—	—	—	—	—	—	—	—
2-Nitrophenol	53	59	57	—	—	—	—	—	—	—	—	—	—	—	—
3-Nitrophenol	7	14	14	—	—	—	—	—	—	—	—	—	—	—	—
4-Nitrophenol	4	9	8	—	—	—	—	—	—	—	—	—	—	—	—
2-Hydroxybenzaldehyde	59	58	57	—	—	—	—	—	—	—	—	—	—	—	—
2-Hydroxybenzoic acid	1	3	2	—	—	—	—	—	—	—	—	—	—	10	bn
3-Hydroxybenzaldehyde	9	17	12	—	—	—	—	—	—	—	—	—	—	—	—
Phenol	26	34	29	—	—	—	—	—	—	—	—	—	—	90	b
2-Ethylphenol	39	49	43	—	—	—	—	—	—	—	—	—	—	—	—
1,2-Dihydroxybenzene	7	12	8	—	—	—	—	—	—	—	—	—	—	—	—
1,3-Dihydroxybenzene	4	4	3	—	—	—	—	—	—	—	—	—	—	—	—
2-Methylphenol	29	33	31	—	—	—	—	—	—	—	—	—	—	—	—
3-Methylphenol	23	29	26	—	—	—	—	—	—	—	—	—	—	—	—
4-Methylphenol	23	31	26	—	—	—	—	—	—	—	—	—	—	—	—
4-Chloro-3-methylphenol	22	30	25	—	—	—	—	—	—	—	—	—	—	—	—
4-Chlorophenol	8	22	20	—	—	—	—	—	—	—	—	—	—	—	—
4-Bromophenol	6	18	18	—	—	—	—	—	—	—	—	—	—	—	—
4-Iodophenol	8	21	20	—	—	—	—	—	—	—	—	—	—	—	—
4-Methoxyphenol	7	15	13	—	—	—	—	—	—	—	—	—	—	—	—
3-Chloroaniline	16	18	16	—	—	—	—	—	—	—	—	—	—	—	—
2,3-Dichloroaniline	30	32	32	—	—	—	—	—	—	—	—	—	—	—	—
2,4,6-Trichloroaniline	66	66	68	—	—	—	—	—	—	—	—	—	—	—	—
Anthraquinone	—	—	—	70	86	70	—	—	—	—	—	—	—	—	—
Anthrone	—	—	—	67	81	68	—	—	—	—	—	—	—	—	—
Benzoquinone	—	—	—	56	71	45	—	—	—	—	—	—	—	—	—
Acetanilide	—	—	—	—	—	—	—	—	—	53	60	66	—	—	—
Benanilide	—	—	—	—	—	—	—	—	—	92	93	22	—	—	—
4-Bromoacetanilide	—	—	—	—	—	—	—	—	—	42	57	24	—	—	—
Sulphanilamide	—	—	—	—	—	—	—	—	—	40	42	12	—	—	—
Sulphadimethoxine	—	—	—	—	—	—	—	—	—	51	48	18	—	—	—
Mefruside	—	—	—	—	—	—	—	—	—	63	73	24	—	—	—
5-Oxo-mefruside	—	—	—	—	—	—	—	—	—	21	34	5	—	—	—
Diethylstilbestol	—	—	—	—	—	—	—	—	—	—	—	—	30	—	—

Layer
- L1 = procoated silica gel 60F254
- L2 = Adsorbosil-5 Prekotes, 250 μm
- L3 = silica gel R

Solvent
- S1 = 20% diethyl ether in hexane
- S2 = 20% methyl-*tert*.-butyl ether in hexane
- S3 = 20% diisopropyl ether in hexane
- S4 = 50% diethyl ether in hexane
- S5 = 50% methyl-*tert*.-butyl ether in hexane
- S6 = 50% diisopropyl ether in hexane
- S7 = 10% diethyl ether in hexane
- S8 = 10% methyl-*tert*.-butyl ether in hexane

Table TLC-12 (continued)
PHENOLS AND ORGANICS

	S9 = 10% diisopropyl ether in hexane
	S10 = diethyl ether
	S11 = methyl-*tert.*-butyl ether
	S12 = diisopropyl ether
	S13 = chloroform-acetonitrile (85:15)
	S14 = benzene-ethyl acetate (1:1)
Detection	D1 = 1% $AgNO_3$ in acetonitrile
	D2 = 10% $FeCl_3$-2% $K_3[Fe(CN)_6]$ (2:1)

REFERENCES

1. **Little, C. J., Dale, A. D., Whatley, J. A., and Wickings, J. A.**, Methyl *tert.*-butyl ether: a new chromatographic eluent, *J. Chromatogr.*, 169, 381, 1979.
2. **Ponder, C.**, Fluorometric determination and thin-layer chromatographic identification of diethylstilbestrol on beef liver, *J. Assoc. Off. Anal. Chem.*, 57, 919, 1974.
3. **Gocan, S., Hodisan, T., Nascu, H., and Georgescu, C.**, Separation and determination of phenols and salicylic acid in industrial wastes, *Rev. Roum. Chim.*, 23, 265, 1978.

Table TLC-13
PHENOLIC COMPOUNDS

Layer	L1	L1	L1
Solvent	S1	S2	S3
Technique			
Detection	D1	D1	D1
Reference	1	1	1
Compound		Rf × 100	
Fucophlorethol-A-octaacetate (2,2′,4′,6,6′-Pentaacetoxy-4-(2,4,6-triacetoxyphenoxy)-biphenyl)	59	—	—
Fucodiphlorethol-A-decaacetate (2,2′4,4′,6,6′-Pentaacetoxy-4-(2,6-diacetoxy-4-(triacetoxy-phenoxy)-phenoxy)-biphenyl)	54	—	—
Fucotriphlorethol dodecaacetate (2,2′4,4′,6,6′-Pentaacetoxy-4-(2,6-diacetoxy-4-(2,4,6-triacetoxy-phenoxy)-phenoxy)-phenoxy)-biphenyl)	44	—	—
Difucophloretholundecaacetate	44	—	—
Phloroglucineacetate	88	—	—
Difucol	67	—	—
Tetrafucol	37	—	—
Trifucol	50	—	—
2,3′,3″,4,5′,5″,6-Heptamethoxy-4-terphenyl	—	18	—
2,3′,4,5′,6-Pentamethoxybiphenyl	—	34	—
4-Bromo-2,6-dimethoxybiphenyl	—	—	52
2,6-Dimethoxy-4-(2,4,6-trimethoxyphenoxy)-biphenyl	—	45	—

Layer	L1 = Kieselgel 60 F254
Solvent	S1 = chloroform-acetone (9:1)
	S2 = dichloromethane
	S3 = *n*-hexane-ethyl acetate (95:5)
Detection	D1 = UV or vanillin-H_2SO_4

REFERENCE

1. **Glombitza, K. W., Rauwald, H. W., and Eckhardt, G.**, Fucophorethole, polyhydroxyoligophenylather aus Fucus vesiculosus, *Planta Med.*, 32, 33, 1977.

Table TLC-14
NATURALLY OCCURRING PHENOLIC COMPOUNDS

Layer	L1		L2		L3											L4	L4
Solvent	S1		S2		S3											S4	S5
Technique																	
Detection	D1	C	D2	C		D3	D4	D5	D6	D7	D8	D9	D10	D11	D12	D13	D13
Reference	1	1	2	2	3	3	3	3	3	3	3	3	3	3	3	4	4
Compound	Rf × 100																
(6)-Gingerol	26	gn-bn	—	—	—	—	—	—	—	—	—	—	—	—	—	—	—
(8)-Gingerol	30	gn-bn	—	—	—	—	—	—	—	—	—	—	—	—	—	—	—
(10)-Gingerol	30	gn-bn	—	—	—	—	—	—	—	—	—	—	—	—	—	—	—
(6)-Gingerol methylether	26	bn-gn	—	—	—	—	—	—	—	—	—	—	—	—	—	—	—
(6)-Gingerol methylether acetate	41	ol-gn	—	—	—	—	—	—	—	—	—	—	—	—	—	—	—
(6)-Gingeryl diacetate	51	gn-bn	—	—	—	—	—	—	—	—	—	—	—	—	—	—	—
(6)-Shogaol	48	gn-bn	—	—	—	—	—	—	—	—	—	—	—	—	—	—	—
(8)-Shogaol	51	gn-bn	—	—	—	—	—	—	—	—	—	—	—	—	—	—	—
(10)-Shogaol	51	gn-bn	—	—	—	—	—	—	—	—	—	—	—	—	—	—	—
Zinzerone	29	Royal-b	—	—	—	—	—	—	—	—	—	—	—	—	—	—	—
Zinzerone acetate	31	Royal-b	—	—	—	—	—	—	—	—	—	—	—	—	—	—	—
Dehydro zinzerone	23	d-p	—	—	—	—	—	—	—	—	—	—	—	—	—	—	—
(6)-Paradol	52	t	—	—	—	—	—	—	—	—	—	—	—	—	—	—	—
(6)-Paradyl monoacetate	53	b-gn	—	—	—	—	—	—	—	—	—	—	—	—	—	—	—
(6)-Paradylbenzoate	60	lt-b	—	—	—	—	—	—	—	—	—	—	—	—	—	—	—
4-Hydroxy-3-methoxybenzaldehyde	3	lt-b-g	—	—	—	—	—	—	—	—	—	—	—	—	—	—	—
Eugenol	55	cr	18	bn	31	bn	y	y	bn	pk-y	r-trace	g-bn-g	g	d-bn	bn	—	—
Isoeugenol	—	—	—	—	27	gn	y	y	y	r-p-p	o-trace	r	g	pk	lt-bn	—	—
Myristicin	—	—	37	bn	71	—	—	—	—	—	—	—	—	—	—	—	—
Safrole	—	—	53	bn	86	—	—	—	—	—	—	—	—	—	—	—	—
Asarone	—	—	8	p	—	—	—	—	—	—	—	—	—	—	—	—	—
3-Methyl-6-isopropylphenol	—	—	22	r	—	—	—	—	—	—	—	—	—	—	—	—	—
Eugenol methylester	—	—	—	—	42	—	—	—	—	—	—	—	—	—	—	—	—
Isoeugenol methylester	—	—	—	—	42	—	—	—	—	—	—	—	—	—	—	—	—
Ellemicin	—	—	—	—	27	—	—	—	—	—	—	—	—	—	—	—	—
Apiole	—	—	—	—	41	—	—	—	—	—	—	—	—	—	—	—	—

Eriodictyol	—	—	—	—	—	—	—	—	—	—	—	—	—	—	—	—	58
Eriodictyol-3′-*O*-β-D-glucoside	—	—	—	—	—	—	—	—	—	—	—	—	—	—	—	26	—
Eriodictyol-3′,5-*O*-β-D-glucoside	—	—	—	—	—	—	—	—	—	—	—	—	—	—	—	45	—
Eriodictyol-4′,5-*O*-β-D-glucocide	—	—	—	—	—	—	—	—	—	—	—	—	—	—	—	46	—
1,2-Dihydroxybenzene	—	—	—	—	—	—	—	—	—	—	—	—	—	—	—	—	52

Layer L1 = silica gel
L2 = silica gel G
L3 = Brinkmann silica gel F254
L4 = polyamide on alumina foil

Solvent S1 = anhydrous diethyl ether-hexane (4:1)
S2 = benzene
S3 = *n*-hexane-chloroform (3:2)
S4 = 96% ethanol-methylethylketone-acetylacetone-water (15:15:5:65)
S5 = benzene-methylethylketone-methanol (50:25:25)

Detection D1 = conc. H_2SO_4 at 140°C
D2 = 1% vanillin in 1% ethanolic H_2SO_4
D3 = CNTNF
D4 = TNB
D5 = DPPH
D6 = Gibb's reagent
D7 = vanillin-H_2SO_4
D8 = conc. H_2SO;
D9 = furfural-H_2SO_4
D10 = chromotropic acid
D11 = $SbCl_3$-$SbCl_5$ (1:1)
D12 = NBDF
D13 = H_2SO_4-30% HCHO or Gibb's reagent

REFERENCES

1. **Connell, D. W, and McLachlan, R.,** Natural pungent compounds. IV. Extraction of the gingerols, shogaols, paradols, and related compounds by thin-layer and gas chromatography, *J. Chromatogr.,* 67, 29, 1972.
2. **Wulf, L. W., Nagel, C. W., and Branes, A. L.,** High-pressure liquid chromatographic separation of the naturally occurring toxicants myristicin, related aromatic ethers and falcarinol, *J. Chromatogr.,* 161, 271, 1978.
3. **Forrest, J. E., Richard, R., and Heacock, R. A.,** Colour reactions of some aromatic ethers found in essential oils, *J. Chromatogr.,* 65, 439, 1972.
4. **Paris, R. R., Alexis, M. N., Faugeras, G., and Jacquemin, H.,** Plates de la Guyane francaise, IV. Sur les polyphenols de l'Helosis guyanesis Rich., Balanophoracees, *Pharm. Acta Helb.* 53, 130, 1978.

Table TLC-15
NATURALLY OCCURRING PHENOLIC COMPOUNDS

Layer	L1	L1	L2	L2									
Solvent	S1	S2	S1	S3									
Technique													
Detection					D1	D2	D3	D4	D5	D6	D7	D8	D9
Reference	1	1	1	1	1	1	1	1	1	1	1	1	1
Compound	Rf × 100												
Apigenin[a] (4′,5,7-Trihydroxyflavone)	0	89	0	70	—	vt—bk	—	y	bk	g	lt-y	g	y
Apigenin[b] (4′,5,7-Trihydroxy flavone)	—	—	—	78	—	vt—bk	—	y	bk	g	lt-y	g	y
Arbutin	22	59	25	53	—	lt-vt	b	—	lt-vt	g	lt-vt	—	w-g
Benzoic acid	—	—	73	—	—	—	w	—	—	—	—	—	—
Catechin	34	59	40	38	bk-tinge	lt-vt	b	bn	lt-vt	g-b-vt	vt	vt	lt-vt
1,2-Dihydroxybenzene	75	89	75	84	—	lt-vt	b	bn-tinge	lt-vt	lt-bn—vt	vt	vt	lt-vt
3,4-Dihydroxycinnamic acid[a]	22	78	21	62	—	b	b	lt-bn	b	b	b	w-b	gn-w—b
3,4-Dihydroxycinnamic acid[b]	58	—	57	66	—	b	b	lt-bn	b	b	b	w-b	gn-w—b
Chlorogenic acid[a]	51	66	52	53	—	b	b	lt-bn	b	b	b	w-b	gn-w
Chlorogenic acid[b]	72	76	71	57	—	b	b	lt-bn	b	b	b	w-b	gn—w
Coumarin	69	96	55	—	—	—	—	—	—	b	—	—	—
Cyanidin chloride	—	—	0	—	—	lt-vt—g	—	—	vt	g—bk	g—b	b	b
2,4-Dimethoxycinnamaldehyde	63	96	66	100	y	gn-y	b	y	d-gn	d—y-gn	pk-y	d-gn	d-y
2,6-Dihydroxybenzoic acid[a]	75	71	72	65	—	gn—lt-y	b	lt-vt	lt-vt	b-vt	b	g	b
2,6-Dihydroxybenzoic acid[b]	—	—	72	72	—	gn—lt-y	y	lt-vt	lt-vt	b-vt	b	g	b
2,5-Dihydroxybenzoic acid	54	86	49	67	—	b	b	bn-tinge	b	g	b	b	intense-b
Ellagic acid	0	40	30	11	—	g-tinge	lt-b	bn	—	bn-y	g	lt-y	g
Esculin hydrate[a]	63	64	49	55	—	b	b	bn	b	b	b—vt	intense-b	b
Esculin hydrate[b]	—	—	—	65	—	b	b	bn	b	b	b—vt	intense-b	b
4-Hydroxy-3-methoxycinnamic acid[a]	24	89	20	91	—	d-b	b	lt-y	b	b	b	b	b-tinge
4-Hydroxy-3-methoxycinnamic acid[b]	—	—	53	95	—	d-b	b	lt-y	b	b	b	b	b-tinge

4-Hydroxy-3-methoxybenzoic acid	48	89	50	91	—	lt-vt	b	tinge-y	lt-vt	lt-vt	lt-vt	—	—
4-Hydroxy-3-methocybenzaldehyde	65	95	63	98	—	—	b	—	b	vt	lt-b	lt-b	g
Vitexin[a]	5	57	2	14	—	lt-vt	—	—	lt-vt	lt-y	w-g—y	lt-y	y
Vitexin[b]	—	—	—	39	—	lt-vt	—	—	lt-vt	lt-vt	w-g—y	lt-y	y
3,4,5-Trihydroxybenzoic acid	33	57	32	37	—	—	b	lt-bn	—	muddy-y	b-g	vt	b-vt
Hesperidin[a]	42	74	42	65	—	—	—	—	—	g	g	g	—
Hesperidin[b]	—	—	—	75	—	—	—	—	—	g	g	g	—
Hesperitin	10	94	8	91	—	g—lt-vt	b	bn	lt-vt	lt—d-g	gn-y	g	y
Hydroquinone	76	84	72	72	—	g—lt-vt	b	—	lt-vt	bn	g—vt	—	—
4-Hydroxycinnamic acid	27	91	27	87	—	b	b	o	b	intense—b	vt	ting-b	—
2-Hydroxycinnamic acid	38	95	35	92	—	w-g	b	w	w-g	lt-gn	b-g	w-g	g
4-Hydroxybenzoic acid	59	—	64	85	—		y	w	—	lt-vt	lt-vt	lt-tinge	—
3-Hydroxybenzoic acid	60	95	65	83	—	—	b—w	lt-y	—	b	lt-vt	—	—
4-Hydroxyphenylacetic acid	85	91	80	82	—	lt-vt	b	lt-y	lt-vt	lt—vt	b	vt-tinge	—
2-Hydroxyphenylacetic acid	87	94	82	86	—	lt-vt	b	lt-y	lt-vt	g—vt	vt	vt-tinge	lt-b
Kaempferol	0	91	0	71	lt—y	muddy-t	—	d-bn	y	o-y	gn-y	y	y
Malvidin chloride	33	43	25	—	vt	pk—bn	—	vt	pk	—	pk	pk	—
Myricetin	s	54	s	22	y	y	—	y	g	bk—vt	y	y	o
Mandelic acid	89	—	90	0	—	—	w-b	—	g	—	—	—	—
Phloridzin	30	75	29	61	—	—	b	bn	—	bn	g—b	—	g
Phloretin	5	94	5	73	—	—	b	d—bn	lt-vt	g	g—lt-y	lt-vt	muddy-y
Quercetin[a]	0	72	0	44	y	muddy-y	b	muddy-y	y	o-bn	y—gn	y	o
Quercetin[b]	—	—	—	40	y	muddy-y	b	muddy-y	y	o-bn	y—gn	y	o
Rutin[a]	30	62	31	40	y	bk-vt	b	y	d-g	y	y	muddy-y	o
Rutin[b]	—	70	—	55	y	bk-vt	b	y	d-g	y	y	muddy-y	o
4-Hydroxy-3,5-dimethoxybenzoic acid	50	88	51	92	—	bk-b	b	lt-y	g	lt-vt	vt	g	b-g
Scopoletin	29	87	25	84	—	b	—	bn	b	lt-b	b	intense-b	d-b
Sakuranetin	5	96	7	99	—	—	lt-b	—	—	g	g—gn	g	muddy-y
Sakuranin[a]	29	79	20	72	—	g	lt-b	—	g	o	g	g	g
Sakuranin[b]	39	—	40	—	—	g	lt-b	—	g	o	g	g	g
4-Hydroxy-3,5-dimethoxycinnamic acid[a]	20	86	18	87	—	b	b	lt-y	b	lt-b—b	d-b—b	b	b
4-Hydroxy-3,5-dimethoxycinnamic acid[b]	49	—	45	—	—	b	b	lt-y	b	lt-b—b	d-b—b	b	b
Taxifolin[a]	29	75	32	58	—	vt-g	b	lt-bn	vt	bn	muddy-y	g	bk-bn—bk
Taxifolin[b]	—	—	—	68	—	vt-g	b	lt-bn	vt	bn	muddy-y	g	bk-bn—bk

Table TLC-15 (continued)
NATURALLY OCCURRING PHENOLIC COMPOUNDS

Layer	L1 = MN-300 cellulose
	L2 = Merck-2330 (Avicel) layer
Solvent	S1 = formic acid-water (2:98)
	S2 = *n*-amyl alcohol-acetic acid-water (10:6:5)
	S3 = benzene-propionic acid-water (20:45:15)
Detection	D1 = daylight
	D2 = UV
	D3 = Turnbull blue
	D4 = Fast Blue RR + NH_3
	D5 = Fuming with NH, UV
	D6 = methanolic NaOH, UV
	D7 = methanolic $A1C1_3$, UV
	D8 = methanolic zinc acetate, UV
	D9 = Flavone, UV

From Dass, H. C. and Weaver, G. M., *J. Chromatogr.*, 67, 105, 1972. With permission.

Table TLC-16
PHTHALIC ACIDS

Layer	L1
Solvent	S1
Technique	
Detection	D1
Reference	1
Compound	Rf × 100
1,2-Benzenedicarboxylic acid dimethylester	20
1,2-Benzenedicarboxylic acid diethylester	30
1,2-Benzenedicarbxoylic acid diisopropylester	41
1,2-Benzenedicarboxylic acid dipropylester	41
1,2-Benzenedicarboxylic acid diisobutylester	52
1,2-benzenedicarboxylic acid dibutylester	50
1,2-Benzenedicarboxylic acid dimethylethylester	0.8
1,2-Benzenedicarboxylic acid di-*n*-pentylester	56
1,2-Benzenedicarboxylic acid diethylethylester	3.8
1,2-Benzenedicarboxylic acid di-*n*-hexylester	60
1,2-Benzenedicarboxylic acid butylbenzylester	45
1,2-Benzenedicarboxylic acid hexylheptylester	61
1,2-Benzenedicarboxylic acid butylnonylester	59
1,2-Benzenedicarboxylic acid dibutylethylester	7.6
1,2-Benzenedicarboxylic acid di-2′-ethylhexylester	68
1,2-Benzenedicarboxylic acid di-*n*-heptylester	63
1,2-Benzenedicarboxylic acid dicyclohexylester	50
1,2-Benzenedicarboxylic acid di-3′,5′,5′-trimethyl hexylester	67
1,2-Benzenedicarboxylic acid diphenylester	33
1,2-Benzenedicarboxylic acid di-*n*-octylester	66
1,2-Benzenedicarboxylic acid di-3′,5′,5′-trimethylcyclohexylester	69
1,2Benzenedicarboxylic acid di-*n*-nonylester	70
1,2-Benzenedicarboxylic acid di-*n*-decylester	74
1,2-Benzenedicarboxylic acid di-*n*-dodecylester	79

Layer L1 = silica gel F254 (Merck)
Solvent S1 = isooctane-isoamyl acetate (85:15)
Detection D1 = UV 254 nm and resorcinol-H_2SO_4-KOH
Color = orange or red

REFERENCE

1. **Bloom, P. J.,** Application des chromatographies sur couche mince et gazliquide à l'analyse qualitative et quantitative des esters de l'acid phthalique, *J. Chromatogr.*, 72, 35, 1972.

Table TLC-17
ORGANIC ACIDS

Layer	L1	L1	L1	L2	L3	L4	L5	L5
Solvent	S1	S2	S3	S4	S5	S6	S7	S8
Technique								
Detection	D1	D1	D1	D2	D3	D4	D5	D5
Reference	1	1	1	2	3	4	5	5
Compound	Rf × 100						[a]	[a]
Formic acid	31	—	—	—	—	—	—	—
Acetic acid	37	—	—	—	58	—	—	—
Propionic acid	45	—	—	—	77	—	—	—

Table TLC-17 (continued)
ORGANIC ACIDS

Compound	Rf × 100						[a]	[a]
n-Butyric acid	52	—	—	—	—	—	—	—
Cinnamic acid	—	54	—	—	—	—	—	—
Benzoic acid	—	41	—	—	—	—	—	—
N-Acetylglycylleucine	49	—	—	—	—	—	—	—
L-Proline	22	—	—	—	—	—	—	—
Citric acid	—	—	39	16	—	—	32	15
Malic acid	—	—	40	29	—	—	37	24
Aconitic acid	—	—	—	35	—	—	—	—
Succinic acid	—	—	—	69	29	—	—	—
Lactic acid	—	—	—	76	—	65	—	—
Fumaric acid	—	—	—	87	—	—	—	—
Methylmalonic acid	—	—	—	—	45	—	—	—
3-Oxobutanoic acid	—	—	—	—	87	—	—	—
Glucose	—	—	—	—	—	3	—	—
Glycerol	—	—	—	—	—	21	—	—
Pyruvic acid	—	—	—	—	—	42	—	—
Cuinic acid	—	—	—	—	—	—	22	35
Ascorbic acid	—	—	—	—	—	—	28	35

[a] Measured on chromatogram.

Layer L1 = crystalline cellulose impregnated with $NaHCO_3$
L2 = cellulose MN 300 HR (Machery-Nagel)
L3 = cellulose MN 300
L4 = Eastman chromatogram (silica gel)
L5 = Avicel microcrystalline cellulose

Solvent S1 = ethanol-water (100:20)
S2 = *n*-butanol-water (100:20)
S3 = ethanol-water (100:60)
S4 = isopropyl ether-formic acid (3:1) + 2-3 ng dichlorofluorescein in 100 mℓ
S5 = water saturated isopropyl ether-formic acid (3:1)
S6 = water saturated diethyl ether-88% formic acid (70:10)
S7 = butanol-formic acid-water (4:1:5)
S8 = ethanol-NH_4OH-water (8:1:2)

Detection D1 = dicyclohexyl carbodiimide
D2 = UV
D3 = $FeCl_3$/acetone
D4 = bromocresol green and radioactivity
D5 = bromophenol blue

REFERENCES

1. **Tanimura, T., Kasai, Y., and Tamura, Z.,** Application of dicyclohexyl carbodiimide for detection of carboxylic acids, *Chem. Pharm. Bull.*, 20, 1845, 1972.
2. **Tokumitsu, Y. and Ui, M.,** Separation and determination of ^{14}C-labeled intermediates of the citric acid cycle and related compounds, *Anal. Biochem.*, 59, 110, 1974.
3. **Tokumitsu, Y. and Ui, M.,** A new simple technique for the determination of radioactivity content of coenzyme A products, *Anal. Biochem.*, 59, 178, 1974.
4. **Tercyak, A., Kuan, H. T., Salcedo, L. L., and Brecher, P.,** Continuous measurement of ^{14}C-labeled carbon dioxide and lactic acid by cultured cells grown in leighton tubes, *Anal. Biochem.*, 86, 407, 1978.
5. **Ryan, J. J.,** Chemical composition of Canadian apple juice, *J. Assoc. Off. Anal. Chem.*, 55, 1104, 1972.

Table TLC-18
ORGANIC ACIDS

Layer	L1	L1	L2	L3	L4	L5	L5	L6
Solvent	S1	S2	S3	S4	S5	S6	S7	S8
Technique								
Detection	D1	D1	D2	D3	D4	D5	D5	D6
Reference	1	1	2	3	4	5	5	6
Compound	Rf × 100 [a]	[a]						
Citric acid	11	18	—	—	15	—	—	34
Tartaric acid	24	19	—	—	—	—	—	—
Malic acid	30	32	—	—	—	—	—	47
Malonic acid	21	52	—	—	—	—	—	—
Aconitic acid	17	60	—	—	—	—	—	—
Glyceric acid	60	32	—	—	—	—	—	—
2-Oxo-glutaric acid	40	51	—	—	—	—	—	69
Succinic acid	37	63	—	79	—	27.2	69.8	74
Methylmalonic acid	26	78	—	—	—	—	—	—
Glycolic acid	67	46	—	—	—	50.3	44.4	—
Pyroglutamic acid	64	49	—	—	—	—	—	—
Glutaric acid	46	71	—	—	—	—	—	—
Adipic acid	55	82	—	—	—	—	—	—
Fumaric acid	51	85	—	91	—	31.4	83.3	85
Phenylacetylglutamic acid	76	67	—	—	—	—	—	—
Lactic acid	72	73	—	—	—	—	—	—
3-Hydroxybutyric acid	77	76	—	—	—	—	—	—
Hippuric acid	84	90	—	—	—	66.8	81.5	—
Hydrochloric acid	70	5	—	—	—	—	—	—
5-Hydroxy-3-methylglutaric acid	—	—	58	—	—	—	—	—
Mevalonic acid	—	—	29	—	—	—	—	—
L-Lactic acid	—	—	—	64	—	58.3	67	—
cis-Aconitic acid	—	—	—	—	40	—	—	—
DL-Mandelic acid	—	—	—	—	—	23.9	30.1	—
Pyrrolidonecarboxylic acid	—	—	—	—	—	46.7	45.6	—
trans-Aconitic acid	—	—	—	—	—	9.3	78.5	—
Isocitric acid	—	—	—	—	—	—	—	34

[a] Measured on chromatogram.

Layer
L1 = Merck No. 5552 or Schleicher & Schill F1440 cellulose
L2 = Eastman silica gel sheet
L3 = silica gel
L4 = MN-cellulose 300
L5 = DC Fertigplatten cellulose with fluorescence indicator

Solvent
S1 = ethanol-0.88 NH_3-water (150:8:40)
S2 = 2-ethyl-1-butanol-formic acid-water (40:12:48)
S3 = ether-formic acid (7:1)
S4 = water saturated ether-formic acid (7:1)
S5 = butanol-acetic acid-water (5:1.5:2)
S6 = ethyl ether-formic acid-water (10:2:1)
S7 = 95% ethanol-25% NH_4OH-water (8:2:1)
S8 = isobutanol-5 *M* formic acid (2:3)

Detection
D1 = bromocresol green or starch-iodine reagent
D2 = radioactivity (Mark II)
D3 = thin-layer chromatography scanner II (radioactivity)
D4 = detect reaction product "citrazinic acid" by fluorescence
D5 = ribose-aniline reagent
D6 = fluorescence

Table TLC-18 (continued)
ORGANIC ACIDS

REFERENCES

1. **Oberholzer, V. G., Wood, C. B. S., Palmer, T., and Harrison, B. N.,** Increased pyroglutamic acid levels in patients or artificial diets, *Clin. Chim. Acta,* 62, 299, 1975.
2. **Murthy, H. R., Moorjani, S., and Lupien, P. J.,** A simplified procedure for the concentration, desalting and thin-layer chromatography of mevalonic acid from reaction mixtures used to assay 3-hydroxy-4-methylglutaryl-coenzyme A reductase activity, *Anal. Biochem.,* 81, 65, 1977.
3. **Matin, A. and Konnings, W. N.,** Transport of lactate and succinate by membrane of Escherichia coli, Bacillus subtilis and a Pseudomonas species, *Eur. J. Biochem.,* 34, 58, 1973.
4. **Massa, V., Susplugas, P., and Salabert, J.,** Determination of citric and cis-aconitic acids by direct fluorimetry of their thin-layer chromatograms, *Trav. Soc. Pharm. Montpellier,* 34, 71, 1974.
5. **Bruinsma, B. L. and Le Tourneau, D.,** Separation of Krebs cycle acids with beaded diethylaminoethyl cellulose, *J. Assoc. Off. Anal. Chem.,* 61, 751, 1978.
6. **Köhler, F.,** A study of the utility of planimetic determinations in the quantification of non-aminated organic acids separated by thin-layer chromatography, *J. Chromatogr.,* 68, 275, 1972.

Table TLC-19
ORGANIC ACIDS

Layer	L1	L2	L3	L4	L5	L5	L6	L6	L6
Solvent	S1	S2	S3	S4	S5	S6	S7	S8	S9
Technique		T1					T1	T1	T1
Detection	D1		D2	D2	D3	D3			
Reference	1	2	3	3	4	4	5	5	5
Compound	Rf × 100								
		[a]			[a]	[a]			
2-Succinylbenzoate	14	—	—	—	—	—	—	—	—
1,4-Dihydroxy-2-naphthoate	22	—	—	—	—	—	—	—	—
Spirodilactone	41	—	—	—	—	—	—	—	—
Oxalacetic acid	—	18	—	—	—	—	—	—	—
2-Oxoglutaric acid	—	36	—	—	—	—	2	0	0
Glyoxylic acid	—	50	—	—	—	—	—	—	—
Pyruvic acid	—	68	50	31	—	—	65	14,21	13
2-Oxobutanoic acid	—	78	—	—	—	—	—	—	—
Isovarglycine	—	—	52	77	—	—	—	—	—
3-Methylcrotonylglycine	—	—	43	72	—	—	—	—	—
Triglylglycine	—	—	43	71	—	—	—	—	—
Propionylglycine	—	—	32	63	—	—	—	—	—
Hippuric acid	—	—	32	69	—	—	—	—	—
3-Hydroxybutyric acid	—	—	52	66	—	—	—	—	—
Adipic acid	—	—	55	37	—	—	—	—	—
Glutaric acid	—	—	47	32	—	—	—	—	—
Succinic acid	—	—	38	27	—	—	—	—	—
Fumaric acid	—	—	49	30	—	—	—	—	—
Methylmalonic acid	—	—	43	22	—	—	—	—	—
Asparagine	—	—	—	—	89	89	—	—	—
Aspartic acid	—	—	—	—	38	50	—	—	—
Malic acid	—	—	—	—	15	36	—	—	—
Citric acid	—	—	—	—	15	26	—	—	—
Fumaric acid	—	—	—	—	—	19	—	—	—
Glutarate semialdehyde	—	—	—	—	—	—	81	71	29
Succunate semialdehyde	—	—	—	—	—	—	73	38	16
Glutamic acid	—	—	—	—	58	58	—	—	—

[a] Measured on chromatogram.

Table TLC-19 (continued)
ORGANIC ACIDS

Layer	L1 = silica gel G
	L2 = silica gel
	L3 = polyamide 6
	L4 = cellulose MN 300
	L5 = Ionex® SB-Ac (anion-exchange resin)
	L6 = Eastman silica gel chromatogram sheet
Solvent	S1 = benzene-ethyl acetate-acetic acid (150:75:2)
	S2 = ethyl formate-petroleum ether (b.p. 60-80)-acetic acid (50:50:7)
	S3 = diethyl ether-light petroleum-carbon tetrachloride-water-formic acid (50:20:20:8:1)
	S4 = ethnol-33% NH_4OH-water (100:16:12)
	S5 = 1% acetic acid
	S6 = 4.4% formic acid
	S7 = hexane-ethyl formate-acetic acid (5:3:1)
	S8 = light petroleum-ethyl acetate-propionic acid (65:36:7.7)
	S9 = hexane-acetic acid-xylene (5:1:3)
Technique	T1 = analyzed as 2,4-dinitrophenylhydrazones
Detection	D1 = Panax TL chromatogram scanner
	D2 = bromocresol purple or aniline-xylose reagent
	D3 = potassium ferricyanide, ferric ammonium sulfate

REFERENCES

1. **Hutson, K. G. and Threlfall, D. R.,** Formation of spirodilactone of 4-(2′-carboxyphenyl)-4,4-dihydroxybutyrate from 2-succinylbenzoate in cell-free extracts of Micrococcus luteus, *Biochim. Biophys. Acta,* 550, 1, 1978.
2. **Kallio, H., Linko, R. R., Pyysalo, T., and Puntari, I.,** Identification of keto acids in arctic bramble, Rubus arcticus L. as methyl esters of their 2,4-dinitrophenylhydrazone, *Anal. Biochem.*90, 359, 1978.
3. **Van den Berg, H. and Hommes, F. A.,** Thin-layer chromatographic determination of some N-acrylglycine conjugates in urine, *J. Chromatogr.,* 104, 219, 1975.
4. **Himoe, A. and Rinne, R. W.,** Separation of citrate, malate, aspartate and glutamate in crude extracts by one-dimensional ion exchange thin-layer chromatography, *Anal. Biochem.,* 88, 634, 1978.
5. **Watson, G. K. and Cain, R. B.,** Microbial metabolism of the pyridine ring. *Biochem. J.,* 146, 157, 1975.

Table TLC-20
ALIPHATIC NONVOLATILE ACIDS

Layer	L1	L1	L1	L1	L1	L1	L1	L1	L1	L1	L1	L1	L1	L1	L1
Solvent	S1	S2	S3	S4	S5	S6	S7	S8	S9	S10	S11	S12	S13	S14	S15
Technique															
Detection	D1	D1	D1	D1	D1	D1	D1	D1	D1	D1	D1	D1	D1	D1	D1
Reference	1	1	1	1	1	1	1	1	1	1	1	1	1	1	1
Compound	Rf × 100														
Oxalic acid	1	1	47	13	16	31	42	24	23	63	22	44	42	40	44
Malonic acid	8	9	63	15	20	40	56	28	29	78	51	70	71	72.5	73.5
Succinic acid	39	40	71	18	25	45	67	32	32	84	60	74	76.5	77.5	78
Glutaric acid	58	56.5	77	23	31	56	81.5	35	36	—	69	79	81	81	83
Adipic acid	71	69	80	27	38	65	90	39	42	—	78	83	85	86	87
3-Hexene-dicarboxylic acid	54	56	—	—	31	59	83	37	37	91	74	82	85	86	86.5
Pimelic acid	84	81	85	38	50	75	94	47	51	—	—	—	—	—	—
Suberic acid	89	90	88	48	58	82	96	56	63	—	—	—	—	—	—
Azelaic acid	95	95	89	—	67	89	—	64	73	—	—	—	—	—	—
Sebacic acid	96	96	91	—	72	91	—	70	82	—	—	—	—	—	—
Undecanic dicarboxylic acid	97	97	92	—	82	95	—	81	91	—	—	—	—	—	—
cis-Aconitic acid	1	2	50	—	11	29	—	26	23	72.5	30	60	62	64	67
trans-Aconitic acid	9	13	58	—	24	56	—	26	23	84	62	78	81.5	83	84.5
Citraconic acid	30	30	69	28	36	61	84	37	34	90	66	79	81	83	83.5
Citric acid	2	3	39	6	5	15	23	23	17.5	55	23.5	56	56	55	57.5
Fumaric acid	32	34	69	—	41	71	90	33	31	90	77	84.5	86.5	88	88.5
Glyceric acid (1)	2	4	53	—	—	7	24	27	22	53.5	15	51	50	49	51
(2)	5	8	—	—	—	17	—	—	—	60	25.5	56	56	55	57
Glycolic acid	14	17	63	12	16	33	50	—	29	74	41	65	67	67.5	70
Itaconic acid	49	50	74	—	40	62	—	37	36	91	70.5	82	83	85	86.5
Lactic acid (1)	32	34	73	22	29	49	78	36	34	85	63	75	79	79	80
(2)	—	—	—	—	34	—	—	—	—	—	—	—	—	—	—
(3)	45	46	—	—	40	63	89	43	43	96	76.5	85	88	89	90
(4)	—	—	—	—	45	72	94	—	—	—	—	91.5	—	96	—
Maleic acid	3	4.5	63	18	25	49	72	35	32	82	48	69	70	71	72.5

Malic acid	4	6.5	50	9	8	20	35	25	22	63	33	60.5	62	61	63
Tartaric acid	1	1.5	32	6	3	8	18	19	17	43	15	45	43	40	42
Tricarvallylic acid	13	16	57	—	14	38	—	27	23	77	46	71	74	75.5	78
1,2-Benzenedicarboxylic acid	43	44	71	—	50	77	93	43	41.5	95	77	85	88	89	89

Layer L1 = Merck 5716 cellulose layer

Solvent S1 = toluene-propionic acid-water (470:470:49)
S2 = benzene-propionic acid-water (470:470:57)
S3 = propionic acid-water (800:200)
S4 = diisopentyl ether-98% formic acid-water (650:250:5.5)
S5 = di-*n*-butyl ether-98% formic acid-water (650:250:22)
S6 = eucaliptol-98% formic acid-water (650:250:74)
S7 = diisopropyl ether-98% formic acid-water (650:250:100)
S8 = 1-nitropropane-98% formic acid-water (650:250:125)
S9 = 2-nitropropane-98% formic acid-water (650:250:130)
S10 = methylethylketone-98% formic acid-water (650:250:200)
S11 = *n*-amyl alcohol-98% formic acid-water (800:150:110)
S12 = *n*-amyl alcohol-absolute ethanol-98% formic acid-water (400:400:150:200)
S13 = 1-butanol-absolute ethanol-98% formic acid-water (400:400:150:200)
S14 = 1-propanol-98% formic acid-water (800:150:200)
S15 = 2-propanol-98% formic acid-water (800:150:200)

Detection D1 = aniline-xylose, furfural

From Munier, R. L., Drapier, A. M., and Faivre, B., *Chromatographia*, 6, 468, 1973. With permission.

Table TLC-21
ALIPHATIC NONVOLATILE ACIDS

Layer	L1	L1	L1	L1	L1	L1	L1	L1	L1	L1	L1	L1	L1	L1
Solvent	S1	S2	S3	S4	S5	S6	S7	S8	S9	S10	S11	S12	S13	S14
Technique														
Detection	D1	D1	D1	D1	D1	D1	D1	D1	D1	D1	D1	D1	D1	D1
Reference	1	1	1	1	1	1	1	1	1	1	1	1	1	1
Compound	Rs	Rs	Rf × 100											
Oxalic acid	0	0	4	4	0	0	0	0	0	2	1	4	11	4.5
Malonic acid	0.055	0.075	4.5	6	0.5	7	13.5	2	12	27	9	7.5	16	49.5
Succinic acid	0.56	0.59	42	44.5	2	26	61	17	48	75	61	47.5	62	72
Glutaric acid	0.71	0.74	54	58	3	30	65	22	53.5	78	71	61	75.5	78
Adipic acid	0.81	0.845	64	67.5	3.5	32	68	26	57	82	79	65.5	81	81
Pimeric acid	0.92	0.94	73	75	4.5	36	72	29	61.5	84.5	86.5	72	85	85
Suberic acid	1	1	80	80	6	40	76	32	65.5	86	92	76	87.5	87

Layer L1 = silica gel (Merck 5721)

Solvent S1 = toluene-propionic acid-water (470:470:49)
S2 = benzene-propionic acid-water (470:470:57)
S3 = cyclohexane-propionic acid-water (470:470:28)
S4 = *p*-xylene-propionic acid-water (470:470:40)
S5 = heptane-acetic acid (900:100)
S6 = heptane-eucalyptol-acetic acid (450:450:100)
S7 = eucalyptol-acetic acid (900:100)
S8 = benzene-acetic acid (900:100)
S9 = benzene-eucalyptol-acetic acid (450:450:100)
S10 = benzene-acetic acid (500:500)
S11 = benzene-propionic acid (500:500)
S12 = eucalyptol-propionic acid-water (750:200:25)
S13 = eucalyptol-1-propanol-propionic acid-water (450:450:200:160)
S14 = eucalyptol-1-propanol-98% formic acid-water (450:450:200:250)

Detection D1 = xylose-aniline reagent, furfural

REFERENCES

Munier, R. L., Drapier, A. M., and Faivre, B., Chromatographie sur couche mince des acides aliphatic non volatile. I. Nouveaux solvants mobiles contenant un acide volatil. Observations prélimi-naires, *Chromatographia,* 6, 468, 1973.

Table TLC-22
AROMATIC AND ALIPHATIC ACIDS

Layer	L1	L2	L3	L3	L4	L5	L6	L7	L8	L8	L8		
Solvent	S1	S2	S3	S4	S5	S6	S7	S8	S9	S10	S11		
Technique													
Detection	D1	D2	D3	D3	D4	D5	D6	D7	D6	D8	D8	D8	C
Reference	1	1	2	2	3	4	4	5	5	6	6	6	6
Compound	Rf × 100							[a]	[a]				
Formic acid	52	—	—	—	—	—	—	8	—	—	—	—	—
Acetic acid	56	—	—	—	—	—	—	19	—	—	—	—	—
Propionic acid	66	—	—	—	—	—	—	28	—	—	—	—	—
Butyric acid	71	—	—	—	—	—	—	37	—	—	—	—	—
Valeric acid	78	—	—	—	—	—	—	48	—	—	—	—	—
Caproic acid	85	—	—	—	—	—	—	59	—	—	—	—	—
Oxalic acid	6	—	—	—	—	—	—	—	0	65	3	13	vt-p
Malonic acid	9	—	—	—	—	—	—	—	35	68	39	19	o&r[b]
Succinic acid	14	—	87	59	—	—	—	—	55	73	52	25	o&r
Glutaric acid	18	—	—	—	—	—	—	—	67	—	—	—	—
Adipic acid	22	—	—	—	—	—	—	—	77	—	—	—	—
Glycolic acid	38	—	—	—	—	35	58	—	—	57	70	15	ro-y
Lactic acid	44	—	82	55	—	—	—	—	—	72	79	32	o&r
2-Hydroxybutyric acid	55	—	—	—	—	—	—	—	—	—	—	—	—
2-Hydroxyvaleric acid	64	—	—	—	—	—	—	—	—	—	—	—	—
Glycine	30	—	—	—	—	—	—	—	—	—	—	—	—
Alanine	38	—	—	—	—	—	—	—	—	—	—	—	—
Norvaline	54	—	—	—	—	—	—	—	—	—	—	—	—
Norleucine	64	—	—	—	—	—	—	—	—	—	—	—	—
Pyruvic acid	25	—	86	59	—	—	—	—	—	—	—	—	—
2-Oxo-valeric acid	45	—	—	—	—	—	—	—	—	—	—	—	—
2-Oxo-isocaproic acid	86	—	—	—	—	—	—	—	—	—	—	—	—
Benzoic acid	—	53	—	—	—	—	—	—	—	—	—	—	—
2-Aminobenzoic acid	—	38	—	—	—	—	—	—	—	—	—	—	—
3-Aminobenzoic acid	—	28	—	—	—	—	—	—	—	—	—	—	—
4-Aminobenzoic acid	—	24	—	—	—	—	—	—	—	—	—	—	—

Table TLC-22 (continued)
AROMATIC AND ALIPHATIC ACIDS

2-Hydroxybenzoic acid	—	55	—	—	—	—	—	—	—	—	—	—	—
3-Hydroxybenzoic acid	—	31	—	—	—	—	—	—	—	—	—	—	—
4-Hydroxybenzoic acid	—	17	—	—	—	—	—	—	—	—	—	—	—
2,4-Dihydroxybenzoic acid	—	19	—	—	—	—	—	—	—	—	—	—	—
1,2-Benzenedicarboxylic acid	—	17	—	—	—	—	—	—	—	84	62	55	r&y
3,4,5-Trihydroxybenzoic acid	—	4	—	—	—	—	—	—	—	—	—	—	—
2-Naphthalene carboxylic acid	—	70	—	—	—	—	—	—	—	—	—	—	—
Diphenyl-2-carboxylic acid	—	75	—	—	—	—	—	—	—	—	—	—	—
Glyoxylic acid	—	—	37	55	—	—	—	—	—	—	—	—	—
Citric acid	—	—	44	14	—	—	—	—	—	42	24	4	y-o&vt
Malic acid	—	—	56	31	—	20	40	—	—	50	47	6	y-o&vt
Isocitric acid	—	—	57	17	—	—	—	—	—	40	23	3	w-o&vt
2-Oxo-glutaric acid	—	—	74	50	—	—	—	—	—	—	—	—	—
Oxaloacetic acid	—	—	86	53	—	—	—	—	—	—	—	—	—
Fumaric acid	—	—	95	60	—	—	—	—	—	82	58	50	y&gn
4-Hydroxyphenoxyacetic acid	—	—	—	—	70	—	—	—	—	—	—	—	—
Mesaconic acid	—	—	—	—	—	—	—	—	—	88	65	72	y&gn
Itaconic acid	—	—	—	—	—	—	—	—	—	79	52	41	y&gn
Methylsuccinic acid	—	—	—	—	—	—	—	—	—	82	59	48	ro-vt&y
Methylmalonic acid	—	—	—	—	—	—	—	—	—	80	43	41	ro-vt&y
trans-Aconitic acid	—	—	—	—	—	—	—	—	—	76	29	14	y&gn
Pyrrolidone carboxylic acid	—	—	—	—	—	—	—	—	—	57	71	16	ro&y
cis-Aconitic acid	—	—	—	—	—	—	—	—	—	39	29	4	y&gn
Citramalic acid	—	—	—	—	—	—	—	—	—	65	52	14	ro&p
Isomalic acid	—	—	—	—	—	—	—	—	—	63	42	11	y-ro&gn
Glyceric acid	—	—	—	—	—	—	—	—	—	36	66	6	r&p
Shikimic acid	—	—	—	—	—	—	—	—	—	32	64	3	y&gn
Quinic acid	—	—	—	—	—	—	—	—	—	18	66	2	ro&p
Tartaric acid	—	—	—	—	—	—	—	—	—	31	40	2	r-p&vt

[a] Measured on chromatogram.
[b] & = color circle.

Layer L1 = rice starch
L2 = rice starch impregnated with fluorescence indicator

L3 = silica gel G
L4 = silica gel
L5 = silica gel G impregnated with 10% aqueous soln. of ethylene glycol
L6 = silica gel G impregnated with 10% aqueous soln. of polyethylene glycol 400
L7 = cellulose and Dowex® 50W

Solvent
S1 = ethanol-water-conc. NH_3 (78:20:13)
S2 = ethanol-butanol-water-conc. NH_3 (40:30:15:15)
S3 = water saturated ether-88% formic acid (7:1)
S4 = chloroform-methanol-formic acid (80:20:1)
S5 = ethyl acetate-chloroform-acetic acid (9:1:0.15)
S6 = butanol-chloroform (20:80)
S7 = 28% NH_4OH-ethanol (1:2)
S8 = petroleum ether (b.p. 40-60°C)-acetone (2:1), 95% saturated with ethylene glycol
S9 = diethyl ether-98% formic acid (50:1), 95% saturated with polyethylene glycol 400
S10 = propanol-methyl benzoate-90% formic acid-water (7:3:2:1)
S11 = propanol-96° ethanol-NH_4OH-water (6:6:6:1.5)
S12 = upper phase of amyl alcohol-90% formic acid-water (4:4:2)

Detection
D1 = 0.5% fluorescein UV254
D2 = UV
D3 = Higgins aniline-ribose reagent
D4 = ferric chloride-potassium ferricyanide
D5 = bromophenol blue
D6 = bromocresol green
D7 = methylred
D8 = Pášková and Munk reagent

REFERENCES

1. **Canić, V. D. and Perišić-Janjić, N. V.**, Separation of aliphatic and aromatic organic acids on starch thin layer chromatographic behaviour and chemical structure, *Z. Anal. Chem.*, 270, 16, 1974.
2. **Beaudoin, A. R., Moorjani, S., and Lemonde, A.**, Thin-layer chromatography of organic acids, *Can. J. Biochem.*, 51, 318, 1973.
3. **Birner, J.**, Determination of penicillin fermentation, *J. Pharm. Sci.*, 61, 933, 1972.
4. **Linder, D. E.**, The separation and determination of glycolic and malic acids by ion-exchange chromatography, *Anal. Chim. Acta*, 75, 355, 1975.
5. **Lupton, C. J.**, Qualitative analysis of carboxylic acids by partition thin-layer chromatography, *J. Chromatogr.*, 104, 233, 1975.
6. **Morot-Gaudry, J. F., Nicol. M. Z., and Jolivet, E.**, Identification, par chromatographie sur couches minces, des acides organiques préalalblement séparés quantitativement sur colonne de gel de silice, *J. Chromatogr.*, 74, 346, 1972.

Table TLC-23
KETO ACIDS

Layer	L1		L1	
Solvent	S1		S2	
Technique	T1		T1	
Detection				
Reference	1		1	
Compound	Rs	Rf×100	Rs	Rf×100
cis-Glyoxylic acid	3.12	—	—	—
trans-Glyoxylic acid	1.21	—	—	—
cis-Pyruvic acid	4.29	—	—	—
trans-Pyruvic acid	1.43	—	1.00	74
cis-2-Oxobutanoic acid	4.18	—	—	—
trans-2-Oxobutanoic acid	1.64	—	—	—
3-Oxobutanoic acid	—	—	0.79	—
cis-2-Oxovaleric acid	5.43	—	—	—
trans-2-Oxovaleric acid	1.84	—	—	—
2-Oxoisovaleric acid	5.46	—	—	—
Levulinic acid	1.00	13.9	0.93	—
cis-2-Oxohexanoic acid	5.40	—	—	—
trans-2-Oxohexanoic acid	2.10	—	—	—
cis-2-Oxoisohexanoic acid	6.36	—	—	—
trans-2-Oxoisohexanoic acid	1.90	—	—	—
2-Oxo-3-methylvaleric acid	6.35	—	—	—
2-Oxo-3-dimethylbutanoic acid	6.08	—	—	—
Oxalacetic acid	—	—	0.44	—
2-Oxoglutaric acid	—	—	0.48	—

Layer L1 = silica gel 0.1 *N* $NaHCO_3$ (1:2 w/v)
Solvent S1 = ethyl acetate saturated with 0.1 *N* $NaHCO_3$-methanol (5:1)
S2 = butanol-ethanol-0.1 *N* $NaHCO_3$ (1:0.3:1), upper phase
Technique T1 = analysed as 2,4-dinitrophenylhydrazones

REFERENCE

1. **Ariga, N.**, Thin-layer chromatography of keto acid 2,4-dinitrophenylhydrazones, *Anal. Biochem.*, 49, 436, 1972.

Table TLC-24
ORGANIC ACID

Layer	L1	L1	L2	L3	L4	L5	L2	L2	L2	L6	L6	L7	L8	L9	L10
Solvent	S1	S2	S3	S4	S4	S5	S6	S7	S8	S9	S10	S11	S12	S13	S14
Technique															
Detection	D1	D1	D2	D3	D3	D4	D5	D5	D5	D6	D6	D7	D8	D9	D10
Reference	1	1	2	3	3	4	5	5	5	6	6	7	8	9	10
Compound	Rf × 100														
	[a]	[a]					Rs	Rs	Rs					[a]	
Hexahydrophippuric acid	53	57	—	—	—	—	—	—	—	—	—	—	—	—	—
3,4,5,6-Tetrahydrohippuric acid	49	57	—	—	—	—	—	—	—	—	—	—	—	—	—
Hippuric acid	41	57	—	0	0	28	—	—	—	—	—	—	—	70	—
3,4-Dihydroxycyclohexanecarboxyic acid	—	20	—	—	—	—	—	—	—	—	—	—	—	—	55
Shikimic acid	—	12	—	—	—	—	—	—	—	—	—	—	—	—	—
Indole-3-acetic acid	—	—	45	—	—	—	—	—	—	—	—	—	—	—	—
Abcisic acid	—	—	57	—	—	—	—	—	—	—	—	—	—	—	—
Benzoic acid	—	—	—	15	72	—	—	—	—	—	—	—	—	—	—
Methylhippuric acid	—	—	—	—	—	32	—	—	—	—	—	—	—	—	—
Urea	—	—	—	—	—	22	—	—	—	—	—	—	—	—	—
3-Hydroxy-4-methoxymandelic acid methylester	—	—	—	—	—	—	0.20	0.37	0.24	—	—	—	—	—	—
4-Hydroxy-3-methoxymandelic acid methylester	—	—	—	—	—	—	0.28	0.58	0.49	—	—	—	—	—	—
4-Phenylbutyric acid	—	—	—	—	—	—	—	—	—	75	95	—	—	—	—
4-Phenylbut-3-enoic	—	—	—	—	—	—	—	—	—	71	95	—	—	—	—
4-Phenylbut-2-enoic acid	—	—	—	—	—	—	—	—	—	—	—	—	—	—	—
trans-Cinnamic acid	—	—	—	—	—	—	—	—	—	67	95	—	—	—	—
Phenylacetic acid	—	—	—	—	—	—	—	—	—	54	95	—	—	—	—
3-Phenylpropionic acid	—	—	—	—	—	—	—	—	—	—	—	—	—	—	—
4-(2′-Hydroxyphenyl)butyric acid	—	—	—	—	—	—	—	—	—	4	59	—	—	—	—
trans-4-(2′-Hydroxyphenyl)but-3-enoic acid	—	—	—	—	—	—	—	—	—	4	37	—	—	—	—
6-(2′-Hydroxyphenyl)hydroxyhexanoic acid	—	—	—	—	—	—	—	—	—	4	18	—	—	—	—
2,5-Dihydroxyphenylacetic acid	—	—	—	—	—	—	—	—	—	—	—	17	—	—	—
cis-3,4-Dihydroxycinnamic acid	—	—	—	—	—	—	—	—	—	—	—	—	18	—	—

Table TLC-24 (continued)
ORGANIC ACID

Compound	Rf × 100														
	[a]	[a]					Rs	Rs	Rs					[a]	
trans-3,4-Dihydroxycinnamic acid	—	—	—	—	—	—	—	—	—	—	—	—	27	—	—
cis-4-Hydroxy-3,5-dimethoxycinnamic acid	—	—	—	—	—	—	—	—	—	—	—	—	43	—	—
trans-4-Hydroxy-3,5-dimethoxycinnamic acid	—	—	—	—	—	—	—	—	—	—	—	—	55	—	—
cis-3-Hydroxy-4-methoxycinnamic acid	—	—	—	—	—	—	—	—	—	—	—	—	56	—	—
3-Dehydroshikimic acid	—	—	—	—	—	—	—	—	—	—	—	—	63	—	—
Shikimate-3-phosphate	—	—	—	—	—	—	—	—	—	—	—	—	—	—	37
3,4-Dihydroxybenzoic acid	—	—	—	—	—	—	—	—	—	—	—	—	—	—	15
2-Methoxyphenol	—	—	—	—	—	1.00	1.00	1.00	—	—	—	—	—	—	95

[a] Measured on chromatogram.

Layer
L1 = silica gel GF254 (Schleicher & Schüll)
L2 = Kieselgel HF254
L3 = SIF silica gel sheet (Riedel de Haen 37350)
L4 = cellulose TLC alumina (Merck AG 5552)
L5 = Schleicher & Schüll G 1500 silica gel
L6 = Eastman silica layer
L7 = silica gel G
L8 = silica gel 60 F254
L9 = Silufol (Kavalier Sklárny, Czec.)
L10 = silica gel 60 F254

Solvent
S1 = acetone-light petroleum-acetic acid (20:40:1)
S2 = *n*-propanol-conc. NH_3 (0.88%) (7:3)
S3 = *n*-propanol-25% NH_3-water (80:10:10)
S4 = *n*-hexane-acetic acid (96:4)
S5 = toluene-acetic acid-water (60:40:2)
S6 = chloroform
S7 = benzene-triethylamine (10:2)
S8 = chloroform-cyclohexane-diethylamine (5:4:1)
S9 = light petroleum-acetic acid (49:1)

S10 = light petroleum-diethyl ether-formic acid (45:5:1)
S11 = upper phase of benzene-acetic acid-water (2:3:1)
S12 = diethyl ether-hexane-chloroform-acetic acid (12:38:50:0.5)
S13 = chloroform-acetone-acetic acid (40:10:5)
S14 = *n*-butanol-acetic acid-water (80:20:20)

Detection
D1 = X ray film
D2 = Ehrlich reagent (red color)
D3 = UV
D4 = dimethylmaminobenzaldehyde/acetic anhydride
D5 = UV or diazotized sulfanilic acid or diazotized 4-nitroaniline
D6 = bromocresol green or Gibb's reagent
D7 = Folin's reagent and sodium carbonate (blue color)
D8 = 360 nm
D9 = dimethylaminobenzaldehyde/acetic acid (red color)
D10 = ammonium molybdate-perchloric acid or 2,4-dinitrophenylhydrazine or alkaline silvernitrate or 2-phenylene diamine or periodic acid-benzidine or periodic acid-aniline or periodic acid-thiobarbituric acid, these reagents were used individually

REFERENCES

1. **Brewster, D., Jones, R. S., and Parke, D. V.**, The metabolism of shikimate in the rat, *Biochem. J.*, 170, 257, 1978.
2. **Duering, H.**, Analysis of abscisic acid and indole-3-acetic acid from fruits of Vitis vinifera L by high pressure liquid chromatography, *Experientia*, 33, 1666, 1977.
3. **Zolg, W. and Ottow, J. C. G.**, Thin-layer chromatography methods for detection of hippurate hydrolase activity among various bacteria (Pseudomonas, Bacillus, Enterobacteriaceae), *Experientia*, 29, 1573, 1973.
4. **Van Kerckhoven, G., Blaton, V., Vandamme, D., and Peeters, H.**, Thin-layer chromatographic evaluation of methylhippuric and hippuric acids in urine, *J. Chromatogr.*, 100, 215, 1974.
5. **Thomas, H., Lax, E. R., and Ockenfels, H.**, Dünnschift- und gaschromatographische Trennung von 4-Hydroxy-3-methoxy- und 3-Hydroxy-4-methoxy-mandelsäure, *J. Chromatogr.*, 84, 187, 1973.
6. **Sariaslani, F. S., Harper, D. B., and Higgins, J. I.**, Microbial degradation of hydrocarbons. Catabolism of 1-phenylalkanes by Nocardia salmonicolor, *Biochem. J.*, 140, 31, 1974.
7. **Fieldman, J. M. and Bowman, J.**, Urinary homogentistic acid; Determination by thin-layer chromatography, *Clin. Chem.*, 19, 459, 1973.
8. **Fenton, T. W., Mueller, M. M., and Clandinin, D. R.**, Isomerization of some cinnamic acid derivatives, *J. Chromatogr.*, 152, 517, 1978.
9. **Gartzke, J. and Weigamn, H. J.**, A simple thin-layer chromatographic method for the estimation of hippuric acid: comparison with a photometric and a gas chromatographic method, *J. Chromatogr.*, 162, 234, 1979.
10. **Ghisalba, O. and Nüesch, J.**, A genetic approach to the biosynthesis of the rifamycin-chromophore in Nocardia mediterranei. II. Isolation and characterization of a shikimate excreting auxotrophic mutant of Nocardia mediterranei with normal rifamycin-production, *J. Antibiot.*, 31, 215, 1978.

Table TLC-25
ORGANIC ACIDS

Layer	L1	L1
Solvent	S1	S2
Technique		
Detection	D1	D1
Reference	1	1
Compound	Rf × 100	
Picric acid	76	74
Hexanoic acid	75	82
2-Hydroxybenzoic acid	72	78
n-Valeric acid	71	77
Benzoic acid	69	79
Phenylacetic acid	68	74
Phenoxyacetic acid	64	63
Puruvic acid	60	46
Mandelic acid	57	44
1,2-Benzenedicarboxylic acid	55	41
Glutaric acid	54	41
Maleic acid	50	30
2-Hydroxybutanoic acid	50	38
Fumaric acid	49	37
Adipic acid	48	36
2-Oxoglutaric acid	45	23
Succinic acid	43	28
Glyoxylic acid	43	21
Malonic acid	40	20
Lactic acid	36	24
Glycolic acid	31	16
Malic acid	26	8
Glyceric acid	24	8
Citric acid	22	4
Tartaric acid	19	3
Ascorbic acid	15	3
Boric acid	7	5
Nicotinic acid	5	2
Oxalic acid	0-30[a]	0-17[a]

[a] Diffuse.

Layer L1 = Sil G25 (Macherey Nagel)
Solvent S1 = *n*-pentyl formate-chloroform-formic acid (70:15:15)
S2 = *n*-pentyl formate-chloroform-formic acid (20:70:10)
Detection D1 = bromocresol green

From Hansen, S. A., *J. Chromatogr.*, 124, 123, 1976. With permission.

Table TLC-26
PHENOLIC COMPOUNDS

Layer	L1	L2
Solvent	S1	S2
Technique		
Detection	D1	D2
Reference	1	2

Table TLC-26 (continued)
PHENOLIC COMPOUNDS

Compound	Rf × 100	
2-Methoxyphenol	88	—
Syringol	90	—
1,2-Dihydroxybenzene	55.5	—
Quinol	52	—
1,3-Dihydroxybenzene	40	—
1,3,5-Trihydroxybenzene	10.5	—
1,2,3-Trihydroxybenzene	30	—
3-Methoxybenzaldehyde	99	—
4-Methoxybenzaldehyde	99	—
3,4-Dimethoxybenzaldehyde	99	—
3,4,5-Trimethoxybenzaldehyde	99	—
Cinnamaldehyde	96.5	—
2-Methoxycinnamaldehyde	96.5	—
2-Hydroxybenzaldehyde	70	—
3-Hydroxybenzaldehyde	66.5	—
4-Hydroxybenzaldehyde	66.5	—
4-Hydroxy-3-methoxybenzaldehyde	72	—
3-Hydroxy-4-methoxybenzaldehyde	70	—
4-Hydroxy-3,5-dimethoxybenzaldehyde	79	—
2,4-Dihydroxybenzaldehyde	55	—
3,4-Dihydroxybenzaldehyde	51	—
Phenylalanine	44	—
Tyrosine	28	—
DOPA	17	—
2-Hydroxybenzoic acid	54	59
3-Hydroxybenzoic acid	54	39
4-Hydroxybenzoic acid	54	38
4-Hydroxy-3-methoxybenzoic acid	59	57
4-Hydroxy-3,5-dimethoxybenzoic acid	59.5	54
2,3-Dihydroxybenzoic acid	33	18
2,4-Dihydroxybenzoic acid	30	—
2,5-Dihydroxybenzoic acid	38.5	14
2,6-Dihydroxybenzoic acid	33	—
3,4-Dihydroxybenzoic acid	34	14
3,5-Dihydroxybenzoic acid	30	8
3,4,5-Trihydroxybenzoic acid	11	4
2-Hydroxycinnamic acid	46	—
3-Hydroxycinnamic acid	46	—
4-Hydroxycinnamic acid	46	41
4-Hydroxy-3-methoxycinnamic acid	59	55
3-Hydroxy-4-methoxycinnamic acid	59	—
4-Hydroxy-3,5-dimethoxy-methoxycinnamic acid	70	—
3-Feruloylquinic acid	45	—
3,4-Dihydroxycinnamic acid	26	16
3-Caffeoylquinic acid (Chlorogenic acid)	21	—
5-Caffeoylquinic acid	17	—
3-Caffeoylquinic acid-potassium-caffeine complex	21.5	—
1,4-Dicaffeoylquinic acid	10	—
4-Hydroxymandelic acid	—	4
2,5-Dihydroxyphenylacetic acid	—	5
4-Hydroxyphenylacetic acid	—	5
4-Hydroxy-3-methoxymandelic acid	—	6
3,4-Dihydroxyphenylacetic acid	—	10
4-Hydroxy-3-methoxyphenyllactic acid	—	12
4-Hydroxy-3-methoxyphenylpyruvic acid	—	13
3,4-Dihydroxymandelic acid	—	14

Table TLC-26 (continued)
PHENOLIC COMPOUNDS

5-Hydroxyindoleacetic acid	—	16
3,4-Dihydroxyphenylpropionic acid	—	16
2-Hydroxyhippuric acid	—	18
3-Hydroxyphenylacetic acid	—	33
4-Hydroxyphenylacetic acid	—	34
2-Hydroxyphenylacetic acid	—	39
3-Hydroxyphenylpropionic acid	—	41
4-Hydroxyphenylpropionic acid	—	41
4-Hydroxy-3-methoxyphenylpropionic acid	—	47

Layer L1 = poly-*N*-vinylpyrrolidone (Polyclar AT by G.A.F. Ltd.)-$CaSO_4 \cdot 1/2H_2O$ (5:1)
L2 = silica gel TLC (Quantum)
Solvent S1 = 2-butanone-methylphenylketone-50% acetic acid (5:5:4)
S2 = upper phase of benzene-acetic acid-water (100:150:50)
Detection D1 = molybdate, diazotized sulfanilic acid, phloroglucinol
D2 = Folin & Ciocalteau's reagent

REFERENCES

1. **Clifford, M. N.,** The use of poly-N-vinylpyrroridones as the adsorbent for the chromatographic separation of chlorogenic and other phenolic compounds, *J. Chromatogr.*, 94, 261, 1974.
2. **Felice, L. J. and Kissinger, P. T.,** Determination of homovanillic acid in urine by liquid chromatography with electrochemical detection, *Anal. Chem.*, 48, 794, 1976.

Table TLC-27
PHENOLIC COMPOUNDS

Layer	L1	L1	L1	L1	L1	L1	L1	L1	L1
Solvent	S1	S2	S3	S4	S5	S6	S7	S8	S9
Technique									
Detection	D1	D1	D1	D1	D1	D1	D1	D1	D1
Reference	1	1	1	1	1	1	1	1	1
Compound	Rf × 100								
3′,4′,5,7-Tetrahydroxyflavone	0	0	0	0	5	0	0	0	0
4′,5,7-Trihydroxyflavone	0	0	0	0	15	0	2	13	10
2-Hydroxybenzoic acid	4	5	5	1	15	8	12	25	30
4-Hydroxycinnamic acid	2	1	4	19	46	1	12	46	30
3,4-Dihydroxybenzoic acid	5	3	12	30	53	4	25	60	15
3,4,5-Trihydroxybenzoic acid	8	5	20	37	53	10	36	70	15
4-Hydroxy-3-methoxycinnamic acid	2	1	8	25	55	4	10	76	55
4-Hydroxybenzoic acid	3	2	10	28	56	5	22	55	35
4-Hydroxy-3,5-dimethoxycinnamic acid	5	3	14	35	60	8	42	75	82
7-Hydroxy-6-methoxycoumarin	0	5	15	45	60	6	20	65	80
4-Hydroxy-3-methoxybenzoic acid	6	3	17	37	63	10	31	70	65

Layer L1 = DC Aluminumfolien Polyamide 11F254 (Merck)
Solvent S1 = water
S2 = water-methanol (75:25)
S3 = water-methanol (50:50)
S4 = water-methanol (25:75)
S5 = methanol
S6 = water-acetone (75:25)

Table TLC-27 (continued)
PHENOLIC COMPOUNDS

	S7 = water-acetone (50:50)
	S8 = water-acetone (25:75)
	S9 = acetone
Detection	D1 = UV

REFERENCE

1. **Mosel, H. -D. and Herrmann, K.,** Chromatographisches Verhalten von Pflanzenphenolen an Polyamid, *J. Chromatogr.*, 87, 280, 1973.

Table TLC-28
ALIPHATIC POLYCARBOXYLIC ACIDS

Layer	L1	L1
Solvent	S1	S2
Technique		
Detection	D1	D1
Reference	1	1
Compound	Rf × 100	
Maleic acid	25	56
Maleic acid monoisopropyl ester	56	—
Maleic acid mono-2-octyl ester	68	—
Fumaric acid	46	60
Fumaric acid monoisopropyl ester	58	—
Terebic acid (DM-telomer, 1:1)	48	62
Terebic acid monoisopropyl ester	53	—
Terebic acid (DM-telomer, 1:2)	27	—
Methylhexylparaconic acid (MH-telomer, 1:1)	60	72
Methylhexylparaconic acid mono-2-octyl ester	71	—
Methylhexylparaconic acid (MH-telomer, 1:2)	30	63
Methylhexylparaconic acid (MH-telomer, 1:2) mono-2-octyl ester	39	—
Methylhexylparaconic acid (MH-telomer, 1:2) di-2-octyl ester	53	—
Methylhexylparaconic acid (MH-telomer, 1:3)	11	47
Methylhexylparaconic acid (MH-telomer, 1:4)	4	40
Methylhexylparaconic aid (MH-telomer, 1:5)	—	34

Layer	L1 = silica gel H (Merck) + 1% $NaHSO_4$
Solvent	S1 = *n*-hexane-acetone (1:1)
	S2 = *n*-hexane-acetone (1:3)
Detection	D1 = dimethyl yellow + pentamethylene red, $KMnO_4$ in H_2SO_4, potassium bichromate in H_2SO_4

From Kučera, J., Pokorný, S., and Čoupek, J., *J. Chromatogr.*, 88, 281, 1974. With permission.

Table TLC-29
PHENOLIC ACIDS

Layer	L1	L1	L2	L3	L4	
Solvent	S1	S2	S3	S4	S5	
Technique						
Detection	D1	D1	D2	D3	D4	C
Reference	1	1	2	3	4	4
Compound	Rf × 100		[a]			
4-Hydroxymandelic acid	8	11	—	—	17	—
Caffeine	18	25	—	—	—	—
4-Hydroxy-3-methoxymandelic acid	17	25	—	54	20	y
2-Hydroxyhippuric acid	42	56	35	—	—	—
4-Hydroxyphenylacetic acid	47	62	—	—	60	—
3-Hydroxybenzoic acid	53	66	—	—	65	—
4-Hydroxybenzoic acid	55	67	80	—	—	—
2-Hydroxybenzoic acid	60	75	—	—	—	—
4-Hydroxy-3-methoxyphenylacetic acid	74	86	—	—	67	y
4-Hydroxy-3-methoxybenzoic acid	80	92	—	87	74	y
2,4,6-Trihydroxybenzoic acid	—	—	6	—	—	—
2,6-Dihydroxybenzoic acid	—	—	10	—	—	—
2,5-Dihydroxyhippuric acid	—	—	16	—	—	—
2,3-Dihydroxybenzoic acid	—	—	56	—	—	—
2,5-Dihydroxybenzoic acid	—	—	67	—	—	—
2-Acetoxybenzoic acid	—	—	86	79	—	—
2,4-Dihydroxybenzoic acid	—	—	90	—	—	—
3,4-Dihydroxymandelic acid	—	—	—	18	—	—
5-Hydroxyindoleacetic acid	—	—	—	48	44	y
2-Hydroxyphenylacetic acid	—	—	—	78	—	—
Normentanephrine	—	—	—	6	—	—
Metanephirine	—	—	—	4	—	—
3,4-Dihydroxyphenylacetic acid	—	—	—	—	35	y
4-Hydroxy-3-methoxycinnamic acid	—	—	—	—	77	y-o
Indole-3-acetic acid	—	—	—	—	72	o

[a] Measured on chromatogram.

Layer
- L1 = microcrystalline cellulose (Merck 5552)
- L2 = silica gel G (Merck)
- L3 = silica gel (Eastman Organic 6061)
- L4 = Kieselgel F254 (Merck)

Solvent
- S1 = benzene
- S2 = toluene-acetic acid-water (70:20:1)
- S3 = diethyl ether-*n*-butyric acid (10:1)
- S4 = toluene-acetic acid-ethyl acetate (50:50:5)
- S5 = benzene-methanol-acetic acid (70:20:10)

Detection
- D1 = alkaline diazotized sulfanilic acid or nitroaniline
- D2 = Scheffel TLC densitometer at 325 nm
- D3 = diazotized 4-nitroaniline
- D4 = reflectance spectroscopy at 400 nm

REFERENCES

1. **Ersser, R. S., Seakins, J. W. T., and Manraj, B.,** Thin-layer chromatography of phenolic acids with use of toluene instead of benzene, *Clin. Chem.,* 24, 836, 1978.
2. **Chrastil, J. and Wilson, J. T.,** Quantitative estimation of salicylic acid and its metabolites by thin-layer densitometry, *J. Chromatogr.,* 152, 183, 1978.
3. **Badella, M., Routj, M. W., Gump, B. H., and Gigliotti, H. J.,** Thin-layer chromatographic method for urinary 4-hydroxy-3-methoxymandelic acid (vanilmandelic acid), *Clin. Chem.,* 22, 2046, 1976.
4. **Huck, H. and Dworzak, E.,** Quantitative Analyse von Katecholamin- und Serotoninmetaboliten auf der Dünnschichtplatte. II. Mitt. Remissionsmessungen nach einer spezifischen photochemischen Reaktion, *J. Chromatogr.,* 74, 303, 1972.

Table TLC-30
GLYCOLIC ACIDS

Layer	L1	L1	L1	L2	L2	L2	L2
Solvent	S1	S2	S3	S4	S5	S6	S7
Detection	D1	D1	D1	D1	D1	D1	D1
Reference	1	1	1	1	1	1	1
Compound	Rf × 100						
Diphenylglycolate-3-quinuclidinyl-HCl	22	74	60	55	5	34	43
Diphenylglycolate-tropinyl-HCl	44	75	73	53	1	50	37
Diphenylglycolate-*N*-methyl-3-piperidinyl and HCl salt	54	74	73	54	0	54	79
Diphenylglycolate-*N*-methyl-4-piperidinyl-HCl	44	76	71	55	1	47	61
Cyclopentyl,phenylglycolate-3-quinuclidinyl	43	73	71	51	5	45	56
Cyclopentyl,phenylglycolate-tropinyl	38	73	71	55	2	47	41
Cyclopentyl,phenylglycolate-*N*-methyl-3-piperidinyl	60	81	77	53	2	56	77
Cyclopentyl,phenylglycolate-*N*-methyl-4-piperidinyl	57	80	75	55	4	49	65
Diphenylglycolate-2-diethylaminoethyl	50	74	69	54	3	44	53
Glycolate-3-quinuclidinyl	10	70	43	43	1	14	40
Diphenylglycolate,1-(2,2,2)-bicyclooctyl	62	79	70	80	60	52	63
Cyclopentyl,phenylglycolate,1-(2,2,2)-bicyclooctyl	65	90	74	83	78	58	73
Methylbenzilate	63	73	74	82	54	61	—
Methylcyclopentylphenylglycolate	68	81	82	78	67	63	—
3-Quinuclidinol	0	25	18	30	2	3	26
Tropine	0	39	28	32	0	8	21
N-methyl-3-piperidinol and HCl salt	7	48	30	26	0	17	59
N-methyl-4-piperidinol	4	47	25	26	1	12	52
2-Diethylaminoethanol	0	70	51	38	33	0	34
1-(2,2,2)-Bicyclooctanol	33	50	55	86	57	45	67
Benzophenone	64	87	75	87	64	59	72
Benzhydrol	32	70	66	86	50	49	71
Cyclopentylphenylcarbinol	36	75	66	86	54	53	71
Cyclopentylphenylketone	64	89	76	87	72	60	72
Mandelic acid	0	0	0	61	5	0	51
Benzilic acid	0	0	3	78	31	0	—
Cyclopentylphenylglycolic acid	0	0	8	81	48	1	—

Layer L1 = alumina (Woelm) + 1% ZS fluorescence indicator (Riedel de Haen)
L2 = silica gel (Merck) + 1% ZS fluorescence indicator
Solvent S1 = chloroform
S2 = chloroform-methanol (9:1)
S3 = chloroform-diethylamine (9:1)
S4 = isobutanol-isopropanol-formic acid-water (15:15:4:10)
S5 = benzene-hexane-acetic acid (2:2:1)
S6 = cyclohexane-chloroform-diethylamine (5:4:1)
S7 = methanol-conc. NH_3 (99:1)
Detection D1 = UV

From Verweij, A., DeJong-De Vos, R., and Teisman, H. G. J., *J. Chromatogr.*, 69, 407, 1972. With permission.

Table TLC-31
METABOLITES OF BENZOIC ACID

Layer	L1	L1
Solvent	S1	S2
Detection	D1	D1
Reference	1	1

Table TLC-31 (continued)
METABOLITES OF BENZOIC ACID

Compound	Rf × 100	
Benzoic acid	77	88
Cyclohexa-1,4-dienecarboxylic acid	77	83
Cyclohexa-2,5-dienecarboxylic acid	77	82
Cyclohexa-1-enecarboxylic acid	91	95
Cyclohexa-3-enecarboxylic acid	91	95
Cyclohxanecarboxylic acid	83	92
2-Hydroxycyclohexanecarboxylic acid	54	21
Heptanedioic acid	60	39
Adipic acid	57	30
6-Hydroxyhexanic acid	43	14

Layer L1 = Kieselgel G
Solvent S1 = benzene-dioxane-acetic acid (90:25:4)
S2 = light petroleum-diethyl ether-acetic acid (50:50:1)
Detection D1 = radioautography on X-ray film

From Williams, R. J. and Evans, W. C., *Biochem. J.*, 148, 1, 1975. With permission.

Table TLC-32
PHENOLIC COMPOUNDS

Layer	L1	L1	
Solvent	S1	S2	
Detection			C
Reference	1	1	1
Compound		Rf × 100	
Salirepin	19	14	y
Salicin	22	18	r
Picein	24	22	r-bn
Salidroside	26	20	y-bn
Triandrin	28	20	d-gn
Grandidetatin	29	21	y
1-*p*-Coumaroyl-*β*-D-glucose	32	25	y-bn
Vimalin	33	20	d-gn
Salicortin	33	23	r
Trichoside	33	18	y-bn
Fragilin	37	32	r
Nigracin	43	37	y
Salireposide	43	36	y
Tricocarpin	43	39	d-gn
Trichocarposide	45	40	o-bn
Populin	47	30	r
Salicyloylsalicin	49	41	r
Tremuloidin	52	48	r
Tremulacin	55	51	r
Salicyloylsalicin-2-*O*-benzoate	67	60	r
Salicyloylsalicin-6-*O*-benzoate	73	70	r
β-Sitosterol	84	80	p
(+)-Catechin	72	72	o

Layer L1 = silica gel
Solvent S1 = ethyl acetate-xylene-formic acid-water (35:1:2:2):
S2 = ethyl acetate-methanol (1:9)

From Steele, J. W., Weitzel, P. F., and Audette, R. C. S., *J. Chromatogr.*, 71, 435, 1972. With permission.

Table TLC-33
ORGANIC ACIDS

Layer	L1	L1	L1	L1	L2	L3	L4	L5
Solvent	S1	S2	S3	S4	S2	S2	S2	S5
Technique								
Detection	D1	D1	D1	D2	D3	D3	D4	D5
Reference	1	1	1	2	3	4	5	6
Compond	Rf × 100			[a]	[a]	[a]		
Resveratrol	71	51	59	—	—	—	—	—
Lecanoric acid	—	—	—	43	—	—	—	—
Orsellinic acid	—	—	—	78	—	—	—	—
Orsinol	—	—	—	91	—	—	—	—
Atranorin	—	—	—	—	85	62	80	—
Chloroatranorin	—	—	—	—	61	—	—	—
Usnic acid	—	—	—	—	50	60	—	—
Evernic acid	—	—	—	—	—	30	—	—
Fumarprotocetraric acid	—	—	—	—	—	3	—	—
Genuine aurintricarboxylic acid	—	—	—	—	—	—	—	28
Aurintricarboxylic acid	—	—	—	—	—	—	—	0

[a] Measured on chromatogram.

Layer
- L1 = silica gel
- L2 = Merck silica gel F254 in 0.2 *N* sodium acetate
- L3 = silica gel (70% ambient moisture)
- L4 = silica gel (Merck 7720) + DC 144 cellulose powder (Schleicher & Schull) + 0.02 *M* oxalic acid
- L5 = silica gel H (Merck)

Solvent
- S1 = chloroform-ethyl acetate-formic aid (5:4:1)
- S2 = benzene-dioxane-acetic acid (90:25:4)
- S3 = toluene-ethyl formate-formic acid (5:4:1)
- S4 = ethyl acetate-methanol-2% acetic acid (10:1:0.15)
- S5 = chloroform-formic acid (1:0.08)

Detection
- D1 = chromatogram scanner (Berthold LB-2723, 14C)
- D2 = blue B salt (Tokyo-Kasei)
- D3 = UV
- D4 = UV
- D5 = UV 350 nm

REFERENCES

1. **Rupprich, N. and Kindl, H.,** Stilbene synthases and stilbenecarboxylate synthases. I. Enzymatic synthesis of 3,5,4′-trihydroxystilbene from p-coumaroyl coenzyme A and malonyl coenzyme A, *Hoppe-Seyler's Z. Physiol. Chem.,* 359, 165, 1978.
2. **Umezawa, H., Shibamoto, N., Naganawa, H., Ayukawa, S., Matsuzaki, M., Takeuchi, T., Kono, K., and Sakamoto, T.,** Isolation of lecanoric acid, an inhibitor of histidine decarboxylase from a fungus, *J. Antibiot.,* 27, 587, 1974.
3. **Ramaut, J. L., Brouers, M., Serusiaux, E., and Corvisier, M.,** Separation of mixtures of atranorin and chloroatronorin by thin-layer chromatography, *J. Chromatogr.,* 155, 450, 1978.
4. **Janssen, A. and Scharf, W.,** Standardisierung der dünnschichtchromatographischen Trennungen von Flechtensäuren zur Chromotaxonomie von Flechten, *Chromatographia,* 12, 106, 1979.
5. **Klee, R. and Streubing, L.,** Quantitative analysis of atranorin by thin-layer chromatography, *J. Chromatogr.,* 129, 478, 1976.
6. **Tsutsui, K., Seki, S., Tsutsui, K., and Oda, T.,** Fractionation of aurintricarboxylic acid and effects of its components on nuclear swelling and nucleic acid synthesis, *Biochim. Biophys. Acta,* 517, 14, 1978.

Table TLC-34
GLYOXYLIC ACIDS

Layer	L1	L1
Solvent	S1	S2
Reference	1	1
Compound	Rf × 100	
Semicarbazone	16	0
Methylglyoxylate(Z)-semicarbazone	34	7
Methylglyoxylate(E)-4-phenylsemicarbazone	62	27
Methylglyoxylate(Z)-4-phenylsemicarbazone	78	62
Methylglyoxylate-2-methylsemicarbazone	35	7
Methylglyoxylate-2-benzylsemicatbazone	46	14
Methylglyoxylate-2-methyl-4-phenylsemicarbazone	82	46
Methylglyoxylate-2-benzyl-4-phenylsemicarbazone	91	70
Metbylglyoxylate-2-ethoxycarbonylmethylsemicarbzone	42	10
1-Methoxycarbonylmethyleneamino-3-ethoxycarbonylmethyl hydrantoin	66	21
Methylglyoxylate-2-methylhydrazone	50	17
Methylglyoxylate-2-benzylhydrazone	71	41
Methylglyoxylate-2-acetyl-2-methylhydrazone	74	42
Methylglyoxylate-2-acetyl-2-benzylhydrazone	84	64
1-Methyl-6-azauracil	52	22
1-Benzyl-6-azauracil	69	43
6-Azauracil	41	11
3-Phenyl-6-azauracil	65	35

Layer L1 = Silufol UV 254 (Kavakier, Czech.)
Solvent S1 = benzene-acetone (3:2)
S2 = benzene-ethyl acetate (2:1)

REFERENCE

1. **Hřebabecký, H. and Beránek, J.**, Isomerisation, alkylation and cyclsation of glyoxylic acid semicarbazone derivatives, *Collect. Czech. Chem. Commun.*, 40, 2364, 1975.

Table TLC-35
PHENOLIC COMPOUNDS (GIBBERELLIN)

Layer	L1	L1	L1
Solvent	S1	S2	S3
Technique	T1	T1	T1
Detection			
Reference	1	1	1
Compound		Rf × 100	
Isosteviol	100	87	92
Steviol	76	68	76
6β,7β-Dihydroxykaurenoic acid	62	64	50
Gibberellin A9	92	69	91
Gibberellin A14	72	59	73
Gibberellin A13	66	50	71
Gibberellin A4	63	56	57
Gibberellin A7	56	51	56
Gibberellin A20	51	46	45
Gibberellin A5	49	41	41
Gibberellin A1	31	23	18
Gibberellin A3	26	19	15

Table TLC-35 (continued)
PHENOLIC COMPOUNDS (GIBBERELLIN)

Layer	L1 = silica gel G impregnated with $AgNO_3$
Solvent	S1 = carbon tetrachloride-acetone (3:1)
	S2 = pentane-acetone (7:3)
	S3 = hexane-ethyl acetate (1:1)
Technique	T1 = analyzed as 4-nitrobenzyl ester
Detection	D1 = UV or FP 365 nm, sulfuric acid

From Heftman, E. and Sauders, G. A., *J. Liquid Chromatogr.*, 1, 333, 1978. With permission.

Table TLC-36
ORGANIC ACIDS

Layer	L1	L1	L1	L2
Solvent	S1	S2	S3	S4
Technique				
Detection				D1
Reference	1	1	1	2
Compund		Rf × 100		
3,4-Epoxy-*trans*-4-phenyl-1(E)-buten-1-carboxylic acid methylester	52	—	—	—
3,4-Epoxy-*cis*-4-phenyl-1(E)-buten-1-carboxylic acid methylester	46	—	—	—
3,4-Epoxy-3-methyl-*trans*-4-phenyl-1(E)-buten-1-carboxylic acid methylester	54	—	—	—
3,4-Epoxy-3-methyl-*cis*-4-phenyl-1(E)-buten-1-carboxylic acid methylester	46	—	—	—
4-Phenyl-1(E),3(E)-butadien-1,3-dicarboxylic acid dimethylester	46	—	—	—
4-Phenyl-1(E),3(Z)-butadien-1,3-dicarboxylic acid dimethylester	40	—	—	—
5c-(2-(methoxycarbonyl)-vinyl)-2c-phenyl-tetrahydro-3γ,4c-furandicarboxylic acid dimethylester	—	53	—	—
5c-(2-(methoxycarbonyl)-vinyl)-2t-phenyl-tetrahydro-3γ,4c-furandicarboxylic acid dimethylester	—	30	—	—
cis-5-(2,2-Dicyanoethyl)-2-phenyl-2,3-dihydro-3,4-furandicarboxylic acid dimethylester	—	—	32	—
trans-5-(2,2-Dicyanoethyl)-2-phenyl-2,3-dihydro-3,4-furandicarboxylic acid dimethylester	—	—	30	—
Chloro-decadiyne(chloro-1-tetraethoxycarbonyl-5,5-10,10-decadiyne-2,7)	—	—	—	42
Cyclodecadiyne(tetraethoxycarbonyl-1,1,6,6-cyclodecadiyne 3,8)	—	—	—	32
Bromodecadiyne(bromo-1-tetraethoxycarbonyl-5,5-10,10-decadiyne 2,7)	—	—	—	43

Layer	L1 = silica gel
	L2 = silica gel GF 254 (Merck)
Solvents	S1 = petroleum ether-ether (2:1)
	S2 = ether-petroleum ether (5:2)
	S3 = benzene-ethyl acetate (8:1)
	S4 = dichloromethane-methanol (0.4-1% methanol)
Detection	D1 = I_2

REFERENCES

1. **Eberbach, W. and Burchardt, B.,** Die Vinyloxiran-Dihydrofuran-Isomerisierung: Mechanismus und präparative Anwendung, *Chem. Ber.*, 111, 3665, 1978.
2. **Etournaud, A. and Wyler, H.,** Synthèse du tétraéthoxycarbonyl-1,1,6,6-cyclodécadiyne-3,8, *Helv. Chim. Acta*, 56, 625, 1973.

Table TLC-37
ORGANIC ACIDS

Layer	L1	L2	L3	L4	L4	L4
Solvent	S1	S2	S3	S4	S5	S6
Technique		T1	T1	T1	T1	T1
Detection						
Reference	1	1	1	1	1	1
Compound	Rf × 100					
DL-*trans*-3-Hydroxy-5-oxo-2,3-tetrahydrofuran dicarboxylate (DL-Hibiscus acid lactone)	**46**	—	—	—	—	—
DL-*trans*-3-Hydroxy-5-oxo-2,3-tetrahydrofuran dicarboxylic acid	25	—	—	—	—	—
2,4-Dinitophenylhydrazine	—	79	—	98	96	—
Methylglyoxal-2,4-dinitrophenylhydrazone	—	13	—	—	—	—
3-Hydroxyquinaldine	—	—	71&70	—	—	—
DL-Isocitrate + DPN^+ + enzyme	—	—	—	57	53	27
2-Oxo-glutaric acid	—	—	—	59	53	28
DL-Homoisocitrate + DPN^+ + enzyme	—	—	—	70	61	45
2-Oxo-adipic acid	—	—	—	68	61	45
Pyruvic acid	—	—	—	34	—	66

Layer L1 = cellulose MN 300G
L2 = Mallinckrodt Chrom AR1000
L3 = Eastman Chromagram cellulose 6065
L4 = Eastman Chromagram 6060 F
Solvent S1 = isopropyl ether-88% formic acid (3:2)
S2 = 0.6 *N* NaOH
S3 = *n*-butanol-ethanol-0.5 *M* NH_4OH (7:1:2)
S4 = benzene-tetrahydrofuran-acetic acid (60:36:4)
S5 = 90% phenol-water-88% formic acid (83:17:1)
S6 = isopropyl ether-88% formic acid-water (7:2:1)
Technique T1 = analyzed as 2,4-dinitrophenylhydrazone

REFERENCE

1. **Plaut, G. W. E., Beach, R. L., and Aogaichi, T.,** Substrate activity of structural analogs of isocitrate dehydrogenases from bovine heart, *Biochemistry,* 14, 2581, 1875.

Table TLC-38
LICHEN PRODUCTS

Layer	L1	L1	L1
Solvent	S1	S2	S3
Detection	D1	D1	D1
Reference	1	1	1
Compound		Rf × 100	
Rhodocladonic acid	0	0	0
Consalazinic acid	2	1	0
Funarprotocetraric acid	1	25	7
Protocetraric acid	3	19	4
Sublimbatic acid	4	23	10
Thamnolic acid	3	25	13
Hypothamnolic acid	4	25	18
Caperatic acid	4	27	6
Constictic acid	7	3	3
Erythrin	6	5	1

Table TLC-38 (continued)
LICHEN PRODUCTS

Compound	Rf × 100		
Porphyrilic acid	9	11	6
Salazinic acid	10	7	4
Connorstictic acid	11	11	2
Variolaric acid	18	13	8
Decarboxythamnolic acid	5	25	18
Squamatic acid	10	25	25
Pannaric acid	8	28	9
Endocrocin	8	33	8
Echinocarpic acid	11	33	11
Siphulin	14	27	4
Pulvinic acid	4	36	7
Barbatolic acid	7	48	19
Conpsoromic acid	15	36	12
Physodalic acid	10	31	19
Haemathamnolic acid	15	35	27
Diploschistesic acid	21	39	13
Haemoventonic acid	33	2	17
Lepraric acid	27	11	21
Galbinic acid	29	12	16
Stictic acid	32	9	18
Arthoniaic acid	31	15	13
Alectoronic acid	33	34	17
Physodic acid	25	35	18
Gyrophoric acid	24	42	24
Hypoprotocetraric acid	25	43	22
Lecamoric acid	28	44	22
Olinetoric acid	34	37	22
Norobtusatic acid	27	50	30
Lividic acid	32	35	31
Rangiformic acid	29	38	33
Lobaric acid	30	46	38
Picrolichenic acid	38	39	36
Protolichesterinic acid	35	47	37
Psoromic acid	36	41	41
4-*O*-methylphysodic acid	37	43	40
Aletorialic acid	39	48	30
4-*O*-methylgyrophoric acid	32	56	46
Virensic acid	26	57	38
Obtusatic acid	35	64	48
Strepsilin	41	21	24
Baeomycesic acid	39	40	42
Anziaic acid	40	59	33
Evernic acid	38	61	43
Grayanic acid	38	62	44
Polyporic acid	43	28	27
Norstictic acid	42	29	77
4-*O*-Demethylplanaic acid	42	32	27
Paludosic acid	29	35	33
Glomelliferic acid	41	43	42
2′-Collatolic acid	40	32	35
Miriquidic acid	41	48	45
4-*O*-demethylbarbatic acid	77	79	78
Roccellic acid	42	60	48
Colensoinic acid	40	73	42
Sekikaic acid	45	57	51
Diffractaic acid	44	64	51

Table TLC-38 (continued)
LICHEN PRODUCTS

Compound	Rf × 100		
Barbatic acid	44	69	52
Divaricatic acid	39	75	51
Didymic acid	44	77	52
Schizopeltic acid	42	27	51
Rhodophyscin	44	32	23
Microphyllinic acid	43	35	42
Homosekikaic acid	47	69	56
Imbricaric acid	43	75	48
Stenosporic acid	45	73	54
Perlatolic acid	44	74	52
Entothein	47	16	25
Norlobaridone	50	36	21
2′-*O*-methylangiaic acid	46	40	34
Planaic acid	52	38	50
Confluentic acid	54	32	48
Zeorin	55	42	43
Cryptochloropbaeic acid	46	46	37
2′-*O*-Methylperlatolic acid	55	50	52
Lichesterinic acid	44	63	43
Thiophanic acid	55	56	52
Merochlorophaeic acid	55	56	61
Ergosterol	56	66	49
Emodin	57	61	39
2-*O*-Methylperlatolic acid	48	60	52
Ursolic acid	59	50	52
Spherophorin	48	76	54
Loxodin	58	40	35
Pseudoplacodiodic acid	60	52	56
Acetylportentol	65	29	45
Methylgyrophorate	60	44	43
Gangaleoidin	64	43	54
Rhizocarpic acid	73	41	66
Ethylorsellinate	62	66	44
4-*O*-Methylcryptochlorophaeic acid	60	61	66
Placodiolic acid	67	63	67
Thiophaninic acid	65	68	65
Scrobiculim	69	55	52
Methyl 3,5-dichlorolecanorate	67	62	66
Usnic acid	70	70	71
Diplocin	64	79	66
Nephroarctin	69	53	66
Epanorin	74	52	66
Vivanicin	73	75	70
Fragilin	73	82	72
Pinastric acid	72	57	76
Pannarin	73	68	76
Leprapinic acid	74	59	81
Lichexanthone	74	72	78
Tenuiorin	77	58	76
Parietin	74	72	77
Vulpinic acid	76	66	82
Atranorin	75	27	78
Chloroatranorin	76	79	80
Calycin	79	81	88
Pulvinin dilactone	77	82	89
Methylhypoprotocetrarate	57	48	41

Table TLC-38 (continued)
LICHEN PRODUCTS

Compound	Rf × 100		
Methyllecanorate	61	49	42
Methylschizopeltate	68	33	56
Methyl 2,2′-*O*-di-*O*-methylevernate	71	33	56
Methylphysodate	60	38	38
Metbyllividate	62	33	48
Dimethylcapratate	71	40	49
Methyl 4-*O*-demethylbarbatate	68	65	53
Methylnorobtusatate	65	57	52
Methyl 2′,4-di-*O*-methylhypoprotocetrarate	73	47	64
Methyl 4-*O*-methylphysodate	72	47	64
Methyl 2,2′,4-tri-*O*-methyldivaricatate	76	43	64
Methyldiffractaiate	76	70	77
Methylsekikaiate	76	65	76
Methylevernate	73	72	76
Methyl 2-*O*-methylperlatolate	75	74	76
Methyldivaricatate	77	77	80
Methylimbricarate	77	80	84
Methylplanaiate	78	50	68
Methyl 2′-*O*-methylperlatolate	76	74	75
Methylbarbatate	78	80	84
Methylstenosporate	77	80	83
Methylsphaerophorate	78	81	85
Methylperlatolate	79	82	85

Layer L1 = silica gel F254 (Merck)
Solvent S1 = benzene-dioxane-acetic acid (180:45:5)
S2 = hexane-diethyl ether-formic acid (130:80:20)
S3 = toluene-acetic acid (200:30)
Detection D1 = UV, H_2SO_4

From Culberson, C. F., *J. Chromatogr.*, 72, 113, 1972. With permission.

Table TLC-39
PHENOLIC COMPOUNDS

Layer	L1	L2	L1	L3	L1	L4
Solvent	S1	S2	S3	S4	S5	S6
Detection	D1	D1	D1	D1	D1	D1
Reference	1	1	1	1	1	1
Compound	Rf × 100					
Ubiquinone-9	31	24	39	51	39	55
Ubiquinone-10	30	14	42	53	42	57
Diacetyldihydro Q-9	31	48	—	—	—	—
Diacetyldihydro Q-10	30	30	—	—	—	—
2-Nonaprenylphenol	50	29	62	76	51	44
6-Methoxy-2-nonaprenylphenol	69	24	81	93	76	36
6-Methoxy-2-phytylprenylphenol	85	89	—	—	—	—
6-Methoxy-2-allylphenol	49	100	—	—	—	—
5-Demethoxy-ubiquinone-9	15	30	24	35	19	53
4-Hydroxybenzoic acid	0	100	0	0	0	0

Table TLC-39 (continued)
PHENOLIC COMPOUNDS

Layer	L1 = silica gel
	L2 = ChromAr® sheet impregnated with light parafin
	L3 = silicAr TLC-7G
	L4 = alumina H
Solvent	S1 = benzene-hexane (50:50)
	S2 = acetone-water (90:10)
	S3 = cyclohexane-diethyl ether (80:20)
	S4 = hexane-diethyl ether-acetic acid (80:20:1)
	S5 = petroleum ether (b.p. 30-60)-diethyl ether (85:15)
	S6 = petroleum ether-diethyl ether (80:20)
Detection	D1 = I_2, Emmerie-Engle reagent, Fast blue salt B, Pauly's reagent, Gibb's reagent

Table TLC-40
LICHEN PRODUCTS

Layer	L1	L1
Solvent	S1	S2
Technique	T1	T1
Detection	D1	D1
Reference	1	1
Compound	Rf × 100[a]	
Anzizic acid	39	72
Divaricatic acid	66	76
Glomellic acid	36	24
Glomelliferic acid	56	83
Loxodellic acid	50	33
4-*O*-Methylolivetoric acid	61	64
Olivetoric acid	23	46
Oxostenosporic acid	53	57
Perlatolic acid	70	83
Stenosporic acid	68	81

[a] Measured on chromatogram.

Layer	L1 = silica gel 60 F254 (Merck)
Solvent	S1 = toluene-acetic acid (2000:30)
	S2 = hexane-diethyl ether-formic acid (120:90:20)
Technique	T1 = two dimensional TLC
Detection	D1 = UV 254 nm, H_2SO_4

REFERENCE

1. **Culberson, C. F. and Johnson, A.,** A standardized two-dimensional thin-layer chromatographic method for lichen products, *J. Chromatogr.,* 128, 253, 1976.

Paper Chromatography Tables

Section I.IV.

PAPER CHROMATOGRAPHY TABLES

Table PC-1
PHENOLS

Paper	P1		P1			P1		P1				P1		P2	P3	P4	P5	P6	P6	P6
Solvent	S1		S2			S2		S1				S3		S4	S5	S4	S6	S7	S8	S9
Technique	T1		T1			T1		T1				T1		T2	T3	T3	T3	T3	T3	T3
Detection	C	C	C	C	C	C		C		C	C	C		CD1						
Reference	1		1			1		1				1		2	3	3	3	3	3	3
Compound	Rf[a] × 100																			
Phenol	1 vt	12 r	13 vt	—	32 r	17 vt	64 r	—	1 vt	—	17 r	72 vt	96 r	—	41	48	38	—	—	—
2-Methylphenol	3 vt	22 r	—	—	—	—	—	—	—	—	—	—	—	—	49	52	51	—	—	—
2-Ethylphenol	10 vt	33 r	—	—	—	—	—	—	—	—	—	—	—	—	—	—	—	—	—	—
2-Isopropylphenol	20 vt	69 r	—	—	—	—	—	—	—	—	—	—	—	—	—	—	—	—	—	—
2-*sec.*-Butylphenol	24 vt	75 r	—	—	—	—	—	—	—	—	—	—	—	—	76	—	—	50	—	—
3-Methylphenol	—	—	13 vt	—	37 r	—	—	—	—	—	—	—	—	—	53	59	54	—	—	—
3-Ethylphenol	—	—	22 vt	—	48 r	—	—	—	—	—	—	—	—	—	—	—	—	—	—	—
3-*tert.*-Butylphenol	—	—	23 vt	—	50 r	—	—	—	—	—	—	—	—	—	—	—	—	—	—	—
3,5-Dimethylphenol	—	—	42 vt	53 o	—	—	—	—	—	—	—	—	—	—	—	—	—	—	—	—
3-Ethyl-5-methylphenol	—	—	51 vt	67 o	—	—	—	—	—	—	—	—	—	—	—	—	—	—	—	—
3-Methyl-5-isopropylphenol	—	—	61 vt	76 o	—	—	—	—	—	—	—	—	—	—	—	—	—	—	—	—
4-Methylphenol	—	—	—	—	—	30 vt	—	—	—	—	—	—	—	34	—	—	—	—	—	—
4-Ethylphenol	—	—	—	—	—	42 vt	—	—	—	—	—	—	—	45	—	—	—	—	—	—
4-Isopropylphenol	—	—	—	—	—	53 vt	—	—	—	—	—	—	—	54	—	—	—	—	—	—
4-*sec*-Butylphenol	—	—	—	—	—	62 vt	—	—	—	—	—	—	—	64	—	—	—	—	—	—
4-*tert.*-Butylphenol	—	—	—	—	—	61 vt	—	—	—	—	—	—	—	58	—	—	—	—	—	—
4-*tert.*-Amylphenol	—	—	—	—	—	65 vt	—	—	—	—	—	—	—	—	—	—	—	—	—	—
4-*tert.*-Octylphenol	—	—	—	—	—	69 vt	—	—	—	—	—	—	—	—	—	—	—	19	55	—
4-Nonylphenol	—	—	—	—	—	72 vt	—	—	—	—	—	—	—	—	—	—	—	13	52	39
2,6-Dimethylphenol	—	—	—	—	—	—	—	—	—	—	63 r	—	—	—	—	—	—	—	—	—
2-*tert.*-Butyl-6-methylphenol	—	—	—	—	—	—	—	—	—	87 o	—	—	—	—	—	—	—	—	—	—

Table PC-1 (continued)
PHENOLS

Compound	Rf[a] × 100																			
2-Methyl-6-*tert*.-octylphenol	—	—	—	—	—	—	—	—	—	90 o	—	—	—	—	—	—	—	—	—	—
2,6-Diethylphenol	—	—	—	—	—	—	—	—	—	—	79 r	—	—	—	—	—	—	—	—	—
2-*tert*.-Butyl-6-ethylphenol	—	—	—	—	—	—	—	—	—	88 o	—	—	—	—	—	—	—	—	—	—
2,6-Di-*tert*.-butylphenol	—	—	—	—	—	—	—	91 y	—	—	—	—	—	—	—	—	—	—	—	—
2,4-Dimethylphenol	—	—	—	—	—	—	—	—	8 vt	—	—	82 vt	—	43	—	—	—	—	—	—
4-Ethyl-2-methylphenol	—	—	—	—	—	—	—	—	11 vt	—	—	—	—	—	—	—	—	—	—	—
4-*tert*.-Butyl-2-methylphenol	—	—	—	—	—	—	—	—	17 vt	—	—	—	—	76	73	66	85	49	—	68
2-Methyl-4-*tert*.-octylphenol	—	—	—	—	—	—	—	—	22 vt	—	—	—	—	—	—	—	—	—	—	—
4-*tert*.-Butyl-2-ethylphenol	—	—	—	—	—	—	—	—	20 vt	—	—	—	—	81	—	—	—	—	—	—
2-*tert*.-Butyl-4-methylphenol	—	—	—	—	—	—	—	—	30 vt	—	—	—	—	76	—	—	—	46	—	—
2,4-Di-*tert*.-butylphenol	—	—	—	—	—	—	—	—	40 vt	—	—	—	—	95	—	—	—	—	—	—
4-Ethyl-3-methylphenol	—	—	—	—	—	—	—	—	—	—	—	90 vt	—	—	—	—	—	—	—	—
4(1,1-Dimethylpropyl)phenol	—	—	—	—	—	—	—	—	—	—	—	—	—	69	—	—	—	—	—	—
4-Benzylphenol	—	—	—	—	—	—	—	—	—	—	—	—	—	25	—	—	—	—	—	—
4-Cyclohexylphenol	—	—	—	—	—	—	—	—	—	—	—	—	—	69	—	—	—	—	—	—
4(1,1,3,3-Tetramethylbutyl)phenol		—	—	—	—	—	—	—	—	—	—	—	—	85	—	—	—	—	—	—
3,4-Dimethylphenol	—	—	—	—	—	—	—	—	—	—	—	—	—	39&49	—	—	—	—	—	—
2,3,4-Trimethylphenol	—	—	—	—	—	—	—	—	—	—	—	—	—	49	—	—	—	—	—	—
2-Bromo-4,6-dimethylphenol	—	—	—	—	—	—	—	—	—	—	—	—	—	43	—	—	—	—	—	—
2-Chloro-4,5-dimethylphenol	—	—	—	—	—	—	—	—	—	—	—	—	—	40&60	—	—	—	—	—	—
2,3-Dimethylphenol	—	—	—	—	—	—	—	—	—	—	—	—	—	—	56	58	70	—	—	—
2-Nitrophenol	—	—	—	—	—	—	—	—	—	—	—	—	—	—	19	11	—	—	—	—
3-Nitrophenol	—	—	—	—	—	—	—	—	—	—	—	—	—	—	23	—	16	—	—	—
4-Methoxyphenol	—	—	—	—	—	—	—	—	—	—	—	—	—	—	25	22	18	—	—	—
2-Chlorophenol	—	—	—	—	—	—	—	—	—	—	—	—	—	—	40	—	30	—	—	—
1-Hydroxynaphthalene	—	—	—	—	—	—	—	—	—	—	—	—	—	—	45	47	—	—	—	—
Octylphenol	—	—	—	—	—	—	—	—	—	—	—	—	—	—	—	—	—	20	60	51
Dodocylphenol	—	—	—	—	—	—	—	—	—	—	—	—	—	—	—	—	—	8	25	20
Cetylphenol	—	—	—	—	—	—	—	—	—	—	—	—	—	—	—	63	—	—	8	3
2-Methyl-6-octylphenol	—	—	—	—	—	—	—	—	—	—	—	—	—	—	62	—	82	17	48	—
2,4-Di-*tert*.-butyl-6-methylphenol		—	—	—	—	—	—	—	—	—	—	—	—	—	—	—	61	—	—	—

[a] Measured on chromatogram.

Paper	P1 = Whatman No. 2 paper impregnated with 20% ethanolic formamide solution
	P2 = Whatman No. 3 paper impreganted with 50% formamide solution
	P3 = Whatman No. 1 paper impregnated with 45% dimethylformamide solution
	P4 = Whatman No. 1 paper impregnated with 30% dimethylformamide solution
	P5 = Whatman No. 1 paper impregnated with 40% dimethylformamide solution
	P6 = Whatman No. 1 paper impregnated with 10% paraffin oil solution
Solvent	S1 = *n*-heptane-benzene (10:1)
	S2 = *n*-heptane-benzene (1:1)
	S3 = benzene
	S4 = *n*-hexane
	S5 = cyclohexane-benzene (5:1)
	S6 = hexane-xylene (10:1)
	S7 = dimethylformamide-methanol-water (8:3:2)
	S8 = dimethylformamide-methanol-water (8:4:1)
	S9 = dimethylformamide-butanol-water (80:5:15)
Technique	T1 = analyzed as 3-methyl-benzthiazolinone-2-hydrazone derivatives
	T2 = analyzed as antipyrine dyes
	T3 = analyzed as 4-(*N,N*- dimethylamino)benzene-4′-azobenzoate
Detection	D1 = diazonized 2,4-dinitroaniline or iron (III) hexacyanoferrate (III)

REFERENCES

1. **Gasparič, J., Svobodová, D., and Pospišilová, M.,** Investigation of the colour reaction of phenols with the MBTH reagent, Identification of organic compounds. LXXXVI, *Microchim. Acta*, 1, 241, 1977.
2. **Gasparič, J., Svobodová, D., and Matysová, A.,** A contribution to the chromatography of phenols as antipyrine dyes, Identification of organic compounds. LXXXI, *J. Chromatogr.*, 88, 364, 1974.
3. **Churáček, J. Pechová, H., and Mareš, V.,** The separation and identification of phenols in the form of p-(N,N-dimethylamino)benzene-p′-azobenzoates by means of paper and thin-layer chromatography, *J. Chromatogr.*, 67, 97, 1972.

Table PC-2
PHENOLIC COMPOUNDS

Paper	P1			ns		ns		
Solvent	S1			S2		S3		
Detection		C1	C2	D1	M.O.P.(V)	D1	M.O.P.(V)	
Reference	1			2		2		
Compound	Rf × 100							
Phenol	0	pk	d-pk	81	0.32	50	0.75	
Quinol	37	nc	g	—	—	—	—	
2-Methylphenol	31	nc	nc	70	0.33	35	0.27[a]	0.84[b]
3-Methylphenol	0	nc	nc	53	0.23	73	0.34[a]	0.98[b]
4-Methylphenol	74	nc	lt-y	—	—	—	—	
1,2-Dihydroxybenzene	65	bf	d-g	—	—	—	—	
1,3-Dihydroxybenzene	77	nc	g	80	0.30	79	0.65	
1,4-Dihydroxybenzene	0	nc	g	73	0.24	52	0.23[a]	0.67[b]
1,2,3-Trihydroxybenzene	79	d-bn	d-bk	58	0.21	60	0.73	
1,3,5-Trihydroxybenzene	—	—	—	84	0.30	81	0.35	
1-Hydroxynaphthalene	45	nc	g	91	0.23	90	0.91	
2-Hydroxynaphthalene	42	nc	bf	85	0.18	83	0.85	
2-Hydroxybenzoic acid	60	lt-y	lt-y	—	—	—	—	
2,4,6-Trinitrophenol	70	pl-y	y	—	—	—	—	
8-Hydroxyquinoline	12	o-y	y	—	—	—	—	
1-Hydroxy-2-nitronaphthalene	33	lt-bn	r-bn	—	—	—	—	
Cyclohexanol-hexahydrophenol	0	nc	nc	—	—	—	—	
4-Hydroxy-3-methoxybenzaldehyde	66	lt-y	lt-y	—	—	—	—	
2-Hydroxy-1-nitrosonaphthalene	30	y	bn	—	—	—	—	
1-Pyridyl-2-azo-naphthol	0	o	o	—	—	—	—	
4′-Nitrophenyl-3,5-dihydroxy-azobenzene	2	y	o	—	—	—	—	
1,2,5,8-Tetrahydroxyanthraquinone	2	pk	vt	—	—	—	—	
Catechin	80	lt-bn-y	bk-gn	—	—	—	—	
Bromocresol blue	35	lt-y	lt-y	—	—	—	—	
Bromothymol blue	15	y-r	d-bn	—	—	—	—	
Phenylfluorone	0	we-r	we-r	—	—	—	—	

[a] First maximum oxidation potential.
[b] Second maximum oxidation potential.

Paper	P1 = Whatman No. 3 paper impregnated with stannic molybdate
	ns = not specified
Solvent	S1 = 1.0 *M* sodium nitrate (pH 6.0)
	S2 = methanol-0.1 *N* lithium chloride solution (80:20)
	S3 = methanol-0.4 *N* lithium chloride solution (80:20)
Detection	D1 = Multi-linear Sweeping Applied Voltage Detector
	M.O.P. = Maximum oxidation potential (volt)
Color	C1 = immediately
	C2 = after 15 min

REFERENCES

1. **Rawat, J. P., Mujitaba, S. Q., and Thind, P. S.,** Chromatographic separation and identification of phenols on paper impregnated with stannic molybdate, *Z. Anal. Chem.,* 279, 368, 1976.
2. **Mori, I. and Chikui, S.,** The quantitative determination of phenols by an anodic oxidation wave using a mercury pool electrode during chromatography, *J. Chromatogr.,* 64, 390, 1972.

Table PC-3
PHENOLIC COMPOUNDS

Paper	P1	P1	P1	P1	P1	P1	P1	P1	P1	P1	P1	P1	P1	P1	P1	P1	P1
Solvent	S1	S2	S3	S4	S5	S6	S7	S8	S9	S10	S11	S12	S13	S14	S15	S16	S17
Reference	1	1	1	1	1	1	1	1	1	1	1	1	1	1	1	1	1
Compound	Rf × 100																
2,4-Dinitrophenol	8	16	24	30	32	34	10	28	56	34	32	30	26	28	40	8	13
1-Hydroxy-2,4-dinitronaphthalene	12	22	26	35	40	44	28	66	80	44	38	26	22	25	62	8	16
2-Aminopyridine	8	17	20	27	29	32	6	10	26	32	28	24	20	22	26	8	14
2-Aminophenol	12	20	22	24	25	26	75	86	90	26	45	60	72	80	90	46	62

Paper	P1	P1	P1	P1	P1
Solvent	S18	S19	S20	S21	S22
Reference	1	1	1	1	1
Compound	Rf × 100				
2,4-Dinitrophenol	17	34	30	24	20
1-Hydroxy-2,4-dinitronaphthalene	20	44	34	24	20
2-Aminopyridine	17	32	26	20	16
2-Aminophenol	72	26	40	50	62

Paper P1 = Whatman No. 4 paper impregnated with McIlvaine's buffer (pH 6.5)

Solvent S1—S6 are mixtures of cyclohexane-chloroform, S1 = 50:50, S2 = 30:70, S3 = 20:80, S4 = 10:90, S5 = 5:95, S6 = 0:100
S7—S9 are mixtures of cyclohexane-tributylphosphate, S7 = 90:10, S8 = 80:20, S9 = 70:30
S10—S15 are mixtures of chloroform-tributylphosphate, S10 = 100:0, S11 = 95:5, S12 = 90:10, S13 = 80:20, S14 = 70:30, S15 = 50:50
S16—S18 are mixtures of cyclohexane-diisobutylketone, S16 = 50:50, S17 = 30:70, S18 = 10:90
S19—S20 are mixtures of chloroform-diisobutylketone, S19 = 100:0, S20 = 90:10, S21 = 70:30, S22 = 30:70

REFERENCE

1. **Matysik, G. and Soczewinski, E.**, Investigations of the relationships between molecular structure and chromatographic parameters. XI. Extraction of solutes containing two or three functional groups using mixtures of chloroform and electron-donner solvents, *J. Chromatogr.*, 160, 29, 1978.

Table PC-4
ORGANIC ACIDS

Paper	P1	P1	P2	P2	P2	P2	P2					P3	P3	P3	P4	P5
Solvent	S1	S2	S3	S4	S5	S6	S7					S8	S9	S10	S11	S12
Technique												T1	T1	T1		T1
Detection	D1	D1						D2	D3	D4	D5				D6	
Reference	1	1	2	2	2	2	2	2	2	2	2	3	3	3	4	5
Compound	Rf × 100															
Maleamic acid	37	53	—	—	—	—	—					—	—	—	—	—
Malic acid	16	57	46	30	20	45	19		1t-y		w	—	—	—	33	—
Fumaramic acid	34	nt	—	—	—	—	—					—	—	—	—	—
Fumaric acid	20	91	78	81	28	60	69	1t-y		y		—	—	—	85	—
Citric aid	—	—	37	23	12	38	11	o	we-r		w	—	—	—	24	—
Lactic acid	—	—	67	55	47	69	50				w	—	—	—	—	—
Succinic acid	—	—	66	63	29	72	49			y		—	—	—	52	—
Glutarate semialdehyde	—	—	—	—	—	—	—					59	54	92	—	—
Succinate semialdehyde	—	—	—	—	—	—	—					52	nd	60	—	—
												52	45	51		
Pyruvate	—	—	—	—	—	—	—					70	70	73	—	—
												10	nd	nd		
3-Methylgarcina lactone	—	—	—	—	—	—	—					—	—	—	—	86
Garcina lactone	—	—	—	—	—	—	—					—	—	—	—	77
DL-4-Isocitricamide	—	—	—	—	—	—	—					—	—	—	—	36
Isocitric lactone	—	—	—	—	—	—	—					—	—	—	—	62
Isocitric acid	—	—	—	—	—	—	—					—	—	—	—	50

Paper
P1 = Whatman No. 1
P2 = Schleicher & Schüll 20
P3 = Whatman No. 1
P4 = Schleicher & Schüll 2043b
P5 = Whatman 3MM

Solvent
S1 = ethanol-conc. NH_4OH-water (20:1:4)
S2 = (diethyl ether-benzene (7:3)) + 3M formic acid/water (organic phase is used)
S3 = *n*-butanol-85% formic acid-water (10:2:5) (organic phase is used)

	S4 = *n*-pentanol-5 *M* formic (1:1), (organic phase is used)
	S5 = 95% ethanol-(200 g/*l*) NH_4OH-water (8:1:1)
	S6 = (phenol-water (3:1))-85% formic acid (100:1)
	S7 = isopentanol-chloroform-85% formic acid (4:1:1), water saturated solution
	S8 = butanol-0.5 *M* NH_4OH-ethanol (7:2:1)
	S9 = isopropanol-conc. NH_4OH-water (20:1:2)
	S10 = butanol-water-ethanol (7:2:1)
	S11 = isopentanol-chloroform-formic acid (4:1:1)
	S12 = *n*-butanol-88% formic acid-water (4:1:2)
Technique	T1 = as analyzed as 2,4-dinitrophenylhydrazone
Detection	D1 = bromophenol blue-citric acid
	D2 = acetic acid-pyridine
	D3* = 3.0 g 4-dimethylaminobenzaldehyde in 7.0 g anhydrous acetic acid (freshly prepared)
	D4* = saturated ammoniummetavanadate aqueous solution
	D5* = color after 2 days (same reagent as D4)
	D6 = aniline-glucose

REFERENCES

1. **Cain, R. B., Houghton, Ch., and Wight, K. A.**, Microbial metabolism of the pyridine ring. Metabolism of 2- and 3-hydroxypyridine by the maleamate pathway in Acromobacter sp., *Biochem. J.*, 140, 293, 1974.
2. **Wollman, H.**, Das Auftreten nichtflüchtiger organischer Säuren in Kulturen von Trichophyton mentagrophytes (Robin) Blanchard, *Pharmazie*, 33, 1, 1978.
3. **Watson, G. K. and Cain, R. B.**, Microbial metabolism of the pyridine ring, *Biochem. J.*, 146, 157, 1975.
4. **Wollman, H.**, Die direktphotometrische Bestimmung organischer Säuren auf Papierchromatogrammen, 29. Beiträge zur Chemie and Physiologic einiger stoffwech selchemisch wichtiger Säuren, *Pharmazie*, 31, 292, 1976.
5. **Plaut, G. W. E., Beach, R. L., and Aogaichi, T.**, Substrate activity of structural analogs of isocitrate dehydrogenases from bovine heart, *Biochemistry*, 14, 2581, 1975.

* Pohloudek-Fabini, Y. R. and Wollmann, H., Über Verfahren zum Nachweis organischen Säuren auf Papier-chromatogrammen, *Pharmazie*, 16, 57, 1961.

Table PC-5
KETO ACIDS

Paper			
Solvent	S1	S2	S3
Technique	T1	T1	T1
Detection	D1	D1	D1
Reference	1	1	1
Compound		Rf × 100	
2-Oxopropanoic acid (pyruvic acid)	40	33	26
2-Oxobutanoic acid	44	37	36
Oxopropanedioic acid (malonic acid)	16	11	—
2-Oxo-5-carbamidopentanoic acid	54	50	—
2-Oxo-4-ethiolbutyric acid	65	60	—
2-Oxopentandioic acid (2-oxoglutaric acid)	40	22	30
2-Oxopentanedioic acid 4-ethylester	59	67	—
Glyoxylic acid	30	25	37
L (and D-)-2-Oxo-3-methylpentanoic acid	60	60	—
2-Oxoisohexanoic acid	58	58	42
Phenylpyruvic acid	55	65	—
Trimethylpyruvic acid	54	55	—
2-Oxoisopentanoic acid	55	54	40
L-4-Glutamyl hydrazide	53	62	—
1-(5-Glutamyl)-1-methylhydrazide	—	90	—
4-*N*-Glutamyl-4-*N*-methyl-*N*′-glyoxylic acid hydrazone	47	48	—
1,4,5,6-Tetrahydro-6-pyridazinone-3-carboxylic acid	53	65	58
3-Hydroxy-tetrahydro-6-pyridazinone-3-carboxylic acid	53	—	48

Solvent S1 = *tert.* butanol-methylethylketone-formic acid-water (40:30:15:15)
S2 = 80% phenol
S3 = *n*-butanol-acetic acid-water (50:25:25)
Technique T1 = analyzed as 2,4-glutamylhydrazones
Detection D1 = ninhydrin

From Cooper, A. J. and Meister, A., *J. Biol. Chem.,* 248, 8489, 1973. With permission.

Table PC-6
ALIPHATIC HYDROXY ACIDS

Paper	P1	P1	P1	P2	P2
Solvent	S1	S2	S3	S4	S5
Technique				T1	T1
Detection	D1	D1	D1	D2	D2
Reference	1	1	1	2	2
Compound			Rf × 100		
(4-^{14}C)keto-2,4-dihydroxy-3,4-dimethyl butanoic acid*	31	44	75	—	—
(4-^{14}C)keto-2,4-dihydroxy-3,4-dimethyl butanoic acid lactone*	82	88	75	—	—
(4-^{14}C)-2,4-dihydroxy-3,4-dimethylbutanoic acid**	71	76	77	—	—
(4-^{14}C)-2,4-dihydroxy-3,4-dimethyl butanoic acid lactone***	83	85	81	—	—
Hydroxyacetic acid	—	—	—	30	60
2-Hydroxypropanoic acid	—	—	—	43.5	72
3-Hydroxypropanoic acid	—	—	—	37	69
3-Hydroxybutanoic acid	—	—	—	48	84
2-Hydroxybutanoic acid	—	—	—	49.5	84
2-Hydroxy-2-methyl propanoic acid	—	—	—	45	90
2(1-Hydroxy-cyclohexyl)-acetic acid	—	—	—	63	94
4(1-Hydroxycyclohexyl)-2-butenoic acid	—	—	—	62	94
2(1-Hydroxycyclohexyl)-2-ethyl acetic acid	—	—	—	67.8	97.8
2(1-Hydroxy-3,3,5-trimethylcyclohexyl)acetic acid	—	—	—	69.3	97.8
Glyceric acid	—	—	—	27	42
2,3-Dihydroxybutanoic acid E	—	—	—	45	58.8
2,3-Dihydroxybutanoic acid F	—	—	—	43.5	57
2,3-Dihydroxy-2-methylbutanoic acid (2RS-3RS)	—	—	—	51	69
2,3-Dihydroxy-2-methylbutanoic acid (2RS-3SR)	—	—	—	48	72
2,3-Dihydroxy-4-methylbutanoic acid	—	—	—	49.5	75
4-Ethoxybutanoic acid	—	—	—	61.5	91.8
4-Benzyloxybutanoic acid	—	—	—	64.5	97.2
Acetoxyacetic acid	—	—	—	45	84
Chloro-acetoxyacetic acid	—	—	—	48	87
2(2-Hydroxy-acetoxy)acetic acid	—	—	—	34.5	63
4-Acetoxybutanoic acid	—	—	—	58.8	91.2
4-Propioxybutanoic acid	—	—	—	62.7	92.4
4-Butyroxybutanoic acid	—	—	—	65.4	96.6
4-Hexanoyloxybutanoic acid	—	—	—	69	98.4

Paper P1 = Whatman 3MM
P2 = Arches 302
Solvent S1 = *sec.*-butanol-propionic acid (95:5), saturated with water
S2 = *n*-butanol-water-propionic acid (10:7:5)
S3 = *n*-butanol-formic acid (95:5), saturated with water
S4 = 95% ethanol-pure-ammonia (95:5)
S5 = *n*-butanol-formic acid-water (4:1:5)
Technique T1 = two dimensional map
Detection D1 * = UV 253.7 nm/ 0.1% semicarbazide in 0.15% sodium acetate solution
D1 ** = 0.04% bromophenol blue in ethanol (pH 7)
D1 *** = iron (III)chloride soln.
D2 = Schweppe reagent for hydroxy acids or 0.1% dichloro-2,6phenol-indophenol in 95% ethanol

REFERENCES

1. **King, H. L., Jr., Dyar, R. E., and Wilken, D. R.,** Ketopantonyl lactone and ketopantonic acid reductases. Characterization of the reactions and purification of two forms of ketopantonyl lactone reductase, *J. Biol. Chem.*, 249, 4689, 1974.
2. **Layole, J., Lattes, A., Battie, B., Zamalik, H., and Carles, J.,** Relations entre la structure et le comportement chromatographique d'acides polyfonctionnels. Constantes de groupe et corrélations Rm-pKa, *J. Chromatogr.*, 76, 441, 1973.

Table PC-7
ORGANIC ACIDS

Paper	P1	P2	P2	P2	P2	P2	P1	P1	P1
Solvent	S1	S2	S3	S4	S5	S6	S7	S8	S9
Technique	T1								
Detection	D1	D2	D2	D2	D2	D2	D3	D3	D3
Reference	1	2	2	2	2	2	3	3	3
Compound	cm	Rf × 100							
4-Hydroxy-3-methoxy mandelic acid	19.5	—	—	—	—	—	—	—	—
3-Hydroxy-4-methoxy mandelic acid	22	—	—	—	—	—	—	—	—
3-Phospha-glycerate	—	40	—	60	—	0	—	—	—
Phospho-enol-pyruvate	—	50	—	70	—	0	—	—	—
Shikimic acid	—	55	40	—	30	0	—	—	—
Shikimic acid methylester	—	70	—	—	—	—	—	—	—
Fumaric acid	—	—	—	—	—	—	93	—	—
Succinic acid	—	—	—	—	—	—	77	—	—
3,4-Dihydroxycinnamic acid	—	—	—	—	—	—	87	93	94
Cinnamic acid	—	—	—	—	—	—	92	—	98
Chlorogenic acid	—	—	—	—	—	—	—	19	—
2-Hydroxybenzoic acid	—	—	—	—	—	—	—	93	—
3,4,5-Trihydroxybenzoic acid	—	—	—	—	—	—	—	—	85
Malic acid	—	—	—	—	—	26	—	—	—
Lactic acid	—	—	—	—	—	60	—	—	—

Paper
P1 = Whatman No. 1
P2 = Whatman 3MM

Solvent
S1 = benzene-acetic acid-water (2:2:1)
S2 = butanol-acetic acid-water (9:4:7)
S3 = butanol-acetic acid-water (100:6:25)
S4 = methanol-conc. ammonia-water (6:1:3)
S5 = propanol-conc. ammonia (7:3)
S6 = ethanol-conc. ammonia-water (80:4:17)
S7 = 85% formic acid-ethyl acetate-*n*-butanol (1:1:1)
S8 = 90% formic acid-water-ethyl acetate (10:5:2)
S9 = benzene-propionic acid-water (2:2:1)

Technique
T1 = 64 hr

Detection
D1 = diazotized 4-nitroaniline and diazotized sulfanilic acid mixture or Roses reagent (β−di-ethylamino-ethyl-4-aminophenyl sulfone)
D2 = potassium permanganate solution of ammonium molybdate solution or bromophenol blue solution or periodate aniline solution and short wavelength UV light
D3 = UV-light or 0.5% bromophenol blue-95% ethanol solution alkalized by 0.1 *N* NaOH solution (pH 9)

REFERENCES

1. **Thomas, H. and Ockenfels, H.,** Papierchromatographische Trennung von 4-Hydroxy-3-methoxy- und 3-Hydroxy-4-methoxy-mandelsäure, *J. Chromatogr.*, 67, 197, 1972.
2. **Floss, H. G., Onderka, D. K., and Carroll, M.,** Stereochemistry of the 3-deoxy-D-arabino-heptulosonate 7-phosphate synthetase reaction and the chorismate sysnthetase reaction, *J. Biol. Chem.*, 247, 736, 1972.
3. **Sovová, M., Nathervá, L., and Poliakova, M.,** Identification by paper chromatographic method of organic acids in crude drugs containing tannin, *Acta Fac. Pharm., Univ. Comeniana*, 24, 53, 1973.

Table PC-8
BENZENE CARBOXYLIC ACIDS

Paper	P1
Solvent	S1
Technique	
Detection	D1
Reference	1

Compound	Rf × 100
1,2-Benzenedicarboxylic acid	27
1,4-Benzenedicarboxylic acid	26
1,3-Benzenedicarboxylic acid	27
2-Hydroxybenzoic acid	72
4-Hydroxybenzoic acid	38
3,4-Dihydroxybenzoic acid	15
2-Carboxybenzenesulfonic acid	27
3-Carboxy-4-hydroxybenzenesulfonic acid	37
1,2,4-Benzenetricarboxylic acid	6
1,2,4,5-Benzenetetracarboxylic acid	1
5-Hydroxy-1,3-benzenedicarboxylic acid	5
2-Aminobenzoic acid	60
3-Aminobenzoic acid	47
4-Aminobenzoic acid	44
4-Amino-2-hydroxybenzoic acid	46
5-Amino-1,3-benzenedicarboxylic acid	11
3-Hydroxy-2-naphthalenecarboxylic acid	80
1,3-Benzenedicarboxylic acid dimethylester	27
1,4-Benzenedicarboxylic acid monomethylester	26
1,4-Benzenedicarboxylic acid dimethylester	26
1,3-Benzenedicarboxylic acid monomethylester	27
4-Carbamino-benzoic acid	26
4-Cyanobenzoic acid	68
3-Cyanobenzoic acid	66
3,5-Dicarboxybenzenesulfonic acid	15
5-Hydroxy-1,3-benzenedicarboxylic acid	5
3-Amino-2-hydroxybenzoic acid	48
2-Nitrobenzoic acid	61
3-Nitrobenzoic acid	69
4-Nitrobenzoic acid	70
3-Nitro-1,2-benzenedicarboxylic acid	27

Paper P1 = Whatman No. 3
Solvent S1 = *n*-propanol-NH_4OH (2:1)
Detection D1 = 2,6-dichlorophenyliodophenol saturated in water for acids 2,4-dinitrophenylhydrazine saturated in 2 *N*HCl for aldehydes

REFERENCES

1. **Franc, J. and Pospišilová, K.,** Reaction paper chomatography of carboxylic acids with the use of electrolytic reduction, *J. Chromatogr.,* 66, 329, 1972.

Table PC-9
CINNAMIC ACIDS

Paper	P1		P1
Solvent	S1		S1
Technique			
Detection	D1		D1
Reference	1		1
Compound	Rf × 100	**Compound**	Rf × 100
4-Nitrocinnamic acid	16	4-Chloro-2′-ethylcinnamic acid	82
3-Nitrocinnamic acid	17	2′-Ethylcinnamic acid	75
3-Chlorocinnamic acid	30	2′-Ethyl-4-isopropylcinnamic acid	92
4-Chlorocinnamic acid	51	2′-Ethyl-4-methylcinnamic acid	83
Cinnamic acid	22	2′-Ethyl-4-methoxycinnamic acid	80
4-Isopropylcinnamic acid	71	2′-Ethyl-4-dimethylaminocinnamic acid	83
4-Methoxycinnamic acid	27	4-Nitro-2′-propylcinnamic acid	62
4-Dimethylaminocinnamic acid	26	3-Nitro-2′-propylcinnamic acid	63
2′-Methyl-4-nitrocinnamic acid	43	3-Bromo-2′-propylcinnamic acid	90
2′-Methyl-3-nitrocinnamic acid	24	3-Chloro-2′-propylcinnamic acid	90
3-Chloro-2′-methylcinnamic acid	70	3-Iodo-2′-propylcinnamic acid	93
4-Chloro-2′-methylcinnamic acid	62	4-Bromo-2′-propylcinnamic acid	90
2′-Methylcinnamic acid	56	4-Chloro-2′-propylcinnamic acid	91
2′-Methyl-4-isopropylcinnamic acid	90	3-Methoxy-2′-propylcinnamic acid	87
2′,4-Dimethylcinnamic acid	78	2′-Propylcinnamic acid	86
4-Methoxy-2′-methylcinnamic acid	64	2′-Propyl-4-isopropylcinnamic acid	95
2′-Methyl-4-dimethylaminocinnamic acid	74	4-Methyl-2′-propylcinnamic acid	92
2′-Ethyl-4-nitrocinnamic acid	42	4-Methoxy-2′-propylcinnamic acid	83
2′-Ethyl-3-nitrocinnamic acid	46	4-Dimethylamino-2′-propylcinnamic acid	90
3-Chloro-2′-ethylcinnamic acid	84	4-Iodo-2′-propylcinnamic acid	93

Paper P1 = Whatman No. 1 impregnated with 5% ammonium formate in formamide
Solvent S1 = benzene
Detection D1 = UV light

REFERENCE

1. **Kuchař, J., Grimová, J., Roubal, Z., Němeček, O., and Kakáč, B.**, Příprava a biologické vlastnosti α-alkylskořicových kyselin, *Česk. Farm.*, 22, 388, 1973.

Section II

Practical Applications

Section II.I.

SAMPLE PREPARATION METHODS FOR ANALYSIS OF PHENOLS BY GAS CHROMATOGRAPHY

Direct Acetylation of Phenols in Water for GC Analysis

A 250 mℓ water sample is added to 10 g of $NaHCO_3$ (this amount has to be excess). After adding 500 μℓ of acetic anhydride to the above mixture, esters are extracted twice with 10 mℓ of CH_2Cl_2. The extracts are combined and concentrated to 20 μℓ and are then injected into a gas chromatograph. Reference: Table GC 3, Reference 4.

Collection of Phenols from Tobacco Smoke

The smoke of 20 weight selected cigarettes or little cigars is led into cold trap (dry ice-acetone), and the remaining particulates are trapped on a Cambridge filter (CM-113A; 92 mm i.d.). The trapped sample is dissolved in peroxide-free ether. The ether is partitioned against 0.1 *N* HCl; the aqueous layer is cross-washed with ether. The pooled ether layers are washed with 0.1% $NaHCO_3$ and then concentrated to 1/3 of their volume. Preequilibrated boric acid (1 *N*) at pH 7 is vigorously shaken with the ether layer. The aqueous layer is then acidified with HCl under cooling to pH 1 and again extracted with ether. The pooled ether layers are washed with a small amount of water to remove excess acid and then dried over Na_2SO_4 and concentrated in vacuo to 1.0 mℓ. Aliquots are injected into a gas chromatograph. Reference: Table GC 8, Reference 2.

Extraction of Total Urinary Phenols

The specific gravity of the urine samples is determined by refractometry, and those of less than 1.010 or more than 1.030 are excluded from analysis. Phenol standards ranging from 0.5 mg% to 10 mg% are prepared in pooled urine. Volumes of 5 mℓ of standard urine blank, standards, or samples are pipetted into 15 mℓ glass-stopped centrifuge tube to which is added 1.0 mℓ of pH 5.0, 2 *M* acetate buffer that contains 1 mg of benzyl alcohol as internal standard and 0.1 mg of amoxycillin as preservative. After thorough mixing, 50 μℓ of Glusulase (Endo Tabs: ca 1500 β-glucuronidase units and 400 arylsulfatase units per mℓ of urine) are added. The tubes are stoppered, thoroughly mixed, and incubated in a water bath at 37°C for 24 hr. After cooling, 0.5 mℓ of concentrated HCl and 1.0 mℓ of diisopropyl ether are added. The stoppered tubes are slowly shaken on a reciprocating mechanical shaker for 20 min. Centrifugation separates the organic and aqueous phases. The diisopropyl ether layer is injected into a gas chromatograph. Reference: Spiegel, H.E. and Opar, G. E., An improved method for determining total urinary phenols, *J. Chromatogr.*, 155, 159, 1978.

Extraction and Derivatization of Chlorophenols in Spent Bleach Liquors

About 1 ℓ of spent liquor is acidified to pH 2 with HCl and continuously extracted with 250 mℓ of diethyl ether for 48 hr. The ether extract is then shaken with two 10 mℓ portions of 0.5 *M* $NaHCO_3$. After separation, the aqueous phase is acidified with 20 mℓ of 1 *N* HCl and extracted with two 10 mℓ volumes of pentane-diethyl ether (1:2). The combined organic phases are transfered to a 50 mℓ centrifuge tube with a screw-cap and PTFE packing, and dried with anhydrous Na_2SO_4 (ca. 5 g). After centrifugation, the sodium sulfate is rinsed with about 5 mℓ of diethyl ether, which is likewise transferred to the test tube by rinsing with a small volume of the same solvent (not more than 0.5 mℓ). After making up to the mark, the solution is transferred to 1 2 mℓ test tube with a screw-cap and alumina foil packing, and then stored in a refrigerator (4 °C) for further investigations.

After separation of the above ether extract from the $NaHCO_3$ phase, the extract is washed with 10 mℓ of ether-saturated water and then shaken with two 10 mℓ portions of 0.5 *N* NaOH solution. The aqueous phase is then acidified with 20 mℓ of 1 *N* HCl, and the procedure is repeated exactly as described above after acidification of the aqueous phase. The phenols are finally dissolved in isooctane-ethanol (9:1). For quantitative analysis, 25 μℓ of the isooctane-ethanol extract are subjected to purification by liquid chromatography. The chromatographic conditions used are: column, 3.2 × 150 mm, Micropak CN, 5 μm; flow rate, 2 mℓ/min; eluent, isooctane-ethanol (9:1); detector, UV flow monitor (280 nm). An elution volume of 2 mℓ is collected.

The isooctane extract is treated with diazomethane generated in a modified apparatus according to the method of Schlenck and Gellerman. When a pale yellow color persists (about 30 sec), the reaction vessel is closed and allowed to stand overnight. The diazomethane reagent is a mixture of 1 mℓ of *N*-nitroso-*N*-ethylurea, 5 mℓ of diethyl ether, 3 mℓ of 10 *N* KOH and 10 mℓ of diethylene glycol monoethyl ether, freshly prepared prior to each derivatization. Reference: Lindstom, K. and Nordin, J., Gas chromatography-mass spectrometry of chlorophenols in spent bleach liquors, *J. Chromatogr.*, 128, 13, 1976.

Extraction and Derivatization of Phenols in Water

A 25 mℓ water sample is acidified to pH 1 with conc. HCl, and 25 mℓ of benzene is added. The mixture is mechanically shaken for 15 min and allowed to stand until the layers separate. A 2 mℓ portion of the benzene extract is dried by passing through a 5 cm × 5 mm glass column packed with anhydrous Na_2SO_4 which has been heated to 600°C for 16 hr. One mℓ of column effluent is collected in a 4 mℓ glass vial and 5 μℓ of heptafluorobutylimidazole reagent are added. The vial is capped and heated at 65°C for 15 min. After cooling, excess of reagent is hydrolyzed by adding 20μℓ of 0.01 *N* HCl and shaking vigorously for 1 min. Prior to gas chromatographic analysis, ca. 250 mg of anhydrous Na_2SO_4 is added to remove excess water. Reference: Table GC 19, Reference 1.

Extraction and Derivatization of Biphenols in Microbiological Media and Homogenates of the Funges

For extraction analysis, 100 μℓ aliquots of a 5 mg/mℓ acetonitrile solution of each hydroxybiphenyl, biphenyl, and 1-naphthol (internal standard) are used to spike 2 mℓ aqueous samples (medium or cell homogenate). After addition of 3 mℓ of 0.2 *M* phosphate buffer (pH 6.0), the mixtures are extracted by rocking with 2 mℓ portions of nitrogen-purged ethyl acetate for 15 min. Following centrifugation, a 200 μℓ aliquot of the organic layer is transferred to a 1 mℓ Reacti-vial, taken to dryness under a stream of nitrogen, and derivatized with heptafluorobutyric anhydride.

Extracts are dissolved in 100 μℓ of acetonitrile and placed in a 1.0 mℓ Reacti-vial. Benzene (100 μℓ) and 25 μℓ aliquots of heptafluorobutyric anhydride are added and heated at 65°C for 90 min. Aliquots of 2 μℓ are injected into a gas chromatograph. Reference: Table GC 12, Reference 1.

Section II.II.

SAMPLE PREPARATION METHODS FOR ANALYSIS OF ORGANIC ACIDS BY GAS CHROMATOGRAPHY

Flash-Heater Trimethylsilylation for GC Analysis

One $\mu\ell$ of a solution of aromatic acid mixture (5 to 300 μg/mℓ in ethyl acetate) is injected together with 1 $\mu\ell$ of *N,O-bis*(trimethylsilyl)trifluoroacetamide into glass capillary GC column. Reference: Table GC 25, Reference 2.

Derivatization of Optical Isomers (Methylsuccinic Acid in Urine)

A 100 mℓ of normal urine is adjusted to pH 12 with NaOH and extracted three times with 100 mℓ portions of diethyl ether to remove neutral and basic compounds. The aqueous layer is adjusted to pH 1 with HCl and saturated with NaCl. This solution is extracted with diethyl ether and the extract is dried with anhydrous Na_2SO_4 and concentrated to 25 mℓ. One mℓ of extract is applied to a 20 × 20 cm glass plate coated with silica gel G (0.5 mm thickness). The plate is developed with an ethanol-water-conc. NH_4OH (35:13:2). After development, the plate is air dried and sprayed with methyl red in ethanol (0.1 g/500 mℓ, adjusted to an orange color with NaOH). The R_f of methylsuccinic acid is 0.57. The area is scraped off the plate and transferred to a screw capped tube and evaporated to dryness. The residue is heated under anhydrous conditions for 1 hr at 105 °C with 0.2 mℓ of 5 *N* HCl in S-(+)-2-butyl alcohol. Excess reagent is evaporated with a stream of dry, filtered N_2, and the residue dissolved in 20 $\mu\ell$ of dichloromethane. Reference: Table GC 44, Reference 1.

Purification of the Carboxylic Acids in Plant Tissues

From 0.125 to 1.5 g of fresh tissue is placed in a small mortar prechilled in dry ice. Liquid nitrogen is added to freeze the tissue. After evaporation of nitrogen, the tissue is ground to a powder with a prechilled pestle. The powder is placed into a 50 mℓ erlenmeyer flask containing 15 mℓ of 95% ethanol. Boiling chips are added and the solution is boiled for 10 min. After cooling, the ethanolic extract is removed with a Pasteur pipette, placed in a 10 × 15 mm test tube, and taken to dryness under filtered compressed air.

The residue is resuspended with a 1 mℓ of water, and lipid is extracted with 0.5 mℓ of chloroform. After centrifugation, the aqueous layer is removed by pipet and placed on ion-exchange columns (a top column contains the formate form of Dowex 1 X 8 and a bottom column contains the H form of Dowex 50 X 8).

After removing amino acids on the top column, organic acids are sorbed on the Dowex 50 column. The combined column is rinsed with distilled water, and the upper column is removed. After eluting neutral or weakly bound acidic compounds with distilled water, organic acids are eluted with 16 *N* formic acid and freeze-dried. After the freeze-drying, the residue is resuspended with distilled water, placed in a Reacti-Vial, and freeze-dried.

Pyridine is added to each sample vial, and the residue is resuspended. Ten minutes before injection into a gas chromatograph, 25 $\mu\ell$ of *N,O-bis*(trimethylsilyl)-acetamide is added, the vial is shaken at room temperature, and left to stand at room temperature until injection. Reference: Table GC 25, Reference 5.

Analysis of α-Keto Acids in Urine as *O*-trimethylsilylquinoxalinol Derivatives

Ten mℓ of urine is mixed with 500 $\mu\ell$ each of acetic acid and toluene. The aliphatic urinary keto acids are then stable for about 5 days without refrigeration. To 2 mℓ of the acidified urine, 50 μℓ of internal standard solution (20 mM α-ketovaleric acid and α-ketocaprylic acid) in water plus acetic acid are added and well mixed. Then 0.5 mℓ of urine plus internal standard are heated for 1 hr at 70°C with 0.5 mℓ of 4 *N* HCl and 1 mℓ of a 1% solution of *o*-phenylenediamine in 2 *N* HCl. The incubation is carried out in vials with PTFE-lined screw-caps using a Reacti-Therm Heating Module.

The samples are then saturated with solid ammonium sulfate and extracted twice with 5 mℓ of chloroform in vials. Vigorous shaking for 1 min each time is performed with a VirTis Model Whirlmix shaker. The extract is dried for 2 hr over anahydrous sodium sulfate and then filtered. After evaporation to dryness in a Heidolph-Elekto Model VVI rotary evaporator at room temperature, the residue is taken up in 50 μℓ of pyridine and incubated for 30 min at 70°C after addition of 50 μℓ of *bis* (trimethylsilyl)trifluoroacetamide. Reference: Table GC 31, Reference 1.

Rapid Tratment (Ethylation) of Oxalic Acid in Urine for Gas Chromatographic Analysis

When necessary, the volume of the 24 hr urine sample is made up to 2 ℓ in order to improve the recovery of the dissolved dry residue. The urine sample is then acidified with 1% HCl, and a 30 mℓ aliquot is evaporated to dryness under reduced pressure at 65° C with a rotating vacuum evaporator. The dry residue is dissolved in 1 mℓ of ethanol, and evaporating vessel is washed twice with 1 mℓ of ethanol. The three extracts are transferred to a graduated test tube and the volume adjusted to 3 mℓ with ethanol. Sulfuric acid (0.1 mℓ) is added to catalyze the ethylation. The test tube is centrifuged at 300 g and the supernatant immediately transferred to a ground-glass stoppered test tube and placed in an incubator at 37° C overnight. Reference: Table GC 24, Reference 7.

Extraction and Derivatization of Citric and Nitrilotriacetic Acid in Water

Formic acid (4 *M*) is added to 50 mℓ of sample to lower the pH to 2.5. The sample is put into a water bath at 60° C, and purified N_2 is bubbled for 30 min. After cooling to room temperature, the sample is transferred to an ion-exchange column (Bio-Rad AG-X2, 50-100 mesh is packed a 2 cm resin bed into 150 × 5 mm i.d.). These acids are eluted with 16 *M* formic acid, and the formic acid is evaporated under a N_2 stream. Dry 3 *N* HCl-butanol is added and the acids are esterified. Reference: Aue, W. A., Hastings, C. R., Gerhardt, K. O., Pierce, II, J. O., Hill, H. H., and Moseman, R. F., The determination of part-per billion levels of citric and nitrilotriacetic acids in tap water and sewage effluents, *J. Chromatogr.*, 72, 259, 1972.

Extraction and *O*-pentafluorobenzyl-methyl Esterification of Phenolic Acids

A 0.5 mℓ of urine sample is adjusted to pH 1 with 6 *N* HCl and saturated with NaCl. The solution is extracted five times with 2 mℓ of diethyl ether by mechanical shaking for 5 min. The interface emulsion is broken by centrifugation for 3 min at 400 g. The combined ether extracts are dried over anhydrous Na_2SO_4, and the solvent is evaporated to dryness under reduced pressure. The residue is dissolved in 1 mℓ of 5% $NaHCO_3$.

The solution is placed into a 10 mℓ glass stoppered test tube. The total volume is made up to 2mℓ with water, and 0.04 mℓ of pentafluorobenzoyl chloride is added. The solution is shaken at 300 rpm for 10 min at room temperature. The reaction mixture is acidified to pH 1 with 10% H_3PO_4(0.05 mℓ) and saturated with NaCl, and 0.5 mℓ of methanolic internal standard solution (DNP-phenylalanine: 100 μg/mℓ; 2,4,6-trichlorophenyl-4′-nitrophenyl ether: 20 μg/mℓ) are added. The solution is extracted three times with 3 mℓ of diethyl ether with vigorous shaking by hand. To the combined ether extracts is added 1 mℓ of methanol, and methylation is carried out by bubbling diazomethane, generated according to the micro-scale procedure, through this solution until the yellow color persists. After standing for 5 min, the solvents are evaporated to dryness at 50° C under a stream of nitrogen. The residue is dissolved in 1 mℓ of ethyl acetate, and the solution is dried over anhydrous Na_2SO_4. This solution is diluted 20 times with *n*-hexane and injected into a gas chromatograph. Reference: Table GC 38, Reference 6.

Extraction and Derivatization of Urinary Acids for GC/MS Analysis

A 3 mℓ urine sample is added the internal standard, 3-chlorophenylacetic acid (0.212 mg) and hydroxylamine hydrochloride (30 mg), and the pH is adjusted to 12 with 2 *N* NaOH. Each sample is heated at 60° C for 30 min to form the oximes of the keto acids. After cooling to room temperature, an aliquot is acidified to a pH below 3 with conc. HCl and transferred directly into a solid phase extraction tube (3mℓ polypropylene tube filled with an inert cellulose gauge matrix (Manhattan Instr., Santa Monica, Calif.)). The organic acids are eluted with a mixture of diethyl ether-ethyl acetate (1:1, 24 mℓ). The extract is dried with $MgSO_4$, evaporated, and transferred to a vial with ethyl acetate-methanol (1:1, 0.5 mℓ). The solvent is removed under a stream of nitrogen and derivatized with *bis* (trimethylsilyl)trifluoroacetamide at 60° C for 30 min. Reference: Table GC 32, Reference 1.

Extraction and Derivatization of Aromatic Acids in Urine

One volume of urine (10 mℓ if the 24 hr volume is greater than 1.5 ℓ, or 5 mℓ if less) is acidified with 0.5 mℓ of 6 *N* HCl, saturated with NaCl, and extracted twice with 5 volumes of ethyl acetate. Aliquots, 4 and 5 volumes of the first and second ethyl acetate extract respectively, are combined and evaporated to dryness under vacuum at 50° C using a rotary evaporator. The residue is esterified at room temperature with 0.5 mℓ of ethanolic-HCl, prepared as required by adding 1.25 mℓ of acetyl chloride dropwise to 10 mℓ of ice-cold anhydrous ethanol. After 30 min, excess ethanolic-HCl is removed under vacuum at 25° C to 30° C, using a water pump. The residue is moistened with 0.1 mℓ of anhydrous ethanol and dissolved in 3 mℓ water, after allowing to soak for a few minutes in the ethanol. The solution is passed through a column (0.55 cm i.d., 5 to 6 cm bed height) of anion-exchange resin (AG 1-X4, acetate form, 100-200 mesh, 0.4 g dry weight). After washing the column with 2 mℓ water, the adsorbed esters are eluted with 6 mℓ of 50% aqueous ethanol. The eluate is evaporated to dryness under vacuum (water pump) at 50° C, and the residue is re-esterified with 0.15 mℓ of ethanolic-HCl at room temperature or 30 min in case some hydrolysis occurred during evaporation. The esterified mixture is transferred with the aid of a Pasteur pipette to a 1 mℓ Reacti-Vial. Solvent is removed under vacuum in a centrifugal freeze-drier modified for use with a water pump, and the remaining liquid adhering to the collecting tube is diluted with 0.15 mℓ of ethanol and combined with the ester residue in the Reacti-Vial. The solvent is again evaporated in the freeze-drier, as above. The contents of the Reacti-Vial are silylated by adding 0.1 mℓ of *N,O-bis* (trimethylsilyl)acetamide, followed by vortex mixing, heating for 10 min at 65° C, vortex mixing again to dissolve completely any undissolved material, and then heating for a further 30 min at 65° C. Reference: Table GC 39, Reference 1.

Extraction and Derivatization of Organic Acids in Urine

Neutralized urine plus 1.0 mℓ of the standard undecanedionic acid solution (0.100 g of undecanedionic acid is dissolved in the minimum volume of 0.1 *N* sodium hydroxide solution and diluted to 100 mℓ with water) is applied to the column and allowed to drain in a 4.5 cm × 7 mm column of A25 DEAE-Sephadex is prepared by pouring an aqueous slurry of gel, previously swollen in water, into the chromatographic column and allowing it to settle. The column is washed with 3 mℓ of water and 20 mℓ of 0.5 *M* aqueous pyridium acetate buffer prior to use. The column is then washed twice with 5 mℓ of water, and the acidic components are eluted with 15 mℓ of 1.5 *M* aqueous pyridium acetate buffer into a 25 mℓ round-bottomed flask containing 50 mg of *O*-ethylhydroxylammonium chloride. The eluate and *O*-ethylhydroxylammonium chloride are mixed and allowed to stand for 4 hr or longer. The eluate is frozen as a film by rotating the flask in a solid carbon dioxide-methanol bath and freeze-dried overnight at 0.5 torr and −10° C as described below.

A 900 μℓ volume of trisil *bis*(trimethylsilyl)acetamide is added to the dry residue together with 50 μℓ of each hydrocarbon internal standard solution, mixed, and the mixture is allowed to stand for at least 4 hr. Additional *bis*(trimethylsilyl)acetamide solution is added, as necessary, so as to give a clear solution before analysis. Reference: Table GC 23, Reference 1′.

Extraction and Derivatization of Organic Acids in Human Saliva

When sample collection is to be performed in the morning, the subjects are required to rinse out the oral cavity with deionized water several times after brushing the teeth upon retiring the previous evening. In the morning, before the breakfast and brushing the teeth, the subjects are instructed to repeat the cleansing of the mouth with deionized water prior to collection. Afternoon sample collection is performed under nonfasting conditions. The oral cavity is rinsed several times with deionized water prior to sample collection. Samples of saliva are obtained by expectoration into glass test tubes.

The saliva is acidified with HCl to pH 1, extracted once with ethyl acetate (50 mℓ), and the organic layer is saved. The remaining aqueous phase is then extracted twice with diethyl ether (25 mℓ), and all the organic phases containing the acidic and neutral metabolites are combined. After backextraction with deionized water and discarding the aqueous layer, the organic solvents are removed using low heat under prepurified nitrogen stream. Portions of methylene chloride (1 mℓ) are used to wash down the aides of the vessel during the solvent removal. The residue is derivatized by adding about 0.1 mℓ *bis*(trimethylsilyl)trifluoroacetamide. The reaction tube with a Teflon® cap liner is then closed and heated to 130° C for 10 min. The sample is concentrated to about 0.1 mℓ and used for direct injection into a gas chromatograph. A blank run is made using artesian water instead of saliva in the above extraction and sample preparation procedure. Reference: Table GC 23, Reference 2.

Section II.III.

DERIVATIZATION OF PHENOLS AND ORGANIC ACIDS

Derivatization of Phenols as Diethylphosphate Esters

The phenol, in solution (0.5 mℓ) of dry benzene and *tert.*-butanol (9:1), is added to NaH pellets (ca. 300 mg) under anhydrous conditions and allowed to stand for 5 min. The solution and subsequent washings (ca. 1 mℓ total) are then dropped directly into two drops of redistilled diethyl phosphorochloride, shaken and kept in a stoppered flask at 50° C. After 30 min the benzene is evaporated and aqueous $NaHCO_3$ solution (5%; 2 mℓ) is added, shaken and then allowed to stand for 2 min. The solution is extracted with benzene, and the benzene is washed thoroughly first with 0.1 *N* HCl and then with water. The benzene solution is dried, evaporated, and the residue made up to a fixed volume for quantification. Reference: Table GC 10, Reference 1.

Derivatization of Phenols as Halogenated Ethers

An excess of NaH is added to a solution of thymol in ether. After the effervescence has subsided, the supernatant solution is added to a stirred solution of the reagent (the solution: 1 *M* thymol (1.1:1)) in anhydrous ether. After stirring for 4 hr the solution is filtered and concentrated to yield the product.

Reagents: monofluoroacetylchloride, monochloroacetylchloride, pentafluoropropionic acid anhydride, heptafluorobutyric acid anhydride, pentafluorobenzyl bromide, pentafluorobenzoylchloride, 1-fluoro-2,4-dinitrobenzene. Reference: Table GC 19, Reference 2.

Preparation of Phenols as *O*-isobutyloxycarbonyl Derivatives

An aqueous aliquot containing up to ca. 2 mg of total phenols is pipetted into a 10 mℓ polyethylene-stoppered vial. To this solution, 0.5 mℓ of 10% sodium carbonate solution is added, and the total volume is made up to 2 mℓ with distilled water. Then 0.1 mℓ of isobutyl chloroformate is added immediately, and the reaction mixture is shaken for 10 min at room temperature. The resulting derivatives are extracted with 0.5 mℓ of diisopropyl ether in alkaline medium. The ethereal solution is dried over anhydrous sodium sulfate and subjected to GC analysis. Reference: Table GC 10, Reference 2.

Flophemesyl Derivatization of Phenols and Acids

The compound (10 mg) is dissolved in toluene (30 μℓ) to which flophemesylamine (30 μℓ, 165 μmoles) is added slowly. If the compound is sparingly soluble in toluene and if reaction is slow, the mixture is heated to 60° C in the closed Reacti-vial.

Phenol ethers are prepared by the general method, or at a faster rate by a catalyzed reaction as follows. Flophemesylamine (15 μℓ, 83 μmoles) and flophemesyl chloride (15 μℓ, 75 μmoles) are added to a solution of phenol (10 μℓ, 110 μmoles) in diethyl ether (30 μℓ). The mixture is centrifuged before withdrawing samples. Reaction is complete in 5 min. Reference: Table GC 34, Reference 4.

Preparation of *O*-acetylated Menthyl Esters of Acids

(-)-Menthol (300 mg) is added to 1 to 5 mg of dry hydroxy acid and esterified at 110° C by bubbling dry HCl gas through the solution for 2 hr. The excess of HCl and menthol are removed under a stream of nitrogen at 110° C. To avoid possible losses of the volatile menthyl ester of lactic acid, it is advisable not to remove menthol completely. The residue is acetylated in 1 mℓ of pyridine-acetic anhydride (1:1) at 100° C for 30 min. Finally, the solvent is evaporated in the presence of absolute ethanol, and the residue is dissolved in chloroform. Reference: Table GC 29, Reference 2.

Derivatization of Carboxylic Acids and Phenols with Pentafluorobenzyl Bromide Using Crown Ether as a Catalyst

To derivatize up to 10 μg of carboxylic acids and/or phenols, 50 μℓ of the derivatizing solution (acetonitrile or benzene or ethyl acetate or dimethylforamide or heptane solution containing pentaluorobenzyl bromide (3.0 μℓ/mℓ) and 18-crown-6 (3.0 mg/mℓ)) is introduced into a 0.3 mℓ Reacti-vial containing the dry acids or phenols, a few grains of powdered K_2CO_3 are added, and the mixture is shaken occasionally. Powdered potassium acetate, bicarbonate, or cyanide are added if one wishes only the carboxylic group of a phenolic acid to be derivatized. Reaction times are 5 to 120 min. For phenolic and catecholic acids, the carboxyic acid group is first esterified (5 min at 25° C) with 20% HCl-methanol (prepared by adding acetyl chloride, 1.0 mℓ, dropwise to methanol, 4.0 mℓ), the methanol is evaporated, and the phenolic groups are derivatized as described above. Reference: Table GC 39, Reference 2.

Preparation of Alkyl Esters of Carboxylic Acids

A 10 mequiv. amount of the acid to be determined is dissolved in 25 mℓ of the alcohol, and 4 mℓ of pyridine is added as catalyst. *N,N'*-dicyclohexylcarbodiimide is added in excess of the theoretical amount (12 mequiv.) or in even larger amounts if the reagents and the acid used are not dry or contain water of crystallization that reacts with *N,N'*-dicyclohexylcarbodiimide. The mixture is stirred gently. If the esterifying alcohol is solid (for example, L-menthol), the amount of pyridine is increased so that it acts as a solvent, reducing the amount of alcohol to 20 meq.

For most of the acids used, esterification with methanol is quantitative at room temperature; but for some acids and alcohols with high molecular weights, it is necessary to heat the mixture at 40 to 80° C for 30 to 120 min. If during the reaction a precipitate of *N,N'*-dicyclohexylurea is formed, it is allowed to settle, and after addition of the internal standard, 0.5 μℓ of the supernatant clear solution is injected into a gas chromatograph.

Reaction:

$$\mathrm{RN{=}C{=}NR} \xrightarrow{\mathrm{H^+}} \mathrm{RN{=}C{=}N^+HR} \xrightarrow{\mathrm{R_1COO^-}} \begin{array}{l}\mathrm{RN{=}C{-}NHR} \\ \quad\;\; | \\ \quad\; \mathrm{O{-}CO{-}R_1}\end{array}$$

$$\xrightarrow[\text{alcohol}]{\mathrm{R_2OH}} \mathrm{R_1COOR_2 + RNH{-}CO{-}NHR}$$

Reference: Table GC 47, Reference 1.

Section II.IV.

DERIVATIZATION AND POST-COLUMN REACTION DETECTOR FOR LIQUID CHROMATOGRAPHY

Phenacyl Ester of Dicarboxylic Acids

A methanol or water solution of the organic acids is neutralized to a phenolphthalein end point with KOH-methanol solution. The solvent is removed under aspirator vacuum to give the potassium salts of the diacids. If the salts look wet, several milliliters of benzene are added to the salts and then removed under vacuum. For every 0.1 m*M* of the acid salt, 1.1 mℓ of alkylating reagent solution A or B is added and the mixture refluxed for 30 min. The solution is injected into a liquid chromatograph.

Reagent A = 0.2 m*M* α-bromoacetophenone and 0.15 n*M* 18-crown-6/mℓ acetonitrile.

B = 0.2 m*M* α-bromoacetophenone and 0.15 n*M* dicyclohexyl-18-acrown-6/mℓ acetonitrile. Reference: Table LC 18, Reference 3.

Nitrobenzyl Ester of Dicarboxylic Acids

The potassium salts of the acids are generated as above for phenacyl esters. For every 0.1 m*M* of acid, 1.2 mℓ of alkylating reagent is added, and the mixture is refluxed for 1.5 hr. The solution is injected into a liquid chromatograph. Reagent: 0.2 m*M* *p*-nitrobenzyl bromide and 0.15 n*M* dicyclohexyl-18-crown-6/mℓ acetonitrile. Reference: Table LC 18, Reference 3.

Quinoxaline Derivatives of 2-oxo-Acids

To 100 μℓ of an aqueous sample solution are added 5 mℓ of 2.5 *N* HCl containing 3 mg of naphthalene-2,3-diamine sulfate, 10 μℓ of 2-mercaptoethanol, and 3 μg of 2-chlorothioxanthone in 100 μℓ of methanol. The mixture is warmed in a water-bath at about 80° C for 2 hr, and then cooled, and the 3-benzyl-2-benzoxybenzoquinoxaline formed is extracted into 10 mℓ of CCl_4. The organic phase is washed with 10 mℓ of 3 *N* HCl, the CCl_4 is evaporated in a rotary evaporator, the residue is dissolved in a few drops of *N,N*-dimethylformamide, and 10 μℓ of this solution are injected into a liquid chromatograph. Reference: Table LC 24, Reference 3.

Selective Post-Column Reaction Detector for 4-hydroxy-3-methoxymandelic Acid

The effluent from the column is mixed with alkaline periodate and heated at 60° C, then oxidized product (vanillin) is detected at 360 nm. Reference: Table LC 30, Reference 1.

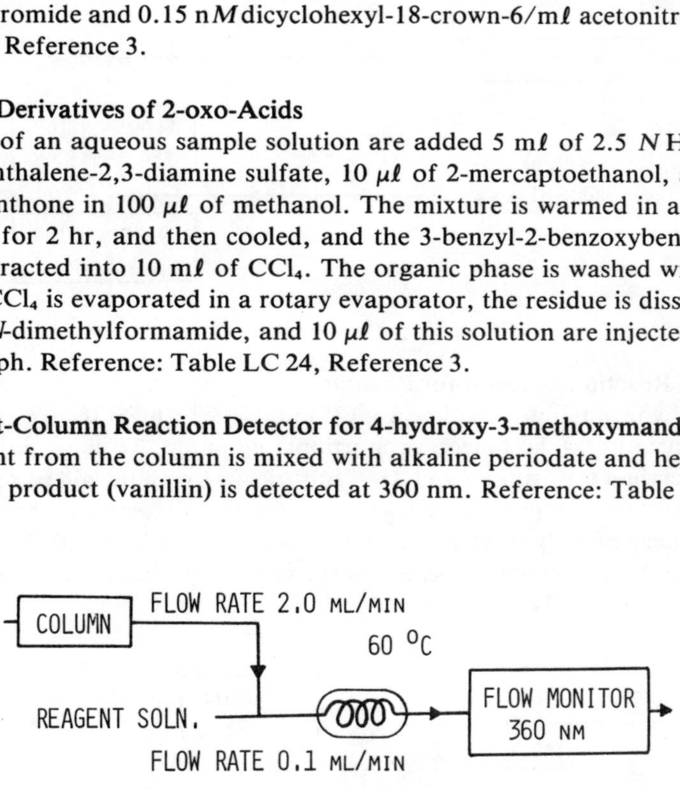

REAGENT SOLUTION

10 mL of 10 mM $NaIO_4$, 9 mL of 2.5 M K_2CO_3 and 10 mL of 5 M NaOH diluted to 50 mL with water

Post-Column Reaction Detector for Carboxylic Acids

Acidic eluent is mixed with reagent solution (0.116 *M* hydroxylamine perchlorate in 0.07 *M* triethylamine) and then added with dicyclohexylcarbodiimide solution (0.13 *M*.) The mixture is heated at 50° C, mixed with 0.005 *M* ferric perchlorate solution, and the reaction products are measured through a spectrophotometer at 530 nm. The diagram is shown below. Reference: Table LC 28, Reference 1.

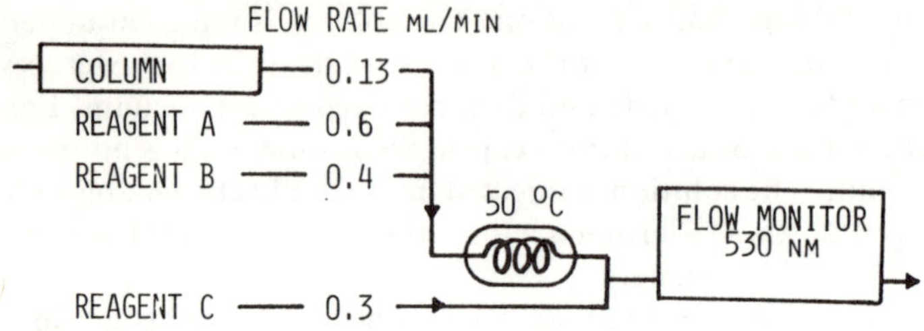

Post-Column Reaction Detector for Carboxylic Acids

The effluent from the column is passed through a UV detector (254 and 313 nm) to monitor aromatic acids, then the effluent is mixed with a solution of sodium 2-nitrophenolate (200 mg of sodium 2-nitrophenolate in 100 mℓ of ethanol) and monitored in second detector (435 nm). Reference: Table LC 18, Reference 1.

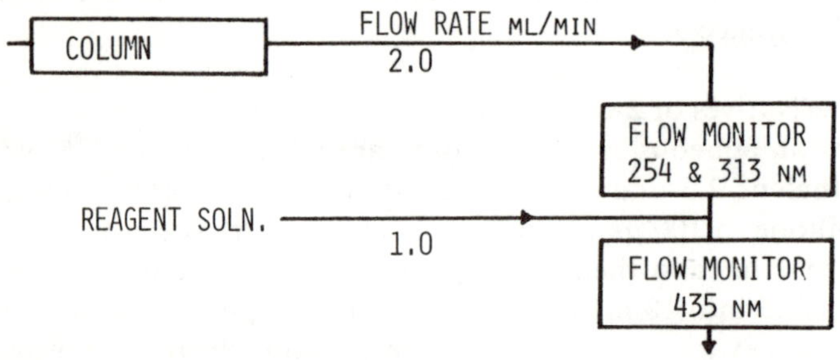

Reagent solution : sodium 2-nitrophenolate soln.

Post-Column Reaction Detector for Phenols

After passing through the UV detector (254 nm), the effluent is segmented into small portions by means of a Technicon proportionating pump. To these portions is added a mixture of cerium (IV) sulfate and 23 *N* sulfuric acid with the aid of the same pump. The combined flows are passed through a short coil maintained at 25° C and into the cell compartment of a fluorescence detector (ex. 260, em. 350 nm). To minimize diffusion effects, the debubbler is located in the cell compartment close to the flow cell. Reference: Table LC 6, Reference 2.

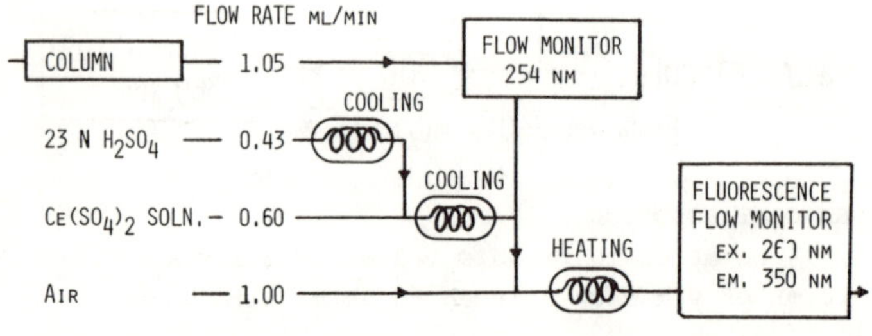

SECTION II.V.

SAMPLE PREPARATION FOR THIN LAYER OR PAPER CHROMATOGRAPHY

Derivatization of Phenols as Antipyrylquinoneimide Dyes

A 2.5 mℓ of 0.01 *M* phenol/ethanol is mixed with 1 mℓ of 4-aminoantipyrine solution (0.5% in buffer solution), 25 mℓ of buffer solution (Britton-Robinson buffer pH 7.5 to 10.0) and 1 mℓ of potassium hexacyanoferrate (III) solution (2% in buffer solution). The mixture is extracted with 1 mℓ of chloroform, and a 5 μℓ aliquot of the separated chloroform is applied for TLC. Reference: TLC 4, Reference 1.

Preparation of *p*-(*N,N*-dimethylamino)benzene-*p*′-azobenzoyl Derivatives of Phenols

0.5 g of phenol and 1.5 g of *p*-(*N,N*-dimethylamino)benzene-*p*′-azobenzoyl chloride are heated with 20 mℓ of pyridine under reflux for 1 hr. After cooling, 20 mℓ of a 5% soln. of sodium hydrogen carbonate are added to the mixture. The solution is filtered, and the separated derivatives are crystallized from dimethylformamide. The esters are red-orange crystals.

For chromatographic purpose, the amount of reagent and phenol are reduced 10 times, and the mixture is heated in a sealed ampoule in boiling water. After cooling, the mixture is applied for TLC. Reference: Table TLC, Reference 1.

Preparation of 2,4-dinitrophenylhydrazone of α-Keto Acids in Berries

Berries are frozen immediately after harvesting and stored at −20° C. α-keto acids are extracted from the berries by blending with 99.5% ethanol followed by centrifuging. The supernatant is passed through a cation-exchange column (Dowex 50W, X8, Cl^- form, 50-100 mesh) to remove anthocyanius and amino acids. The acids are eluted with distilled water and applied to an anion-exchange column (Dowex 1 X8, formate form, 100-200 mesh). After remove sugars with water, the keto acids are eluted with 6 *N* formic acid, and the solvent is removed in a vacuum evaporator. The 2,4-dinitrophenylhydrazone of keto acids are prepared and purified in an alumina column. Reference: Table TLC 19, Reference 2.

Extraction of Salicylic Acid and Metabolites in Urine or Plasma

A 10 mℓ sample of urine is alkalinized to pH 8.5 with 0.2 *N* NaOH soln. The urine is shaken with 14 mℓ of ethyl acetate and 1 mℓ of ether for 2 min. The lower layer is transferred, and 1 mℓ of conc. HCl is added. Acids are extracted with 14 mℓ of ethyl acetate and 1 mℓ of ether. After centrifuging, the ethyl acetate layer is applied for TLC.

Hydrolyzed urine is treated in the same way, except that before the second extraction the acidified urine is boiled for 2 hr under reflux. After cooling, the hydrolyzed urine is extracted with an ethyl acetate and ether mixture.

One mℓ of blood plasma is deproteinized by 5% trichloroacetic acid (1 mℓ), centrifuged, and the supernatant is alkalinized to pH 8.5 with 1 *N* NaOH soln. This alkaline solution is then treated in the same manner as urine. Reference: Table TLC 29, Reference 2.

Extraction of Shikimate from Urine or Feces

For analysis of metabolites: 5 mℓ of diluted urine (1:1) is acidified to pH 1 with 5 *N* HCl and extracted with 5 × 6 mℓ portions of ethyl acetate. The combined extract is evaporated to dryness and dissolved in 200 μℓ of methanol for analysis by TLC.

For analysis of whole urine: 5 mℓ of diluted urine (1:1) is freeze-dried, and the residue suspended in 10 mℓ of methanol and transferred to a silica gel column (15 × 1 cm i.d. containing 1 g of Kieselgel 60). The clear methanolic eluate is collected, and the silica gel further eluted with 10 mℓ of methanol. Eluates are combined, evaporated, and dissolved in 300 μℓ of methanol for TLC.

For analysis of feces: 10 mℓ of aqueous 5% homogenates of 24 hr fecal samples are freeze-dried and suspended in 10 mℓ of methanol. The suspensions are transferred to a silica gel column (15 × 1 cm i.d., containing 1 g of Kieselgel 60). The eluates are collected, and the silica gel is further eluted with 20 mℓ of methanol. The methanolic eluates are combined and decolorized with 100 mg of activated charcoal. After filtration, the extracts are evaporated to dryness and dissolved in methanol for TLC analysis. Reference: Table TLC 24, Reference 1.

Section II.VI.

DETECTION REAGENT FOR THIN LAYER AND/OR PAPER CHROMATOGRAPHY

1. Aminoantipyrine

   Preparation: solution a = 2 g aminoantipyrine in 100 mℓ of ether
   solution b = 8 g $K_3Fe(CN)_6$ in 100 mℓ of water
   solution c = 25% aq. ammonia

   Procedure: after spraying solution a, respray solution b and placed in ammonia vapor (solution c) for 10 min.

   Reference: Table TLC 1, Ref. 1.

2. Anilium chloride-sodium nitrite

   Preparation: Prepare the thin layer with 5 g of anilium chloride, 100 g of silica gel, and 200 mℓ of water.

   Spray reagent = 5% sodium nitrite aqueous solution

   Procedure: after development, the plate is air dried and sprayed with the sodium nitrite solution.

   Reference: Table TLC 5, Ref. 1.

3. Cerium (IV) ammonium nitrate

   Preparation: 1.0 g cerium (IV) ammonium nitrate in 2 *N* nitric acid (250 mℓ).

   Procedure: After development the plate is air dried, examined in UV light, and sprayed with the chromogenic reagent.

   Reference: Table TLC-7 and Volume II, of this Handbook.

4. Ceric ammonium nitrate-hydroxylamine

   Preparation: soln. a = 5% $(NH_4)_2Ce(NO_3)_6$ in acetone, freshly prepared.
   soln. b = 5% $NH_2OH \cdot HCl$ in 80% acetone

   Procedure: The plate is sprayed with a followed by b and dried with hot air. Brown spots are observed in daylight. Heating of plates at 110° C for 5 min can increase sensitivity.

   Result: The spots and colorless background are stable for a long time (at least 1 year). The reaction under these conditions is specific for aromatic hydroxy compounds and indole derivatives. Aliphatic alcohols or amines, sugars, saturated cycloalcohols, carboxylic acids, aldehydes, and ketones do not react. Some aromatic amines in high concentrations can also give color, especially with heat.

   Reference: Chrastil, J., A method for detecting free phenolic groups and indole derivatives in thin-layer or paper chromatography, *J. Chromatogr.*, 115, 273, 1975.

5. Chromic acid

   Preparation: 0.6% aqueous solution of chromic acid

   Procedure: After development the plate is dried with a hot air blower and sprayed with the chronic acid solution.

   Reference: Table TLC-8, Ref. 1.

6. Chromotropic acid

   Preparation: Conc. sulfuric acid (15 mℓ) is slowly added to solution of chromotropic acid (4,5-hydroxy-2,7-naphthalenedisulfonic acid, 1 g in 15 mℓ of water).

   Procedure: After spraying, plates are heated for 15 to 20 min at 110 to 120° C.

   Reference: Table TLC-14, Reference 3.

7. Diazotized *p*-nitroaniline
   Preparation: soln. a = 0.5 g *p*-nitroaniline in 100 mℓ of 2 *N* HCl.
   soln. b = 5 g $NaNO_2$ in 100 mℓ water.
   soln. c = 20 g sodium acetate in 100 mℓ of water or 10% of aqueous potassium carbonate*.
   Procedure: The reagent is prepared immediately before spraying the plate by mixing three solutions a, b, and c (5mℓ + 0.5mℓ + 15mℓ).
   Reference: Table TLC-1, Ref. 1. *Table TLC-29, Ref. 3.
8. Diazotized sulfanilic acid
   Preparation: soln. a = 4.5 g sulfanilic acid + 45 mℓ of 12 *N* HCl in 500 mℓ water.
   soln. b = 5 g $NaNO_2$ in 100 mℓ water
   soln. c = 10 g Na_2CO_3 in 100 mℓ water
   Procedure: A mixture of solutions a and b (10 mℓ + 10 mℓ) is freshly prepared; then after 10 min, 15 mℓ of solution c is added.
   Reference: Table TLC-1, Reference 1. and Volume II, of this Handbook.
9. Dichloroflurescein
   Preparation: 0.1% dichlorofluorescein in ethanol
   Procedure: Spots are visualized by spraying with the reagent.
   Reference: Table TLC-10, Reference 1.
10. 9-Dicyanomethylene-2,4,7-trinitrofluorene
   Preparation: 2% 9-dicyanomethylene-2,4,7-trinitrofluorane in acetone (prepare immediately before use)
   Procedure: After drying in air the plate is sprayed with the reagent.
   Reference: Table TLC-14, Reference 3.
11. Dicyclohexyl carbodiimide
   Preparation: 10% methylene chloride solution of dicyclohexyl carbodiimide and 1% of acidic alcoholic solution of $FeCl_3 \cdot 6H_2O$ saturated with hydroxylamine hydrochloride.
   Procedure: The plate is sprayed with the reagent. The color develops in a minute.
   Reference: Table TLC-17, Reference 1.
12. Fremy's salt
   Preparation: 1% potassium nitrosodisulfonate in water
   Procedure: After development, the plate is examined in UV light and sprayed with the reagent.
   Reference: Table TLC-7, Reference 1.
13. Furfural-sulfuric acid
   Preparation: 2% of furfural solution in 62.5% sulfuric acid.
   Procedure: After spraying the solution, plates are heated 15 to 30 min at 110 to 120° C.
   Reference: Table TLC-14, Reference 3.
14. Genuine blue
   Preparation: solution a = 0.15 g genuine blue salt B in 100 mℓ water
   solution b = 0.1 *N* NaOH
   Procedure: solution a is sprayed, followed by solution b.
   Reference: Table TLC-1, Reference 1.
15. Isoniazid (isonicotinic acid hydrazide)-sodium hydroxide
   Preparation: solution a = 0.5% isoniazid in water
   solution b = 0.5 *N* NaOH
   Procedure: The plate is sprayed with solution a, partly dried for 10 min, and resprayed solution b so as to prevent spreading of the spots. The color is noted after 10 min.

Result: All *o*-dihydroxyphenolic compounds give a pinkish orange or red color against a clear background. Monohydroxy, *m*-dihydroxy, *m*-trihydroxy, vicinal hydroxymethoxy, methylated compounds, and amino compounds do not respond.

Reference: Bajaj, K. and Bhatia, M. S., Structure-specific reagents for the detection of phenolic compounds, *J. Chromatogr.*, 117, 445, 1976.

16. Periodic acid-benzidine

Preparation: solution a = 0.1% HIO_4
solution b = a mixture of 0.37 g benzidine in 20 mℓ of 50% ethanol, 4 mℓ of acetone and 2 mℓ of 0.2 *N*HCl.

Procedure: The plate is sprayed with solution a followed in 10 min by solution b.

Reference: Table TLC-24, Reference 10.

17. Periodic acid-aniline

Preparation: solution a = 0.16 g $NaIO_4$ in 25 mℓ of 1 *M* acetate buffer
solution b = 3% aniline in ethanol

Procedure: The plate is sprayed with solution a followed in 15 min by solution b.

Reference: Table TLC-24, Reference 10.

18. Periodic acid-thiobarbituric acid

Preparation: solution a = 0.02 *M* $NaIO_4$
solution b = ethylene glycol-acetone-95% H_2SO_4 (50:50:0.3)
solution c = 1.2 g thiobarbituric acid in 8.3 mℓ of 1 *N* NaOH and 11.7 mℓ of water

Procedure: The plate is sprayed with solution a, after 15 min with b, and after 10 min with c; heat 5 min at 100° C.

Reference: Table TLC-24, Reference 10.

19. Potassium metaperiodate

Preparation: 0.5% aqueous solution of potassium metaperidate

Procedure: The paper is sprayed with the solution.

Result: *o*- and *p*-dihydroxyphenols and their monomethyl ethers are detected as yellow-orange spots. This response is also positive to vanillin, vanillic acid and methyl vanilloyl ketone, negative to monohydroxy (2-hydroxybenzoic acid, 3-hydroxybenzoic acid, 4-hydroxybenzoic acid, methyl phenyl ketone, tyrosine, etc.), *m*-dihydroxy and trihydroxy (2,4-dihydroxybenzene, 2,4-dihydroxybenzoic acid, 2,4,6-trihydroxybenzene, etc.) compounds.

Reference: Clifford, M. N. and Wight, J., Metaperiodate a new structure specific locating reagent for phenolic compounds, *J. Chromatogr.*, 86, 222, 1973.

20. Potassium persulfate-silver nitrate

Preparation: 1% $K_2S_2O_8$ and $AgNO_3$ in water

Procedure: After development the plate is dried with a hot air blower and sprayed the solution.

Reference: Table TLC-8, Reference 1.

21. Sodium tungstate-sodium trichloroacetic acid-sodium anitrite

Preparation: reagent a = mixture of 6 mℓ of a 10% solution of sodium tungstate, 6 mℓ of a 5% solution of trichloroacetic acid, and 3 mℓ of 0.5 *N* HCl to which is added 6 mℓ of freshly prepared 5% sodium nitrite solution.
reagent b = 0.5 *N* sodium hydroxide solution.

Procedure: After spraying with chromogenic reagent a, see color below, then after 3 min, respray with reagent b.

	Color	
Compound	Reagent a	Reagents a and b
4-hydroxy-3-methoxybenzoic acid	y	y
3,4-dimethoxybenzoic acid	y	y
4-hydroxy-3,5-dimethoxybenzaldehyde	bn	bn
4-hydroxy-3-methoxybenzaldehyde	y	y
vanillal	y	y
4-hydroxy-3,5-dimethoxybenzoic acid	y	y
4-hydroxy-3-methoxycinnamic acid	y	y
2-methylphenol	g	g
4,4′-dihydroxy-3,3′-dimethoxystilbene	pk	pk
hesperdin	—	—
4-hydroxy-3,5-dimethoxycinnamic acid	pk	pk
naringin	—	—
phenol	—	—
5-methoxy-2-isopropyl-1-phenol	y	y
3-hydroxybenzoic acid	ft-bn	ft-bn
4-hydroxybenzaldehyde	y	y
4-hydroxybenzoic acid	ft-y	ft-y
4-aminophenol	y	y
1-hydroxynaphthalene	ft-y	bn
4-hydroxycinnamic acid	y	y
coumarin	ft-y	ft-y
3-hydroxybiphenyl	ft-bn	ft-bn
2-hydroxybenzoic acid	pk	pk
3-hydroxyanthranilic acid	y	pk
2-aminophenol	y	y
4-hydroxyphenylpyruvic acid	g	g
tyrosine	y	y
2-nitrophenol	y	intense-y—r-y
2,4,6-trinitrophenol	y	r-y
3,5-dihydroxybenzoic acid	ft-y	ft-y
2,5-dihydroxybenzoic acid	r-y	p
2,5-dihydroxyphenylacetic acid	g	g
2,4-dihydroxyresorcylaldehyde	ft-pk	ft-pk
2,4-dihydroxybenzoic acid	bn-y	bn
1,3,5-trihydroxybenzene	intense pk-y	r
rutin	y	r
3,4-dihydroxycinnamic acid	y-bn	r-bn
1,2-dihydroxybenzene	y	intense-r
dihydroquercetin	intense-y	y-r
D(+)-catechin	y	y-r
3,4-dihydroxybenzoic acid	y	d-r
3,4-dihydroxybenzaldehyde	lt-y	lt-r
2,3-dihydroxybenzoic acid	y	d-r
chlorogenic acid	y	r
ellagic acid	g	r
adrenaline	r	r
quecetin	bn-y	bn-y
tannic acid	y	bn-y
1,2,3-trihydroxybenzene	pk-bn	bn
3,4,5-trihydroxybenzoic acid	bn	bn
3,4,5-trihydroxybenzoic acid methylester	lt-bn	bn-y—bn

Reference: Bhatia, I. S., Singh, J., and Bajaj, L. K., A new chlomogenic reagent for the detection of phenolic compounds on thin-layer plates, *J. Chromatogr.*, 79, 350, 1973.

22. Titanium chloride
Preparation: 20% $TiCl_4$ in conc. HCl
Procedure: after spraying the reagent, color is immediately given.
Result: The color complex is stable for up to 4 to 6 hr, after which the spots tend to darken.

Compound	Color	Compound	Color
4-hydroxy-3-methoxybenzoic acid	y	3,4-dihydroxycinnamic acid	lt-bn
4-hydroxy-3-methoxycinnamic acid	y	3,4-dihydroxybenzoic acid	bn
naringenin	y	1,2-dihydroxybenzene	bn
phenol	—	rutin	d-bn
4-hydroxybenzoic acid	lt-y	chlorogenic acid	d-bn
4-hydroxycinnamic acid	y	quercetin	d-bn
2-hydroxybenzoic acid	y	catechin	bn
4-hydroxy-3,5-dimethoxybenzoic acid	y	1,3-dihydroxybenzene	o-y
4-hydroxy-3,5-dimethoxycinnamic acid	o-y	1,3,5-trihydroxybenzene	y

Reference: Eskin, N.A.M. and Frenkel, C., Titanium chromogenic reagent for phenolic compounds in thin-layer plates, *J. Chromatogr.*, 150, 293, 1978.

Section II.VII.

SOURCES OF CHROMATOGRAPHIC MATERIALS AND PRODUCTS

Gas Chromatography

Most of packings and liquid phases are supplied by following companies. Special materials are noted in footnotes to each table. (See also Volume II of the Handbook, p. 255 to 283).

U.S.

Alltech Associates
2051 Waukegan Road, Deerfield, Ill. 60015

Analabs
80 Republic Drive, North Haven, CT 06473

Applied Science
P.O. Box 440, State College, Penn. 16801

Chrompack
P.O. Box 4131, Whittier, Calif. 90607

Regis Chemical Company
8210 Austin, Morton Grove, Ill. 60053

Supelco Inc.
Supelco Park, Bellefonte, Penn. 16823

Canada

Chromatographic Specialties Limited
P.O. BAG 1150, 300 Laurier Blvd., Brockville, Ont. K6V 5W1

Mandel Scientific Company
395 Norman St. Ville St. Pierre, Montreal, Quebec H8R 1A3

Europe

Chrompack
P.O. Box 3, 4330 AA Middeburg, The Netherlands

Japan

Gasukuro Kogyo Co., Ltd.
6-12-18 Nishishinjuku, Shinjuku-ku, Tokyo 160

Liquid Chromatography

Packings for liquid chromatography are listed in following tables (See also Volume II of the Handbook, p. 284-303).

Table 1
COLUMNS AND PACKINGS FOR GEL PERMEATION CHROMATOGRAPHY

So	Name	No	PS μm	ps nm	MW	plate N 1000/m	Column size cm × mm i.d.	Remarks	Mfgr. & supplier
Glass									
AO	Bio-Glass	5	44	20-250	10^3-9×10^6	—	—		BR
AO	CPG-10	10	36-75	25-200	2000-10^7	—	—		EN
AO	CPG-40,100,250,550	4	5-177	4-55	1000-35×10^4			sa = 70-190	Pi
Silica gel									
AO	BioSil® GFC 10-100	3	10	10-100		—	—		BR
AO	LiChrospher SI 100-4000	4	5,10,30	10-400		—	—		EM,AS,At,BD, Ch,Ra,Su
AO	Silarex CPS 75,100,150	3		7.5-15	-3×10^4	28	25 × 4.6	pv = 1.3-1.6	Li
AO	Porasil®	6	37-75	6-200	-2×10^6	—	—		Wa
AO	μPorasil® GPC 60A	1	10	6	-10^4	—	—		Wa
AO	Spherosil XOA-C		100-200	8-300		—	—	sa = 10-400	Su
Surface treated silica gel									
AO	Aquapore OH 100-4000	5	10	10-400	5×10^4-8×10^6	12-20	25 × 4.6	Glycerylpropylsilane	CX
AO	μBondagel E 125-1000	4	10	12.5-100	2000-2×10^6	12		-Si(RO)nCH_3	Wa
AO	SizeExclusion SE60-1000	4	10	6-100	1000-10^7		25 × 6.2		DP
AO	SynChropak GPC 100-4000		10	10-400	-6×10^5		25 × 4.6		SC,AS,Re
AO	TSK G2000-4000SW	3	10 ± 2		-6×10^5	20	30 × 7.8		TS,BR,Ax,Va
AO	Zorbax PSM 60,1000	2	10	6-100	100-10^4		25 × 6.2		DP
Polystyrene gel									
O	BioBeads S	6	37-75	—	400-14×10^3	—	—		BR
O	Microgel®	7	20	5-10^5	-4×10^7	6	60 × 7.7		Ch
O	Microgel®	7	10	5-10^5	-4×10^7	15	25 × 7.7		Ch
O	PL gel	7	10	5-10^5	20-10^7	25	30,60 × 7.7		AS
O	PL gel	7	5	5-10^5	20-10^7	40	30,60 × 7.7		AS
O	Poragel	4	37-75	6-200	100-2×10^4	—	—		Wa
O	Shodex A801-806	6	—	—	-5×10^7	26	30,50 × 7.8		SD,PE

O	Shodex H201-203	3	—	—	—	20	—		SD
O	μSherogel	7	—	$5\text{-}10^5$	-10^6	18	30 × 8		Ax
O	Styragel®	8	37-75	$100\text{-}10^8$	-4×10^8	(2.1)			Wa
O	μStyragel®	6	10	$100\text{-}10^6$	$50\text{-}2 \times 10^8$	12	30 × 7.8		Wa
O	TSK G1000H-7000H	8	—	$4\text{-}10^7$	-4×10^8	40	30 × 7.8	(THF)	TS, Va
O	TSK G1000HG-7000HG	8	—	$4\text{-}10^7$	-4×10^8	27	30 × 7.8	($CHCl_3$)	TS, Va
O	TSK S1000S-7000S	8	—	$4\text{-}10^7$	-4×10^8	—	—		TS
Sulfonated polystyrene gel (controlled pore-sized ion-exchange resin)									
A	Aquapack A440		37-75	—	-4×10^8				Wa
A	Ionpak S801-805	5	—	—	-10^7	14	50 × 7.8		SD
A	TSK LS211-217	7	6,12,18	—	-4×10^8	—	—		TS
Other semi-hard organic gels									
A	Merckogel OR-PGM 2000	1	63-100	—	−2500	—	—	Polyethylenedi-methylacrylate	EM
O	Merckogel OR-PVA	7	32-63	—	$500\text{-}10^6$	—	—	Polyvinylacetate	EM
A	OH Pak B804	1	—	—	-4×10^5	12	50 × 7.8	Hydroxypolyester	SD
A	PVA Pak Q801-803	3	—	—	-10^4	—	50 × 7.8	Polyvinylalcohol	SD
A	Sephasorb HP ultrafine		10-23	—		—	—	Hydroxypropylated dextran	Ph
A	Toyo Pearl HW 40-75	6	20-40,30-60,50-100		-5×10^7	—	—	Polyvinyl	TS
A	TSK G 1000W	1			60-600	20	30 × 7.8	Dextran	TS
A	TSK G 2000W	1			150-3000	20	30 × 7.8	Starch	TS
A	TSK G3000W-6000W	4			-10^7	20	30 × 7.8	Ethylene glycol	TS
A	TSK 2000PW-5000PW	4	10-17		-10^7		30,60 × 7.5	Hydroxylated polyether	TS, BR, Va
Soft organic gel									
A	Biogel® A 0.5-150m	6	45-850		$500\text{-}15 \times 10^7$			Agarose	BR
A	Biogel® P2-300	10	45-850		$2000\text{-}3 \times 10^6$			Polyacrylamide	BR
A	Sephadex G10-200	8	10-300	(4)	-6×10^5			Dextran sa = 10-400, bv = 2-40	Ph
A	Sephacryl S 200,300	2			$5000\text{-}15 \times 10^7$			Allyldextran with *N,N'*-methylene bis acrylamide, pH 3-11	Ph

Table 1 (continued)
COLUMNS AND PACKINGS FOR GEL PERMEATION CHROMATOGRAPHY

So	Name	No	PS μm	ps nm	MW	plate N 1000/m	Column size cm × mm i.d.	Remarks	Mfgr. & supplier
A	Sepharose 2B,4B,6B	3			10^4-4×10^7			Agarose, pH 4-9, 0-40°C	Ph
AO	Sepharose CL 2B,4B,6B	3			10^4-4×10^7			Agarose, pH 3-14, −70°C	Ph
AO	Sephadex LH20,60	2			10^3-7×10^4			Hydroxypropylated Sephadex pv = 1.6-13.6, pH 2-10	Ph

Abbreviations (I)
So = solvent, (A = aqueous, O = organic solvent)
No = Number of types
PS = particle size (μm)
ps = pore size (nm)
MW = molecular weight (generally it means exclusion limit of packing)
pv = pore volume (mℓ/g)
sa = surface area (m^2/g)
Abbreviations of mfd. and supplier are shown at the end of liquid chromatography with their address.

Table 2
INORGANIC POROUS PACKINGS

Name	Sa	PS μm	ps nm	sa m^2/g	pv mℓ/g	Plate N 1000/m	Column size cm × mm i.d.	Remarks	Mfgr. & supplier
Alumina									
BioRad® AG Alumina	I	−74		200					BR
Alox® 60D	I	5,10,20	6	60				Basic	MN,At,Ch,Ra,Su
LiChrosorb® AloxT 5,10	I	5,10,30		70-90	0.25-0.30			Basic (Mercksorb AloxT)	EM,At,AS,Ch,Ra, Re,Su.

Micropak®-Al	I	5,10		70				Prepacked	Va
Spherisorb A5Y,10Y,20Y	S	5,10,20		95					PS,At,AS,Ch,Ra, Re,Su.
Woelm Alumina A18	I	18-30		250				Acidic	Wm
Woelm Alumina B18	I	18-30		250				Basic	Wm
Woelm Alumina N18	I	18-30		250				Neutral	Wm
AL-10A	I	10		70-90	0.25-0.30	14	25 × 4.6	(LiChrosorb AloxT 10)	BL, Rh
AL-MP	I			70-90	0.25-0.30		5,10 × 4.6	(LiChrosorb AloxT)	BL,Rh
Diatomite									
Chromosorb® LC-6	I	37-44							JM,AL,AS,At,CS, Su.
Silica gel (see also columns and packings for gel permeation chromatography)									
Biogel® 4		37-44							BR
Biosil® HP-10		10		340		25	25 × 4		BR
Chromosorb® LC-6	I	10	13	300-350					JM,AL,AS,At,Ch, CS,Su
Chrompack	I	10	6	500		22	25 × 4.6	(LiChrosorb Si 60)	Ch
Chrompack	I	5	6	500		40	25 × 4.6	(LiChrosorb Si 60)	Ch
HI-Chrom reversible Si	S	5				45	25 × 4.6		Re
Hitachi 3040,1,3	S	30,10,5							Hi
Hypersil	S	5-7	10-3	200					SN,Ch
Chromegasorb R	I	5,10	6	500					ES
Chromegasorb 100R	I	5,10	10	—					ES
LiChrosorb® Si-60	I	5,7,10	6	500	0.75			(Mercksorb Si60)	EM,AS,At,Ch,BD,Ra,Su.
LiChrosorb® Si-100	I	5,10	10	400	1.0			(Mercksorb Si 100)	EM,AS,At,Ch,BD,Ra,Su.
LiChrospher Si 100	S	5,10	10	250	1.2				EM,AS,At,Ch,BD,Ra,Su.
LiChrospher Si 300	S	5,10	10-50	250	2.0				EM,AS,At,Ch,BD,Ra,Su.
LiChrospher Si 500	S	10	50	50	0.8				EM,AS,At,Ch,BD,Ra,Su.
LiChrospher Si 1000	S	10	100	20	0.8				EM,AS,At,Ch,BD,Ra,Su.
LiChrospher Si 4000	S	10	400	6	0.8				EM,AS,At,Ch,BD,Ra,Su.
LiChroprep Si 60	I	15-25,25-40	6	500	0.75				EM,AS,At,Ch,BD,Ra,Su.
Micropak SI	I	5,10	10	400					Va
Microsil		4.2	62.5	445		27	30 × 5		MM

Table 2 (continued)
INORGANIC POROUS PACKINGS

Name	Sa	PS μm	ps nm	sa m²/g	pv mℓ/g	Plate N 1000/m	Column size cm × mm i.d.	Remarks	Mfgr. & supplier
Nucleosil 50	S	5,7,10	5	500	0.8				MN,At,Ch,Ra,Su.
Nucleosil 100, 100V	S	5,10	10	400	1.0,1.5				MN,At,Ch,Ra,Su.
Partisil® 5,10,20	I	5,10,20	5.5-6	400	0.7				Wh,At,Ch,Ra.
Partisil® 5	I	5	5.5-6	400	0.7	40	25 × 4.6		Wh,At
Partisil® 10	I	10	5.5-6	400	0.7	24	25 × 4.6		Wh,At
Polygosil (Sil 60D)	I	5-30 (6)	6	500	0.75				MN,At,Ch.
Porasil A-F (6)	S	35-75	10-150	500-8					Wa
Porasil T	I	15-25	15	300					Wa
μPorasil	I	10	15	300-350		9	30 × 3.9		Wa
SI-5A	I	5	10	400	1.0	36	25 × 4.6	(LiChrosorb Si 100)	BL,Rh
SI-10A	I	10	10	400	1.0	20	25 × 4.6	(LiChrosorb Si 100)	BL,Rh
SO1-10A	S	10	10	250	1.2	20	25 × 4.6	(LiChrospher Si 100)	BL,Rh
SO3-10A	S	10	10-50	250	2.0	20	25 × 4.6	(LiChrospher Si 300)	BL,Rh
SO5-10A	S	10	50	50	0.8	16	25 × 4.6	(LiChrospher Si 500)	BL,Rh
S10-10A	S	10	100	20	0.8	12	25 × 4.6	(LiChrospher Si 1000)	BL,Rh
S40-10A	S	10	400	6	0.8	8	25 × 4.6	(LiChrospher Si 4000)	BL,Rh
SI-MP	I			400	1.0		5,10 × 4.6		BL,Rh
Spherosil	S	7	3.5,9	580,860	0.78	1.3			Su
Sil X-I,II	I	13,35		400,12					PE
Silarex SR 1		10	6	500		24	25 × 4.6		Li
Silarex SR II		10	10	350		28	25 × 4.6		Li
Spherisorb S5W,10W	S	5,10	8 (5.5-11)	200-220					PS,AS,At,Ch,Ra,Re,Su.
Supelcosil LC-Si	S	5				65-75	15,25 × 4.6		Su
TSK LS 310	S	5,10,15							TS
Ultrashere Si 5	S	5				50	15,25 × 4.6		Ax,BM
Ultrasil Si 10	I	10					25 × 4.6,10		Ax
Vydac TP Adsorbent	S	10	33	100	0.6				SG,AL,AS,At,Ch.
Zorbax SIL	S	5-6	7	400			15,25 × 4.6, 25 × 6.2		DP

Table 3
INORGANIC BONDED PACKINGS AND COLUMNS

Organic phase	Name	Basic support or packing	Plate N 1000/m	Column size cm × mm i.d.	Remarks	Mfd. & supplier
Dimethyl	Chromegabond C2	Chromegasorb 100R				ES
	CP 10A	LiChrosorb RP2	16	25 × 4.6		BL,Rh
	CP MP	LiChrosorb RP2		5,10 × 4.6		BL,Rh
	LiChroprep RP2	LiChroprep Si 60				EM,AS,At,Ch,BD, Ra,Su.
	LiChrosorb® RP2	LiChrosorb Si 60				EM,AS,At,Ch,BD, Ra,Su.
Trimethyl	Supelcosil LC-1	Supelcosil LC-Si	62-70	15,25 × 4.6		Su
	Zorbax TMS	ZorbaxSIL	32	15,25 × 4.6		DP
Hexyl	Spherisorb C6	Spherisorb S5W				PS,AS,At,Ch,Ra, Re,Su.
	HI-CHROM reversible C6	HI-CROM reversible Si	45	25 × 4.6	σ = 0.6 mM/g	Re
Cyclohexyl	Chromegabond C6H11	Chromegasorb 100R				ES
Phenyl	Allylphenyl Sil X-I (phenyl Sil X-1)	Sil X-I				PE
	Bondapak phenyl/Porasil B	Porasil B			o = 10%	Wa
	μBondapak phenyl	μPorasil	9	30 × 3.9	o = 10%	Wa
Octyl	Chromegabond C8	Chromegasorb 100R				ES
	Chrompack RP8	LiChrosorb RP8 (10 μm)	18	25 × 4.6		Ch
	Chrompack RP8	LiChrosorb RP8 (5 μm)	30	25 × 4.6		Ch
	LiChroprep RP8	LiChroprep Si 60				EM,AS,At,BD,Ch, Ra,Su.
	LiChrosorb RP8	LiChrosorb Si 100				EM,AS,At,BD,Ch,Ra,Su.
	Microsil C8	Microsil	27	30 × 5	sa = 272, PS = 4.4	MM
	Nucleosil C8	Nucleosil 100				MN,At,Ch,Ra,Su.
	Polygosil C8(Sil 60D C8)	Polygosil				MN,At.
	RP 10A	LiChrosorb RP8 (10 μm)	18	25 × 4.6		BL,Rh
	RP 5A	LiChrosorb RP8 (5 μm)	36	25 × 4.6		BL,Rh
	RP MA	LiChrosorb RP8				BL,Rh
	Sil 60D C8	Sil 60D				MN
	Silarex FNP8	Silarex SR 1	16	25 × 4.6		Li

Table 3 (continued)
INORGANIC BONDED PACKINGS AND COLUMNS

Organic phase	Name	Basic support or packing	Plate N 1000/m	Column size cm × mm i.d.	Remarks	Mfgr. & supplier
	Supelcosil LC-8	Supelcosil LC-Si	65-80	15,25 × 4.6	Octyldimethyl	Su
	Ultrasil octyl	Ultrasil Si 10		25 × 4.6,10		Ax,BM
	Ultrasphere octyl 5	Ultrashere Si 5	50	15,25 × 4.6		Ax,BM
	Zorbax C8	Zorbax SIL		15,25 × 4.6		DP
Octadecyl	18-10A	LiChrosorb RP18 (10 μm)	12	25 × 4.6		BL,Rh
	18-5A	LiChrosorb RP18 (5 μm)	24	25 × 4.6		BL,Rh
	18-MP	LiChrosorb RP18		5,10 × 4.6		BL,Rh
	BioSil® ODS 10	BioSil® HP 10	20	25 × 4		BR
	Bondapak C18/Porasil B	Porasil B				Wa
	μBondapak C18	μPorasil	9	30 × 3.9	o = 10%	Wa
	Chromegabond C18	Chromegasorb 100R				ES
	Chrompack RP18	LiChrosorb® RP18 (10 μm)	18	25 × 4.6		Ch
	Chrompack RP18	LiChrosorb® RP18 (5 μm)	30	25 × 4.6		Ch
	Chromosorb LC-7	Chromosorb LC-6			o = 15%	JM,AL,AS,At,Ch,CS,Su.
	HI-CROM reversible ODS	HI-CROM reversible Si	45	25 × 4.6	σ = 0.3mM/g	Re
	Hitachi 3050,3051,3053	Hitachi 3040,3041,3043				Hi
	Hypersil ODS	Hypersil			Si-C	SH,Ch
	LiChroprep RP18	LiChroprep Si 60				EM,AS,At,Ch,BD,Ra,Su.
	LiChrosorb RP18	LiChrosorb Si 100			Si-O-Si-C	EM,AS,At,Ch,BD,Ra,Su.
	Micropak CH	Micropak SI				Va
	Microsil C18	Microsil	27	30 × 5	sa = 272, PS = 4.4	MM
	Nucleosil C18	Nucleosil 100				MN,At,Ch,Ra,Su.
	ODS Sil X-I,II	Sil X-I,II				PE
	Partisil® 5-ODS	Partisil® 5	35	25 × 4.6	o = 5%	Wh,At,Ch,Ra.
	Partisil® 10-ODS	Partisil® 10	15	25 × 4.6	o = 5%	Wh,At,Ch,Ra.
	Partisil® 10-ODS-2	Partisil® 10	25	25 × 4.6	o = 15%	Wh,At,Ch,Ra
	Partisil® 10-ODS-3	Partisil® 10	20	25 × 4.6		Wh,At,Ch,Ra.
	Polygosil C18	Polygosil				MN,At,Ch
	Sil 60D C18	Sil 60D				MN,Ch
	Sherisorb ODS	Spherisorb S5W,10W				PS,AS,At,Ch,Ra,Re,Su.
	Shodex ODS pack	Shodex Silicapak				SD

	Silarex FNP18	Silarex SR II	16	25 × 4.6		Li
	Supelcosil LC-18	Supelcosil LC-Si	65-80	25 × 4.6	octadecyldimethyl	Su
	TSK LS 410	TSK LS 310				TS
	Ultrasil ODS 10	Ultrasil 10		25 × 4.6,10		Ax,BM
	Ultrasphere ODS 5	Ultrasphere Si 5	50	15,25 × 4.6		Ax,BM
	Vydac TP reverse phase C18	Vydac TP				SG,Al,AS,At,Ch.
	Zorbax ODS	Zorbax SIL	30	15,25 × 4.6	o = 25%	DP
Diol	Chromegabond R(OH)2	Chromegasorb 100R				ES
	LiChrosorb® DIOL	LiChrosorb® Si 100				EM,AS,At,BD,Ch,Ra,Su.
	OH-10A	LiChrosorb® DIOL	16	25 × 4.6		BL,Rh
	OH-MP	LiChrosorb® DIOL		5,10 × 4.6		BL,Rh
Fluoro	Fluoro ether Sil X-I	Sil X-I				PE
Polystyrene	TSK LS 510	TSK LS 310				TS
Nitrile	μBondapak CN	μPorasil®	9	30 × 3.9	o = 9%	Wa
	Chromosorb LC-8	Chromosorb LC-6				JM,AL,AS,At,Ch,CS,Su.
	Cyano Sil X-I	Sil X-I				PE
	HI-CROM reversible CN	HI-CROM reversible Si	45	25 × 4.6	σ = 0.6mM/g	Re
	Micropak CN	Micropak SI				Va
	Microsil CN	Microsil (10 μm)		30 × 5	PS = 10	MM
	Nucleosil CN	Nucleosil 100				MN,At,Ch,Ra,Su.
	Polygosil CN(Sil 60D CN)	Polygosil				MN,At,Ch.
	Sil 60D CN	Sil 60D				MN
	Partisil® PAC	Partisil® 10	15	25 × 4.6		Wh,At,Ch,Ra.
	Spherisorb CN	Spherisorb S5W, 10W				PS,AS,At,Ch,Ra,Re,Su.
	Silarex CN I	Silarex SR I	16	25 × 4.6		Li
	Vydac TP bonded phase	Vydac TP			Si-R-O-CH_2CH_2CN	SG,AL,AS,At,Ch.
	Zorbax CN	Zorbax SIL				DP
Nitro	Nucleosil NO_2	Nucleosil 100				MN,At,Ch,Ra,Su.
	Polygosil NO_2(Sil 60DNO_2)	Polygosil				MN,At,Ch.
	Sil 60D NO_2	Sil 60D				MN
Primary-amine (NH_2)						
	Amino Sil-X-I	Sil-X-I				PE
	μBondapak NH_2	μPorasil	9	30 × 3.9	o = 9%,pH2-8,ec = 0.3-0.4meq/g	Wa
	Chromegabond NH_2	Chromegasorb 100R				ES
	Chromosorb LC-9	Chromosorb LC-6				JM,AL,AS,At,Ch,CS,Su.

Table 3 (continued)
INORGANIC BONDED PACKINGS AND COLUMNS

Organic phase	Name	Basic support or packing	Plate N 1000/m	Column size cm × mm i.d.	Remarks	Mfgr. & supplier
	H2 10A	LiChrosorb® NH_2	18	25 × 4.6		BL,Rh
	H2 MP	LiChrosorb® NH_2		5,10 × 4.6		BL,Rh
	HI-CROM reversible NH_2	HI-CROM reversible Si	45	25 × 4.6	$\sigma = 0.6$mM/g	Re
	Hypersil APS	Hypersil				SN,Ch
	LiChrosorb® NH_2	LiChrosorb® SI 100				EM,AS,At,BD,Ch,Ra,Su.
	Micropak NH_2	Micropak SI				Va
	Microsil NH_2	Microsil (10 μm)		30 × 5		MM
	Nucleosil NH_2	Nucleosil 100				MN,At,Ch,Ra,Su.
	Spherisorb S5 NH_2	Sherisorb S5W			Aminopropyl	PS,AS,At,Ch,Ra,Re,Su.
	Sil 60D NH_2	Sil 60D (Polygosil NH_2 - Polygosil)				MN,At,Ch
	Ultrasil NH_2	Ultrasil		25 × 4.6,10		Ax,BM
Dimethylamine -$N(CH_3)_2$	Nucleosil $N(CH_3)_2$	Nucleosil 100				MN,At,Ch,Ra,Su.
	Polygosil $N(CH_3)_2$	Polygosil				MN,At,Ch.
	Sil 60D $N(CH_3)_2$	Sil 60 D (Polygosil $N(CH_3)2$)				MN
Trialkylamine (strong anion-exchanger)	AX 10A	LiChrosorb® AX	16	25 × 4.6		BL,Rh
	AX MP	LiChrosorb® AX		5,10 × 4.6		BL,Rh
(-$N^+Me_3CL^-$)	LiChrosorb® AX	LiChrosorb® Si 100			ec = 1meq/g	EM,AS,At,BD,Ch,Ra,Su.
	Microsil SAX	Microsil (10 μm)		30 × 5	NMe_3Cl, pH 1-9	MM
	Nucleosil SB	Nucleosil 100			ec = 1meq/g, pH 1-9	MN,At,Ch,Ra,Su.
	Partisil® 10 SCX	Partisil® 10	15	25 × 4.6	ec = 1meq/g, pH 1.5-10, $NMe_3H_2PO_4$	Wh,At,Ch,Ra.
	Polygosil SB	Polygosil			NMe_3Cl	MN,At,Ch.
	Ultrasil AX	Ultrasil		25 × 4.6		Ax,BM
	Vydac TP anion exchange	Vydac TP			NMe_3Cl	SG,AL,AS,At,Ch.
DEAE	Synchropak AX 300			25 × 4.6		SC,AS,Re.
	TSK IEX 540 DEAE	TSK 2000 SW				TS
Sulfonate	Chromegabond SO_3H (CAT-SCX)	Chromegasorb 100R				ES
	CX 10A	LiChrosorb® CX	16	25 × 4.6		BL,Rh
	CX MP	LiChrosorb® CX		5,10 × 4.6		BL,Rh

LiChrosorb® CX (KAT)	LiChrosil Si 100			SO_3H, ec = 1meq/g	EM,AS,At,BD,Ch,Ra,Su.
Microsil SCX	Microsil (10 μm, 10 nm)		30 × 5	SO_3H, pH 1-9	MM
Nucleosil SA	Nucleosil 100			ec = 1meq/g, pH 1-9	MN,At,Ch,Ra,Su.
Partisil® SCX	Partisil® 10	10	25 × 4.6	ec = 1meq/g, pH 1.5-10, SO_3NH_4	Wh,At,Ch,Ra.
Polygosil SA	Polygosil			SO_3Na	MN,At,Ch.
Ultrasil CX	Ultrasil		25 × 4.6	Si-$(CH_2)_2$Ph-SO_3H, pH 2-7.5	Ax,BM
Vydac TP cation exchange	Vydac TP			SO_3H	SG,AL,AS,At,Ch.

Table 4
INORGANIC PELLICULAR PACKINGS

Material	Name	Sa	PS μm	ps nm	sa m^2/g	Remarks	Mfgr. & supplier
Alumina	Chromosorb LC-3	S	37-44				JM,AL,AS,AT,Ch,CS,Su.
	HS Pellumina	S	37-44		4	Mono layer	Wh,At,Ch,Ra.
	HC Pellumina	S	37-44		8	Double layer	Wh,At,Ch,Ra.
Diatomite	Diachrom	I	37-44		10		AS
Silica	Corasil	S	37-50	5	25-30	Capacity = 1-2 mg/column	Wa
	Chromosorb LC-2	S	37-44		15-30		JM,AL,AS,At,Ch,CS,Su.
	HS Pellosil	S	37-44		4	Mono layer	Wh,At,Ch,Ra.
	HC Pellosil	S	37-44		8	Double layer	Wh,At,Ch,Ra.
	Perisorb A	S	30-40	6	14		EM,AS,At,BD,Ch,Ra,Su.
	Vydac 101 Si	S	30-44	5.7	12	Density = 1.4g/mℓ	SG,AL,AS,At,Ch.
	Zipax	S	25-37	100	0.8-1	Density = 1.5g/mℓ, thickness = 1-2μm	DP
Silicaglass	Liqua-Chrom	I	49-53		10		AS

Table 5
PELLICULAR BONDED PACKINGS

Organic	Name	Basic support	PS μm	sa m^2/g	ec μeq/g	Remarks	Mfgr. & supplier
Carbon	Pellicarb	Pellosil					Wh,At,Ch,Ra.
Dimethyl	Perisorb RP2	Perisorb A	30-40				EM,AS,At,BD,Ch,Ra,Su.
Ether	Permaphase ETH	Zipax	25-37	1			DP
Phenyl	Bondapak phenyl/Corasil	Corasil	37-50			o = 0.8%, diphenyldichlorosilane	Wa
	Chromosorb LC-5	Chromosorb LC-2	37-44				JM,AL,AS,At,Ch,CS,Su.
Octyl	Perisorb RP8	Perisorb A	30-40				EM,AL,AS,At,Ch,CS,Su.
Octadecyl	BondapakC18/Corasil	Corasil	37-60	7		o = 1%, octadecyltrichlorosilane	Wa
	Chromosorb LC-4	Chromosorb LC-2	37-44				JM,AL,AS,At,Ch,CS,Su.
	Co: Pell ODS	Pellosil	37-44				Wh,At,Ch,Ra.
	Perisorb RP18	Perisorb A	30-40				EM,AS,At,BD,Ch,Ra,Su.
	Permaphase ODS	Zipax	25-37	1			DP
	Vydac 201 RP	Vydac 101 Si	30-44				SG,AL,AS,At,Ch.
Nitrile	Co: Pell PAC	Pellosil	37-44				Wh,At,Ch,Ra.
	Vydac 501 PP	Vydac 101 Si	30-44				SG,AL,AS,At,Ch.
Polyamide	Perisorb PA6	Perisorb A	30-44				EM,AS,At,BD,Ch,Ra,Su.
Polycaprolactam	Pellamidon	Pellosil	37-44				Wh,At,Ch,Ra.
Primary amine	AL-Pellionex WAX	Pellosil	44-53	1			Wh,At,Ch,Ra.
	HL-Pellionex WAX	Pellosil	44-53				Wh,At,Ch,Ra.
Trialkylamine	AS-Pelliones SAX	Pellosil	44-53	1	10		Wh,At,Ch,Ra.
	Bondapak AX/Corasil	Corasil	37-50	7	10	o = 1.3%, pH = 2-7.5, $SiONMe_3Cl$	Wa
	Pellicular anion		-40		10		Va
	Perisorb AN	Perisorb A	30-40		30		EM,AS,At,BD,Ch,Ra,Su.
	Permaphase AAS	Zipax	25-37		12		DP
	Vydac 301 SB	Vydac 101 Si	30-44		100		SG,AL,AS,At,Ch.
Sulfonate	Bondapak CX/Corasil	Corasil	37-50	7	35	o = 1.1%, $SiO\text{-}Ph\text{-}SO_3Na$	Wa
	HC Pelliones SCX	Pellosil	44-53		60	SO_3H	Wh,At,Ch,Ra.
	Pellicular cation		-40		10		Va
	Perisorb KAT	Perisorb A	30-40		50	SO_3H	EM,AS,At,BD,Ch,Ra,Su.
	Vydac 401 SA	Vydac 101 Si	30-44		100	SO_3H	SG,AL,AS,At,Ch.

Table 6
ORGANIC POLYMER PACKINGS (EXCLUDING PACKINGS FOR GPC AND ION-EXCHANGE)

Polyer	Name	PS (μm)	Remarks	Mfgr. & supplier
Polystyrene	DN-X12-11	11 ± 1		Du
	DN-X8-11	11 ± 1		Du
	DN-X2-11	11 ± 1		Du
	HITACHI 3010,3011,3013,3019	20,10,5,35	ps = 4, sa = 300	Hi
	HN-X2.00 - X35.00 (24 types)	50,30,7-10		Ha
	Shodex polymer HP125,225	12.5, 25	sa = 500	SD
	TSK-LS-110 P5,P10,P15	5,10,15	sa = 250	TS
	TSK-LS-111 P5,P10,P15	5,10,15	sa = 250	TS
Polyacrylate	HITACHI 3030	20		Hi
	HN-X20-11	11 ± 1		Du
	TSK-LS-120, 130	5,10,15	sa = 250	TS
Polyvinylacetate	TSK-LS-140	5,10,15	sa = 250	TS
Polystyrene-benzyl-alcohol	HITACHI 3011-0	10		Hi
	TSK-LS-150	5,10,15	sa = 250	TS
Polyglycol	HITACHI 3020	25		Hi
	TSK-LS-160	5,10,15	sa = 250	TS
Starch	TSK-LS-170	5,10,15	sa = 250	TS
Hydroxypolyester	Shodex OH pak	15-20,20-25		SD
Polyvinylalcohol	Shodex PVA pak			SD
Polyamide 6	PA-6	5-20,20-32		MN
Acetylated cellulose	CEL AC 30X, 40X	15-35	acetyl content = 30 & 40%	MN

Note: See also packings for gel permeation chromatography.

Table 7
ORGANIC ION-EXCHANGERS

Exchange group	Name	ec (μeq/g)	PS (μm)	Remarks	Mfgr. & supplier
Sulfonate (strong cation)	Aminex A4-A7 (4 types)	1700	7-24 (4)*	X = 8	BR
	Aminex A8,A9	2000	5-8,11.5-0.5		BR
	Aminex Q150S, Q15S	1700	28±7,22±3	X = 8	BR
	Aminex 50W X4	1200	25±5	X = 4	BR
	Aminex A9 packed column, 25 cm long × 4 mm, N = 8000/m				BR
	AG MP-50	1700	38-850 (4)*	sa = 35	BR
	Dowex® AG 50W-X2-X16 (6)*	700-2600	38-1180 (5)*	X = 2,4,8,10,12,16	Do, BR
	Durrum DC1A-6A (4)*		6-21 (4)*		Du
	DC-X12-11	5000	11±1	X = 12	Du
	DC-X8-11	5000	11±1	X = 8	Du
	HC X2-X35 (24)*		7-25 (4)*	X = 2-35	Ha
	Hitachi 2611		13-17	X = 8	Hi
	Hitachi 3011-S		10		Hi
	Shodex HC125,HC175	4000	12.5,17.5		SD
	TSK LS 211	4600	6,12,18	ps = 10	TS
	TSK LS 212,213	4200	6,12,18	ps = 100,1000	TS
	TSK LS 214-217(4)*	2700	6,12,18	ps = 10^4-10^7(4)*	TS
	10 NEX SA-10,15,20,052	5000	5,10,20		BR
Carbonate (weak cation)	BioRex 70	3300	75-1180(5)*		BR
	HITACHI 3011-C		10-15		Hi
	TSK LS 230	7500	6,12,18		TS
Carboxymethyl	CM-Cellulose				Wh
	CM Sephadex®	4500±500			Ph
	CM Sepharose	120±20			Ph
	Cellex CM	700±100		Cellulose	BR
Sulfopropyl	SP Sephadex®	2300±300			Ph
Phosphate	Cellex P	850±100		Cellulose	BR

Quaternary amine	Aminex A,14	1200	20±3	X = 4	BR
(strong anion)	Aminex A25-29	1400	5-20(4)*	X = 8	BR
	Aminex A27 packed column, 25 cm long × 4mm i.d., N = 8000/m				BR
	Dowex AG1-X2,X4,X8	700-1400	45-850(5)*	X = 2,4,8	Do,BR
	AG MP-1	1200	38-850(4)*	sa = 23	BR
	Durrum DAX2,X4,X6	2000	20±5	X = 2,4,6	Du
	DA-X12-11,X4-11	4000	10±1	X = 4,8,12	Du
	HA X4-X10		7-25(2)*	X4,6,8,10	Ha
	HITACHI 2630,2632		16-20	X = 4,8	Hi
Starch	TSK-LS-170		5,10,15	sa = 250	TS
Hydroxypolyester	Shodex OH pak		15-20,20-25		SD
Polyvinylalcohol	Shodex PVA pak				SD
Polyamide 6	PA-6		5-20,20-32		MN
Acetylated cellulose	CEL AC 30X, 40X		15-35	acetyl content = 30 & 40%	MN
	HITACHI 3011-N,3013-n		10,5		Hi
	MCI GEL CDR-10		6±1		Mi
	TSK LS 221	4600	6, 12, 18	ps = 10	TS
	TSK LS 222, 223	4200	6, 12, 18	ps = 100, 1000	TS
	TSK LS 224-227 (4)*	2700	6, 12, 18	ps = 10^4-10^7 (4)*	TS
Tertiary amine	AG2-X8, X10	1400,1500	45-850 (4)*	X = 8, 10 (strong anion)	BR
	Dowex AG3-X4A	1900	45-1180 (3)*	X = 4 (weak anion)	Do, BR
	TSK LS 240	4300	6, 12, 18	(weak anion)	TS
Pyridium	BioRex 9	1300	45-1180 (4)*		BR
Quat. + Tert. amine	BioRex 5	2800	45-1180 (4)*		BR
Imminodiacetate	Chelex 100	900	45-850 (4)*		BR
base + acid	AG3-X4A			Ph-$CH_2N^+Me_3$ + R-CH_2-COO^-	BR
Diethylaminoethyl	DEAE Cellulose				Wh
	DEAE Sephacel	1400 + 100	40-160		Ph
	DEAE Sephadex A-25, 50	3500 + 500			Ph
	DEAE Sepharose CL-6B	130 + 20		pH 3-10	Ph
	Cellex D	400-900		cellulose	BR
2-Hydroxypropylamino	QAE Sephadex	3000 + 400		pH 2-12	Ph

Table 7 (continued)
ORGANIC ION-EXCHANGERS

Exchange group	Name	ec (μeq/g)	PS (μm)	Remarks	Mfgr. & supplier
	Cellex QAE	800 + 100		cellulose	BR
Triethylaminoethyl (TEAE)	Cellex T	500 + 100		cellulose	BR
ECTEOA (mixed amine)	Cellex E	300 + 50		cellulose	BR
Benzoylated DEAE	Cellex BD	380 + 80		cellulose	BR

Note: See also packings for gel permeation chromatography.

Table 8
CONTROLLED PORE GLASS PACKINGS WITH ORGANIC PHASE

Name	Organic group	PS (μm)	ps (nm)	pv (mℓ/g)	ec (meq/g)	Mfgr.
DEAE Glycophase	Diethylaminoethyl	37-74	20	1.0	0.1	Pi
CM Glycophase	Carboxymethyl	37-74	20	1.0	0.1	Pi
QAE Glycophase	Triethylammonium chloride	37-74	20	1.0	0.05	Pi
SP Glycophase	Sulfonic	37-74	20	1.0	0.05	Pi
Long chain alkylamine	$Si-(CH_2)_6-NH_2$	125-177	50	1.0	0.17	Pi
Aminopropyl	$Si-(CH_2)_3-NH_2$	125-177	50	1.0	0.1	Pi
Carboxy		125-177	50	1.0	0.1	Pi
NHS Glycophase	*N*-hydroxysuccinimide	125-177	50			Pi
Diazonium Salt		125-177	50			Pi

Abbreviations (II)

σ = surface concentration ($\mu M/m^2$ or mM/g)

o = organic weight percent

ec = exchange capacity (meq/g or μeq/g)

Sa = shape of packings (S = spheroidal, I = irregular)

MANUFACTURERS AND SUPPLIERS

Symbol	Company and address
AL	Analabs, 80 Republic Drive, North Haven, CT 096475
AS	Applied Science, P.O. Box 400, State College, PA 16801
At	Alltech Associates, 2051 Waukegan Road, Deerfield, Il 60015
BD	BDH Chemicals, 350 Evans Avenue, Toronto, Ont. M8Z 1K5
BL	Brownlee Labs, 2065 Martin Avenue (107), Santa Clara, CA 95050
BR	BioRad Laboratories, 2200 Wright Avenue, Richmond, CA 94804
Ch	Chrompack, P.O. Box 4131 Whittier, CA 90607
CS	Chromatographic Specialties Limited, P.O. BAG 1150, 300 Laurier Blvd. Brockville, Ont. K6V 5W1
CX	Chromatix Inc., 560 Oakmead Parkway, Sunnyvale, CA 94086
Do	Dow Chemical Co., Midland, MI
DP	DuPont Company Scientific & Process Instruments Divisions, Concord Plaza,McKean Building, Wilmington, DE 19898
Du	Durrum, 1228 Titan Way, Sunnyvale, CA 94086
ES	ES Industries, 8 S Maple Avenue, Marlton, NJ 08053
Ge	Gelman Instrument Company, 600 South Wagner Road, Ann Arbor, MI 48106
Ha	Hamilton Company, P.O. Box 10030, Reno, NV 89510
Hi	NSI Hitachi Scientific Instrument, 450 East Middle Field Road, Mountain View, CA 94043
JM	Johns-Manville Sales Corporation, Ken-Caryl Ranch, Denver, CO 80217
Li	Lion Technology Inc., P.O. Drawer 898, Dover, NJ 07801
MM	Micrometrics Instrument Corporation, 5680 Goshen Springs Road, Norcross, GA 30093
MN	Macherey-Nagel Co. GmbH, Werkstr 6-8, Postfach 307, D-5160, Düren, Germany
PE	Perkin Elmer Corporation, Norwalk, CT 06856
Ph	Pharmacia Fine Chemicals AB, Box 175, S-75104 Uppsala 1, Sweden
Pi	Pierce Chemical Company, Rockford, Il 61105
PS	Phase Separation Limited, Deeside Industrial Estate, Queensferry Clwyd, GB
Ra	Rainin Instrument Co. Inc., Mack Road, Woburn, MA 01801
Re	Regis Chemical Company, 8210 Austin, Morton Grove, Il 60053
Rh	Rheodyne Incorporated, 2809 Tenth Street, Berkeley, CA 94710
SD	The Separation Group, P.O. Box 867, 16640 Spruce Street, Hespecia, CA 92345
SL	Separation Labs Inc., P.O. Box 1744, Bloomfield, NJ 07003
SN	Shadon Southern Products Ltd., 93/96 Chadwick Road, Astwoor Industrial Estate, Runcorn, Cheshire, WA7 1PR
SS	Schleicher & Schüll Inc., Keene, NH 03431
Su	Supelco Inc., Supelco Park, Bellefonte, PA 16823
TS	Toyosoda Manufacturing Co. Ltd., 762 Marietta Blvd. N.W., Atlanta, GA 30318
Va	Varian Instrument Group, 611 Hansen Way, Palo Alto, CA 94303
Wa	Waters Associates Inc., Milford, MA 01757
Wh	Whatman Inc., 9 Brindewell Place, Clifton, NJ 07014
Wm	Woelm

Thin-Layer Chromatography

Thin-layers and adsorbents are supplied by following companies (See also Volume II of the Handbook, p. 310-317).

Adsorbent	Mfd.	Name
Alumina	Analabs	Anasil AG, AGF
	E. Merck	Aluminum oxide 60F 254, 150F 254, Preparative Aluminum oxide
	Whatman	K3/K3F
Cellulose	Analabs	Anasil C, CF
	Applied Sci.	Permakotes microcrystalline cellulose
	E. Merck	Cellulose, Cellulose F, PEI Cellulose
	Schleicher & Schüll	G1440/G1440 LS254 Cellulose, G1800/21 or 45 acetylated-cellulose, G1440/A5, A10 Cellulose W/Dowex 2-X8, G1440/K5, K10 Cellulose W/Dowex 50W-X8, G1440 PEI/G1440 PEI LS254 PEI cellulose.
	Whatman	K2/K2F
Silica	Analabs	Anasil O, OF, G, GF, H, HF
	Applied Sci.	Permakotes Adsorbosil 1, 1-P, 5, 5-P Permakotes Silicagel G, GF, H, HF Safety-Kotes, Silufol, Thick layer.
	E. Merck	Silica gel 60, 60F-254, RP-2, RP-2F254, RP-8 F254, RP 18 F254. (Kieselguhr F254, Silica gel 60/Kieselguhr F254) Preparative Silica gel 60, 60F254. Preconcentrating Zone TLC (Silica gel 60, 60F254).
	Gelman	ITLC SA/SAF, ITLC SG.
	Macherey-Nagel	Sil 20 Nano, Sil 20 UV Nano.
	Schleicher & Schüll	G1500/G1500 LS254, G1505, G1510.
	Whatman	LK6/LK6F, LK5/LK5F, MKC 18F, KC18F/KC18D8, HP-KF, MK6F, K6/K6F, K4/K4F, K1/K1F, MK1F
Polyamide	Schleicher & Schüll	G1600/G1600 LS254, F1700.

Paper Chromatography

See Volume II of the Handbook p. 304-309 for chromatographic papers.

Section II VIII

REVIEWS AND BOOKS FOR CHROMATOGRAPHY OF PHENOLS AND ORGANIC ACIDS

1. **Blau, K. and King, G.,** *Handbooks of Derivatives for Chromatography,* Heyden & Son Ltd., 1977.
2. **Churáček J. and Coupek, J.,** Phenols, in *Liquid Column Chromatography,* Deyl, Z., Macek, K., and Janák, J., Eds., Elsevier, Amsterdam, 1975.
3. **Harborne, J. B.,** Chromatography of phenolic compounds, in *Chromatography,* Heftmann, E., Ed., Van Nostrand, New York, 1975.
4. **Jandera, P. and Churáček, J.,** Ion-exchange chromatography of carboxylic acids, *J. Chromatogr.,* 86, 351, 1973.
5. **Jandera, P. and Churáček, J.,** Ion-exchange chromatography of sulphur compounds, phenols, phosphorus compounds and esters of carboxylic acids, *J. Chromatogr.,* 86, 423, 1973.
6. **Khym, J. X.,** Analytical ion-exchange procedures, in *Chemistry and Biology,* Prentice-Hall, Englewood Cliffs, N.J., 1974.
7. **Knapp, D. R.,** *Handbook of Analytical Derivatization Reactions,* John Wiley & Sons, New York, 1979.
8. **Sergeeva, U. N.,** *Chromatographic Analysis in the Chemistry of Wood,* Zinatne, 1975.

Section III

Indexes

Compound	Synonym	Number of tables and/or figures
Abietic acid		GC-25,46,
Abscisic acid		LC-21
		TLC-24
Acetic acid		GC-3,24,28
		LC-14,17,18,28
		TLC-17,22
Acetoacetic acid	—3-Oxobutanoic acid	
Acetobenzenepentacarboxylic acid		GC-41
Acetoxyacetic acid		PC-6
2-Acetoxybenzoic acid		GC-25
		LC-30
		TLC-29
4-Acetoxybutanoic acid		PC-6
4-Acetyl-1,2-dimethoxybenzene		GC-6
2-Acetylmethoxybenzene		GC-6
3-Acetylmethoxybenzene		GC-6
4-Acetylmethoxybenzene		GC-6
Acetylportentol		TLC-38
2-Acetylsalicylic acid	—2-Acetoxybenzoic acid	
Aconitic acid	—1,2,3-propanetricarboxylic acid	
cis-Aconitic acid		GC-22,28
		LC-17,18,28
		TLC-18,20
trans-Aconitic acid		GC-25,29,32
		LC-18,28
		TLC-18,20
Acrylic acid	—propenoic acid	
Adipic acid	—hexanedioic acid	
Alectoronic acid		TLC-38
Aletorialic acid		TLC-38
Allaric acid		GC-48
Allonic acid		GC-48
Allonic acid, 1,4-lactone		GC-48
4-Allyl-1,2-dimethoxybenzene		GC-6
4-Allylmethoxybenzene		GC-6
4-Allyl-2-methoxyphenol		GC-10
		LC-6,14
		TLC-1
Altraric acid		GC-48
Altronic acid		GC-48
Altronic acid 1,4-lactone		GC-48
5-Amino-1,3-benzenedicarboxylic acid		LC-30
		PC-8
2-Aminobenzoic acid		GC-45
		LC-15,19,20
		TLC-22
		PC-8
3-Aminobenzoic acid		LC-19,30
		TLC-22
		PC-8
4-Aminobenzoic acid		GC-25
		LC-19,20,30
		TLC-22
		PC-8
6-Aminocaproic acid	—6-Aminohexanoic acid	
3-Amino-2,5-dicolorobenzoic acid (chloramben)		LC-25
2-Amino-4,6-dinitrophenol		TLC-3
2-(4′-Amino-3′,5′-dinitrophenyl)-2-(4″-nitrophenyl)propane		LC-7
2-(4′-Amino-3′,5′-dinitrophenyl)-2-phenylpropane		LC-7

Compound	Synonym	Number of tables and/or figures
6-Aminohexanoic acid		GC-25
3-Amino-4-hydroxybenzenesulfonic acid		TLC-3
2-Amino-3-hydroxybenzoic acid		TLC-3
3-Amino-2-hydroxybenzoic acid		PC-8
4-Amino-2-hydroxybenzoic acid		LC-3
		TLC-3
		PC-8
5-Amino-2-hydroxybenzoic acid		TLC-3
1-Amino-7-hydroxynaphthalene		TLC-3
5-Amino-1-hydroxynaphthalene		TLC-3
7-Amino-2-hydroxynaphthalene		TLC-3
8-Amino-1-hydroxy-3,6-naphthalenedi-sulfonic acid		LC-30
2-(3′-Amino-4′-hydroxy-5′-nitrophenyl)-2-phenylpropane		LC-7
4-Amino-methylbenzene		LC-11
2-Amino-4-methylphenol		LC-11
2-Amino-5-methylphenol		LC-11
3-Amino-6-methylphenol		LC-11
1-Amino-7-naphthol	—1-Amino-7-hydroxynaphthalene	
5-Amino-1-naphthol	—5-Amino-1-hydroxynaphthalene	
7-Amino-2-naphthol	—7-Amino-2-hydroxynaphthalene	
8-Amino-1-naphthol-3,6-disulsulfonic acid	—8-Amino-1-hydroxy-3,6-naphthalene disulphonic acid	
2-Amino-4-nitrobenzoic acid		LC-15
2-Amino-4-nitrophenol		TLC-3
2-Amino-5-nitrophenol		TLC-3
4-Amino-2-nitrophenol		TLC-3
2-Aminophenol		LC-6
		TLC-3
		PC-3
3-Aminophenol		TLC-3,12
4-Aminophenol		LC-3,6,11
		TLC-3,12
2-Aminophenol-4-sulfonic acid	—3-Amino-4-hydroxybenzene sulfonic acid	
2-Aminopyridine		PC-3
4-Aminosalicylic acid	—4-Amino-2-hydroxybenzoic acid	
5-Aminosalicylic acid	—5-Amino-2-hydroxybenzoic acid	
2-Amino-3,4,6-trichlorophenol		TLC-3
4-*tert.*-Amylphenol		LC-12
		PC-1
4-*tert.*-Amylpyrocatechol	—3,4-Dihydroxy-*tert.*-pentylbenzene	
Anacardic acid		TLC-10
1,4-Anhydro-3- deoxypentitol-2-carboxylic acid		GC-48
2,5-Anhydro-3,4-dideoxypentonic acid		GC-48
2,5-Anhydrogluconic acid		GC-48
2,5-Anhydromannonic acid		GC-48
2,5-Anhydrotalonic acid		GC-48
Anisol	—Methoxybenzene	
Anthranilic acid	—2-Aminobenzoic acid	
Anthraquinondicarboxylic acid		GC-41
Anziaic acid		TLC-38
Apigenin	—4′,5,7-Trihydroxyflavone	
Arabinaric acid		GC-48
Arabinonic acid		GC-48
Arabinonic acid 1,4-lactone (1,4-Arabinolactone)		GC-36,48
Arabonic acid		GC-36
D-Arabonic acid lactone		GC-43
Arabonic acid 1,4-lactone (1,4-Arabono lactone)		GC-36

Compound	Synonym	Number of tables and/or figures
Arbutin	—4-Hydroxy-β-D-glucopyranoside	
Arthoniaic acid		TLC-38
Asarone		TLC-14
Ascorbic acid		GC-22,36,40
		TLC-17,25
Atranorin (Usnarin, Parmelin, Atranoric acid)		TLC-33,38
Atropinesulfate		LC-31
Aurintricarboxylic acid		TLC-33
Azelacnic acid		GC-24
Azelic acid	—Nonanedioic acid	
Baeomycesic acid		TLC-38
Barbatic acid		TLC-38
Barbatolic acid		TLC-38
1,2-Benzenedicarboxylic acid		GC-20,21,24,25,41
		TLC-20,22,25
		PC-8
1,2-Benzenedicarboxylic acid butylbenzylester		GC-20
		TLC-16
1,2-Benzenedicarboxylic acid butylnonylester		GC-20
		TLC-16
1,2-Benzenedicarboxylic acid dibutylester		GC-20
		LC-16
		TLC-16
1,2-Benzenedicarboxylic acid dibutylethylester		GC-20
		TLC-16
1,2-Benzenedicarboxylic acid dicyclohexylester		GC-20
		TLC-16
1,2-Benzenedicarboxylic acid didecylester		GC-20
		TLC-16
1,2-Benzenedicarboxylic acid di-*n*-dodecylester		GC-20
		LC-16
		TLC-16
1,2-Benzenedicarboxylic acid diethylester		GC-20
		LC-16
		TLC-16
1,2-Benzenedicarboxylic acid diethylethylester		GC-20
		TLC-16
1,2-Benzenedicarboxylic acid di-2′-ethylhexylester		GC-20,36
1,2-Benzenedicarboxylic acid diheptylester		GC-20
		LC-16
		TLC-16
1,2-Benzenedicarboxylic acid di-*n*-hexylester		GC-20
		LC-16
		TLC-16
1,2-Benzenedicarboxylic acid diisobutylester		GC-20
		LC-16
1,2-Benzenedicarboxylic acid diisodecylester		TLC-16
		LC-16
1,2-Benzenedicarboxylic acid diisopropylester		GC-20
		LC-16
		TLC-16
1,2-Benzenedicarboxylic acid dimethylester		GC-20
		LC-16
		TLC-16
1,2-Benzenedicarboxylic acid dimethylethylester		GC-20
		TLC-16
1,2-Benzenedicarboxylic acid di-*n*-nonylester		GC-20
		LC-16
		TLC-16

Compound	Synonym	Number of tables and/or figures
1,2-Benzenedicarboxylic acid di-*n*-octylester		GC-20
		LC-16
		TLC-16
1,2-Benzenedicarboxylic acid dipentylester		GC-20
		LC-16
		TLC-16
1,2-Benzenedicarboxylic acid diphenylethylester		GC-20
1,2-Benzenedicarboxylic acid diphenylester		GC-20
		TLC-16
1,2-Benzenedicarboxylic acid dipropylester		GC-20
		LC-16
		TLC-16
1,2-Benzenedicarboxylic acid hexylheptylester		GC-20
		TLC-16
1,2-Benzenedicarboxylic acid di-3′,3′,5′-trimethylcyclohexylester		GC-20
		TLC-16
1,2-Benzenedicarboxylic acid di-3′,5′,5′-trimethylhexylester		GC-20
		TLC-16
1,2-Benzenedicarboxylic acid di-2′,2′,4′-trimethylpentylester		GC-20
1,3-Benzenedicarboxylic acid (Isophthalic acid)		GC-21,24,41
		PC-8
1,4-Benzenedicarboxylic acid (Terephthalic acid)		GC-21,24,36
		LC-20
		PC-8
1,4-Benzenedicarboxylic acid diethylester		LC-16
1,4-Benzenedicarboxylic acid dimethylester		LC-16
Benzenehexacarboxylic acid		GC-41
Benzenepentacarboxylic acid		GC-41
Benzenetetracarboxylic acid		GC-41
1,2,3,4-Benzenetetracarboxylic acid (Prehnitic acid)		GC-21
1,2,3,5-Benzenetetracarboxylic acid		GC-41
1,2,4,5-Benzenetetracarboxylic acid (Pyromellitic acid)		PC-8
1,2,3-Benzenetricarboxylic acid		GC-41
		LC-30
1,2,4-Benzenetricarboxylic acid (Trimellitic acid)		GC-21,41
		LC-24
		PC-8
1,3,5-Benzenetricarboxylic acid		GC-21,41
Benzenesulfonic acid		LC-15
Benzhydrol (Diphenylmethanol)		TLC-30
Benzilic acid (Diphenyl-(hydroxy)-actic acid)		TLC-30
Benzoic acid		GC-9,21,23,25,28,33,34, 36,42
		LC-2,14,15,19,20,21, 30,32
		LC-2,14,15,19,20,21, 30,32
		TLC-15,17,22,24,25,31
Benzoylformic acid (methylester)		GC-34
3-Benzyl-2-hydroxybenzoquinoxaline		LC-24
2-Benzyl-4-methylphenol		TLC-1
2-Benzylphenol		TLC-1
4-Benzylphenol		TLC-1
		PC-1
Biphenyl-2,2′-dicarboxylic acid		GC-41
Biphenylhexacarboxylic acid		GC-41
Biphenyltetracarboxylic acid		GC-41

Compound	Synonym	Number of tables and/or figures
2-Bromobenzoic acid		GC-34
3-Bromobenzoic acid		GC-34
4-Bromobenzoic acid		GC-34
		LC-19
4-Bromo-2-*tert*.-butyl-6-methylphenol		LC-5
6-Bromo-2-*tert*.-butyl-4-methylphenol		LC-5
2-Bromo-4-*sec*.-butylphenol		LC-5
2-Bromo-4-*tert*.-butylphenol		LC-5
4-Bromo-2,6-di-*sec*.-butylphenol		LC-5
4-Bromo-2,6-di-*tert*.-butylphenol		LC-5
2-Bromo-6-chlorophenol		TLC-14
2-Bromo-4-cyclohexylphenol		LC-5
2-Bromo-5,4′-dihydroxybiphenyl		LC-6
5-Bromo-2,4′-dihydroxybiphenyl		LC-6
4-Bromo-2,6-dimethoxybiphenyl		TLC-13
2-Bromo-4,6-dimethylphenol		PC-1
4-Bromo-2,6-dimethylphenol		LC-5
6-Bromo-2,4-dimethylphenol		LC-5
6-Bromo-2,5-dimethylphenol		TLC-4
2-Bromo-4-ethylphenol		LC-5
6-Bromo-4-*tert*.-hexyl-2-methylphenol		LC-5
1,3-Bromomethyl-5-hydroxymethylcyclo-hexaneacetate		GC-41
1-Bromomethyl-3,5-hydroxymethylcyclo-hexanediacetate		GC-41
2-Bromo-4-methylphenol		LC-5
6-Bromo-2-methylphenol		TLC-4
4-Bromo-2-methyl-6-propylphenol		LC-5
2-Bromo-4-octylphenol		LC-5
2-Bromo-4-*tert*.-pentylphenol		LC-5
2-Bromophenol		GC-11,19
		LC-6,13
		TLC-3,4
3-Bromophenol		GC-11
		LC-6
		TLC-4
4-Bromophenol		GC-11,19,34
		LC-6,13
		TLC-3
3-Bromo-2′-propylcinnamic acid		PC-9
4-Bromo-2′-propylcinnamic acid		PC-9
2-Bromo-4-propylphenol		LC-5
4-Bromo-2,3,5,6-tetramethylphenol		LC-5
6-Bromo-2,4,5-trimethylphenol		LC-5
Butylbenzylphthalate	—1,2-Benzenedicarboxylic acid butylbenzylester	
4-*tert*.-Butylcatechol	—4 *tert*.-Butyl-1,2-dihydroxybenzene	
4-*tert*.-Butyl-2-chlorophenol		TLC-1
o-*tert*.-Butyl-*o*-cresol	—2-*tert*.-Butyl-6-methylphenol	
o-*tert*.-Butyl-*p*-cresol	—2-*tert*.-Butyl-4-methylphenol	
p-*tert*.-Butyl-*o*-cresol	—4 *tert*.-Butyl-2-methylphenol	
2-*tert*.-Butyl-6-ethylphenol		TLC-4
		PC-1
3-*n*-Butyl-6-ethylphenol		GC-5
4-*tert*.-Butyl-2-ethylphenol		PC-1
2-*tert*.-Butylhydroquinone	—2-*tert*.-Butyl-1,4-dihydroxybenzene	
3-*tert*.-Butyl-2,5-dihydroxymethylbenzene		LC-12
3-*tert*.-Butyl-4,5-dihydroxymethylbenzene		LC-12
4-*tert*.-Butyl-2,3-dihydroxymethylbenzene		LC-12
4-*tert*.-Butyl-2,5-dihydroxymethylbenzene		LC-12
2-*tert*.-Butyl-1,4-dihydroxybenzene		LC-12

Compound	Synonym	Number of tables and/or figures
3-*sec*.-Butyl-1,2-dihydroxybenzene		LC-12
4-*n*-Butyl-1,2-dihydroxybenzene		LC-12
4-*sec*.-Butyl-1,2-dihydroxybenzene		LC-12
4-*tert*.-Butyl-1,2-dihydroxybenzene		GC-8 LC-12 TLC-1
2-*tert*.-Butyl-4,6-dimethylphenol		LC-8,12 TLC-1
4-*tert*.-Butyl-2,6-dimethylphenol		LC-12
2-*sec*.-Butyl-4,6-dinitrophenol (Dinoseb)		LC-25
2-*tert*.-Butyl-4,6-dinitrophenol (Dinoterb)		LC-25 TLC-1
2-*sec*.-Butyl-4,6-dinitrophenylisopropyl-carbonate (Dinobuton)		LC-25
2-*tert*.-Butyl-4-hydroxyanisole	—2 *tert*.-Butyl-4-methoxyphenol	
3-*tert*.-Butyl-4-hydroxyanisole	—3-*tert*.-Butyl-4-methoxyphenol	
1-*sec*-Butyl-2-hydroxybenzene	—2-*sec*.-Butylphenol	
1-*tert*.-Butyl-2-hydroxybenzene	—2-*tert*.-Butylphenol	
1-*tert*.-Butyl-3-hydroxybenzene	—3-*tert*.-Butylphenol	
1-*sec*.-Butyl-4-hydroxybenzene	—4-*sec*.-Butylphenol	
1-*tert*.-Butyl-4-hydroxybenzene	—4-*tert*.-Butylphenol	
3-*tert*.-Butyl-2-hydroxy-5-methylbenzoic acid		TLC-1
1-(3′-*tert*.-Butyl-2′-hydroxy-5′-methylbenzyl)-3-*tert*.-butyl-2-hydroxy-5-methylbenzene		LC-4
1-(5′-*tert*.-Butyl-4′-hydroxy-3′-methylbenzyl)-5-*tert*.-butyl-4-hydroxy-3-methylbenzene		LC-4
1-(5′-*tert*.-Butyl-2′-hydroxy-3′-methylbenzyl)-5-*tert*.-butyl-2-hydroxy-3-methylbenzene		LC-4
3-*tert*.-Butyl-4-hydroxytoluene	—2-*tert*.-Butyl-4-methyphenol	
3-*tert*.-Butyl-6-hydroxytoluene	—4-*tert*.-Butyl-2-methylphenol	
2-*tert*.-Butyl-4-methoxyphenol		LC-8,26
3-*tert*.-Butyl-4-methoxyphenol		LC-8,26 TLC-1
3-*tert*.-Butyl-5-methyl-2-hydroxybenzoic acid	—3-*tert*.-Butyl-2-hydroxy-5-methylbenzoic acid	
2-*tert*.-Butyl-4-methylphenol		GC-4,5 LC-1,5,12 TLC-1,2 PC-1
2-*tert*.-Butyl-5-methylphenol		GC-4,5 LC-1,12 TLC-1
2-*tert*.-Butyl-6-methylphenol		GC-5 LC-5,6,12 TLC-1,4 PC-1
4-*tert*.-Butyl-2-methylphenol		GC-5 LC-12 TLC-1,2 PC-1
4-*tert*.-Butyl-3-methylphenol		GC-5
3-*tert*.-Butyl-5-methylpyrocatechol	—3-*tert*.-Butyl-4,5-dihydroxymethylbenzene	
3-*tert*.-Butyl-5-methylsalicylic acid	—3-*tert*.-Butyl-2-hydroxy-5-methylbenzoic acid	
2-*n*-Butylphenol		GC-5
2-*sec*.-Butylphenol		GC-5,19 LC-11 TLC-2,4 PC-1

Compound	Synonym	Number of tables and/or figures
2-*tert*.-Butylphenol		GC-4,5
		LC-1,11,12
		TLC-1,4
3-*n*-Butylphenol		GC-5
3-*tert*.-Butylphenol		GC-5
		LC-11,12
		TLC-1,4
		PC-1
4-*n*-Butylphenol		GC-5
4-*sec*.-Butylphenol		GC-4,5
		LC-1,5,12
		PC-1
4-*tert*.-Butylphenol		GC-4,5,19
		LC-1,5,11,12
		TLC-1,5
		PC-1
3-*sec*.-Butylpyrocatechol	—3-*sec*.-Butyl-1,2-dihydroxybenzene	
4-*n*-Butylpyrocatechol	—4-*n*-Butyl-1,2-dihydroxybenzene	
4-*sec*.-Butylpyrocatechol	—4-*sec*.-Butyl-1,2-dihydroxybenzene	
4-*tert*.-Butyl-pyrocatechol	—4-*tert*.-Butyl-1,2-dihydroxybenzene	
4-Butyroxybutanoic acid		PC-6
Caffeic acid	—3,4-Dihydroxycinnamic acid	
cis-Caffeic acid	—*cis*-3,4-Dihydroxycinnamic acid	
trans-Caffeic acid	—*trans*-3,4-Dihydroxycinnamic acid	
Caffeic acid-tartaric acid-glucose ester	—3,4-Dihydroxycinnamic acid-tartaric acid-glucose ester	
3-Caffeoylquinic acid	—Chlorogenic acid	
4-*O*-Caffeoylquinic acid		LC-10
5-Caffeoylquinic acid		TLC-26
3-Caffeoylquinic acid potassium-caffeine complex		TLC-26
Caffeoyltartaric acid		LC-24
Calycin		TLC-38
Camphor		LC-9
Caperatic acid		TLC-38
Caproic acid (Hexanoic acid)		TLC-22
Cardanol		GC-13
		TLC-10
Cardanol acetate		GC-13
		TLC-10
Cardanol dimethylether		GC-13
		TLC-10
Cardanol methylether		GC-13
		TLC-10
Cardol		GC-13
		TLC-10
Cardol diacetate		GC-13
		TLC-10
D-Catechin		LC-21
Catechin		GC-17
		LC-12
		TLC-15
(+)-Catechin		LC-10
		TLC-32
Catechol	—1,2-Dihydroxybenzene	
Cetylphenol	—Hexadecylphenol	
Chloroacetic acid		LC-17
Chloroacetoxyacetic acid		PC-6
Chloroatranorin		TLC-33,38

Compound	Synonym	Number of tables and/or figures
2-Chlorobenzoic acid		GC-34
		LC-2
3-Chlorobenzoic acid		GC-34
		LC-2
4-Chlorobenzoic acid		GC-34
		LC-2
4-Chlorobenzoylformic acid (methyl ester)		GC-34
3-Chlorocinnamic acid		PC-9
4-Chlorocinnamic acid		PC-9
3-Chloro-*o*-cresol	—3-Chloro-2-methylphenol	
4-Chloro-*o*-cresol	—4-Chloro-2-methylphenol	
4-Chloro-*m*-cresol	—4-Chloro-3-methylphenol	
2-Chloro-*p*-cresol	—2-Chloro-4-methylphenol	
3-Chloro-2-cyclopentylphenol		TLC-1
3-Chloro-4,4′-dihydroxybiphenyl		TLC-7
5-Chloro-2,3-dihydroxymethylbenzene		TLC-6
2-Chloro-3,5-dimethylphenol		TLC-4
2-Chloro-4,5-dimethylphenol		PC-1
2-(4′-Chloro-3′,5′-dinitrophenyl)-2-(4″-nitrophenyl)propane		LC-7
2-(4′-Chloro-3′,5′-dinitrophenyl)-2-phenylpropane		LC-7
3-Chloro-2′-ethylcinnamic acid		PC-9
4-Chloro-2′-ethylcinnamic acid		PC-9
5-Chloro-2-hydroxybenzoic acid		LC-23
2-Chloro-4-hydroxybiphenyl		TLC-7
3-Chloro-4-hydroxybiphenyl		TLC-7
4′-Chloro-4-hydroxybiphenyl		TLC-7
5-Chloro-3-methylcatechol	—5-Chloro-2,3-dihydroxy-methylbenzene	
3-Chloro-2′-methylcinnamic acid		PC-9
4-Chloro-2′-methylcinnamic acid		PC-9
2-Chloro-4-methylphenol		TLC-6
2-Chloro-5-methylphenol		TLC-1
3-Chloro-2-methylphenol		TLC-6
4-Chloro-2-methylphenol		TLC-1,6
4-Chloro-3-methylphenol		GC-10
		TLC-1,12
4-Chloro-2-methyl-phenoxyacetic acid (MCPA)		LC-25
2-Chloro-4-nitrophenol		TLC-1
4-Chloro-2-nitrophenol		TLC-1
2-Chlorophenol		GC-9,11,19
		LC-2,6,13
		TLC-1,2,3,4,5,8,11
3-Chlorophenol		GC-9,11
		LC-2,6
		TLC-1,3,4,5,8,11
4-Chlorophenol		GC-9,10,11,19,34
		LC-2,6,9,13
		TLC-1,3,5,8,11,12
2-Chlorophenoxyacetic acid		LC-19
4-Chlorophenoxyacetic acid		LC-19
3-Chloro-2′-propylcinnamic acid		PC-9
4-Chloro-2′-propylcinnamic acid		PC-9
5-Chlorosalicylic acid	—5-Chloro-2-hydroxybenzoic acid	
2-Chlorothymol (2-Chloro-5-methyl-6-isopropylphenol)		TLC-1
Chlorogenic acid (3-*O*-Caffeoylquinic acid)		GC-17
		LC-10,20,21
		TLC-15,26
		PC-7
Cinnamaldehyde		TLC-26

Compound	Synonym	Number of tables and/or figures
Cinnamic acid (3-Phenylpropenoic acid)		GC-17,24
		LC-20,30,31
		TLC-17
		PC-7,9
trans-Cinnamic acid		GC-40
		TLC-24
Citraconic acid		GC-25
		LC-17
		TLC-20
Citramalic acid		GC-25
		LC-18
Citric acid (2-Hydroxypropane-1,2,3-tricarboxylic acid)		GC-22,23,24,25,27,28, 32,36,48
		LC-17,18,28
		TLC-17,18,19,20,22,25
		PC-4
Colensoinic acid		TLC-38
2′-Collatolic acid		TLC-38
Confluentic acid		TLC-38
Connorstictic acid		TLC-38
Conpsoromic acid		TLC-38
Consalazinic acid		TLC-38
Constictic acid		TLC-38
Coumaric acid	—Hydroxycinnamic acid	
cis-Coumaric acid	—*cis*-Hydroxycinnamic acid	
trans-Coumaric acid	—*trans*-Hydroxycinnamic acid	
o-Coumaric acid	—2-Hydroxycinnamic acid	
p-Coumaric acid-tartaric acid-glucose ester	—4-Hydroxycinnamic acid-tartaric acid-glucose ester	
1-*p*-Coumaroyl-*β*-D-glucose		TLC-32
p-Coumaroyltartaric acid (4-Coumaroyltartaric acid)		LC-24
o-Cresol	—2-Methylphenol	
m-Cresol	—3-Methylphenol	
p-Cresol	—4-Methylphenol	
Crotonic acid		GC-25
trans-Crotonic acid		LC-28
Cryptochlorophaeic acid		TLC-38
Cyanadin		GC-17
Cyanidin chloride		TLC-15
Cyanin		GC-17
3-Cyanobenzoic acid		PC-8
4-Cyanobenzoic acid		PC-8
Cyclohexa-1,4-dienecarboxylic acid		GC-42
		TLC-31
Cyclohexa-2,5-dienecarboxylic acid		GC-42
		TLC-31
3-Cyclohexane-1-carboxylic acid		GC-42
Cyclohexanecarboxylic acid		GC-42
		TLC-31
cis-1,3,5-Cyclohexanetricarboxylic acid		GC-41
trans-1,3,5-Cyclohexanetricarboxylic acid		GC-41
Cyclohexanol-hexahydrophenol		PC-2
Cyclohex-1-enecarboxylic acid		GC-42
		TLC-31
Cyclohex-3-enecarboxylic acid		GC-42
		TLC-31
4-Cyclohexylphenol		LC-5,12
		PC-1

Compound	Synonym	Number of tables and/or figures
2-Cyclohexyl-2-phenylacetic acid		GC-42
2-Cyclohexyl-2-phenylglycolic acid		GC-42
2-Cyclopentyl-4-chlorophenol		TLC-1
Cyclopentylphenylcarbinol		TLC-30
Cyclopentylphenylglycolate-1-(2,2,2)bicyclooctyl		TLC-30
Cyclopentylphenylglycolate-*N*-methyl-3-piperidinyl		TLC-30
Cyclopentylphenylglycolate-*N*-methyl-4-piperidinyl		TLC-30
Cyclopentylphenylglycolate-3-quinuclidinyl		TLC-30
Cyclopentylphenylglycolate-tropinyl		TLC-30
Cyclopentylphenylglycolic acid		TLC-30
Cyclopentylphenylketone		TLC-30
Decandioic acid		GC-26
Decarboxythamnolic acid		TLC-38
Dehydroabietic acid		GC-46
3-Dehydroshikimic acid		TLC-24
Dehydrozingerone		TLC-14
Delphindin		GC-17
Delphinin		GC-17
4-*O*-Demethylbarbitaric acid		TLC-38
4-*O*-Demethylplanaic acid		TLC-38
2-Deoxy-*arabino*-hexaric acid		GC-48
3-Deoxy-*arabino*-hexaric acid		GC-48
2-Deoxy-*arabino*-hexonic acid		GC-48
2-Deoxy-*arabino*-hexonic acid 1,4-lactone		GC-48
3-Deoxy-*arabino*-hexonic acid		GC-48
3-Deoxy-*arabino*-hexonic acid 1,4-lactone		GC-48
2-Deoxy-2,4-dioxo-tetrahydrofuran		GC-36
3-Deoxy-2,4-dioxo-tetrahydrofuran		GC-36
4-Deoxyerythronic acid		GC-36,48
2-Deoxy-*erythro*-pentaric acid		GC-48
3-Deoxy-*erythro*-pentaric acid		GC-48
2-Deoxy-*erythro*-pentonic acid		GC-36,48
3-Deoxy-*erythro*-pentonic acid		GC-48
2-Deoxy-*erythro*-pentonic acid 1,4-lactone		GC-48
3-Deoxy-*erythro*-pentonic acid 1,4-lactone		GC-48
6-Deoxygalactonic acid		GC-48
6-Deoxygalactonic acid 1,4-lactone		GC-48
3-Deoxy-2-*C*-hydroxymethyl-*erythro*-pentaric acid		GC-48
3-Deoxy-2-*C*-hydroxymethyl-*erythro*-pentonic acid		GC-48
3-Deoxy-2-*C*-hydroxymethyl-*erythro*-pentonic acid 1,4-lactone		GC-48
2-Deoxy-3-*C*-hydroxymethyl-tetraric acid		GC-48
3-Deoxy-2-*C*-hydroxymethyl-tetronic acid		GC-48
3-Deoxy-2-*C*-hydroxymethyl-tetronic acid 1,4-lactone		GC-48
3-Deoxy-2-*C*-hydroxymethyl-*threo*-pentaric acid		GC-48
3-Deoxy-2-*C*-hydroxymethyl-*threo*-pentonic acid		GC-48
3-Deoxy-2-*C*-hydroxymethyl-*threo*-pentonic acid 1,4-lactone		GC-48
2-Deoxy-*lyxo*-hexaric acid		GC-48
3-Deoxy-*lyxo*-hexaric acid		GC-48
2-Deoxy-*lyxo*-hexonic acid		GC-48
3-Deoxy-*lyxo*-hexonic acid		GC-48
2-Deoxy-*lyxo*-hexonic acid 1,4-lactone		GC-48
3-Deoxy-*lyxo*-hexonic acid 1,4-lactone		GC-48
6-Deoxymannoic acid		GC-48
6-Deoxymannonic acid 1,4-lactone		GC-48
2-Deoxy-3-C-methyltetraric acid		GC-48
3-Deoxy-*ribo*-hexaric acid		GC-48
3-Deoxy-*ribo*-hexonic acid		GC-48
3-Deoxy-*ribo*-hexonic acid 1,4-lactone		GC-48

Compound	Synonym	Number of tables and/or figures
2-Deoxytetronic acid		GC-48
3-Deoxytetronic acid		GC-48
4-Deoxytetronic acid		GC-48
2-Deoxytetronic acid 1,4-lactone		GC-48
3-Deoxytetronic acid 1,4-lactone		GC-48
Deoxythreonic acid		GC-36
4-Deoxythreonic acid		GC-48
2-Deoxy-*threo*-pentaric acid		GC-48
3-Deoxy-*threo*-pentaric acid		GC-48
2-Deoxy-*threo*-pentonic acid		GC-48
3-Deoxy-*threo*-pentonic acid		GC-48
2-Deoxy-*threo*-pentonic acid 1,4-lactone		GC-48
3-Deoxy-*threo*-pentonic acid 1,4-lactone		GC-48
3-Deoxy-*xylo*-hexaric acid		GC-48
3-Deoxy-*xylo*-hexonic acid		GC-48
3-Deoxy-*xylo*-hexonic acid 1,4-lactone		GC-48
2,6-Dibromo-4-*sec*.-butylphenol		LC-5
2,6-Dibromo-4-*tert*.-butylphenol		LC-5
2,6-Dibromo-4-cyclohexylphenol		LC-5
2,6-Dibromo-3,5-dimethylphenol		TLC-4
2,6-Dibromo-4-ethylphenol		LC-5
2,6-Dibromo-4-methylphenol		LC-5
2,6-Dibromo-4-octylphenol		LC-5
2,6-Dibromo-4-*tert*.-pentylphenol		LC-5
2,4-Dibromophenol		GC-11,19 LC-13
2,6-Dibromophenol		GC-11 TLC-4
2,6-Dibromo-4-propylphenol		LC-5
4,6-Di-*tert*.-butyl-*o*-cresol	—2,4-Di-*tert*.-butyl-6-methylphenol	
2,5-Di-*tert*.-butyl-1,4-dihydroxybenzene		LC-12
2,6-Di-*tert*.-butyl-1,4-dihydroxybenzene		LC-12
3,5-Di-*tert*.-butyl-1,2-dihydroxybenzene		LC-12
3,6-Di-*tert*.-butyl-1,2-dihydroxybenzene		LC-12
4,6-Di-*tert*.-butyl-1,3-dihydroxybenzene		TLC-1
3,3′-Di-*tert*.-butyl-2,2′-dihydroxy-5,5′-dimethylbiphenyl		LC-4
5,5′-Di-*tert*.-butyl-2,2′-dihydroxy-3,3′-dimethylbiphenyl		LC-4
5,5′-Di-*tert*.-butyl-4,4′-dihydroxy-3,3′-dimethylbiphenyl		LC-4
Dibutylethylphthalate	—1,2-Benzenedicarboxylic acid dibutylethylester	
2,5-Di-*tert*.-butylhydroquinone	—2,5-Di-*tert*.-butyl-1,4-dihydroxybenzene	
2,6-Di-*tert*.-butylhydroquinone	—2,6-Di-*tert*.-butyl-1,4-dihydroxybenzene	
3,5-Di-*tert*.-butyl-4-hydroxybenzoic acid		LC-14
1-(3′,5′-Di-*tert*.-butyl-2′-hydroxybenzyl)-3,5-di-*tert*.-butyl-2-hydroxybenzene		LC-4
1-(3′,5′-Di-*tert*.-butyl-4′-hydroxybenzyl)-3,5-di-*tert*.-butyl-4-hydroxybenzene		LC-4
3,5-Di-*tert*.-butyl-4-hydroxy-methybenzene		LC-26
2,6-Di-*tert*.-butyl-4-hydroxy-methylphenol		LC-26
1,1-Di-(3′-*tert*.-butyl-2′-hydroxy-5′-methylphenyl)butane		LC-4
1,1-Di-(5′-*tert*.-butyl-4′-hydroxy-3′-methylphenyl)butane		LC-4
2,2-Di-(3′-*tert*.-butyl-2′-hydroxy-5′-methylphenyl)butane		LC-4
2,2-Di-(5′-*tert*.-butyl-4′-hydroxy-3′-methylphenyl)butane		LC-4
1,1-Di-(5′-*tert*.-butyl-4′-hydroxy-3′-methylphenyl)cyclohexane		LC-4
1,1-Di-(3′-*tert*.-butyl-2′-hydroxy-5′-methylphenyl)ethane		LC-4
1,1-Di-(5′-*tert*.-butyl-2′-hydroxy-3′-methylphenyl)ethane		LC-4
1,1-Di-(5′-*tert*.-butyl-4′-hydroxy-3′-methylphenyl)ethane		LC-4
2,2-Di-(3′-*tert*.-butyl-2′-hydroxy-5′-methylphenyl)propane		LC-4
2,2-Di-(5′-*tert*.-butyl-2′-hydroxy-3′-methylphenyl)propane		LC-4
2,2-Di-(5′-*tert*.-butyl-4′-hydroxy-3′-methylphenyl)propane		LC-4

Compound	Synonym	Number of tables and/or figures
Di-(3-*tert*.-butyl-2-hydroxy-5-methylphenyl)sulfide		LC-4
Di-(5-*tert*.-butyl-2-hydroxy-3-methylphenyl)sulfide		LC-4
Di-(5-*tert*.-butyl-4-hydroxy-3-methylphenyl)sulfide		LC-4
Di-(3-*tert*.-butyl-2-hydroxy-5-methylphenyl)disulfide		LC-4
Di-(5-*tert*.-butyl-4-hydroxy-3-methylphenyl)disulfide		LC-4
Di-(5-*tert*.-butyl-4-hydroxy-3-methylphenyl)trisulfide		LC-4
1,1-Di-(3′,5′-di-*tert*.-butyl-2′-hydroxyphenyl)ethane		LC-4
1,1-Di-(3′,5′-di-*tert*.-butyl-4′-hydroxyphenyl)ethane		LC-4
2,2-Di-(3′,5′-di-*tert*.-butyl-2′-hydroxyphenyl)propane		LC-4
2,2-Di-(3′,5′-di-*tert*.-butyl-4′-hydroxyphenyl)propane		LC-4
Di-(3,5-di-*tert*.-butyl-2-hydroxyphenyl)sulfide		LC-4
Di-(3,5-di-*tert*.-butyl-4-hydroxyphenyl)sulfide		LC-4
2,4-Di-*tert*.-butyl-6-methylphenol		LC-12 TLC-2 PC-1
2,6-Di-*tert*.-butyl-4-methylphenol		GC-4 LC-1,12 TLC-1
2,4-Di-*tert*.-butylphenol		LC-12 TLC-1 PC-1
2,5-Di-*tert*.-butylphenol		GC-4 LC-1
2,6-Di-*sec*.-butylphenol		GC-4 LC-1,5
2,6-Di-*tert*.-butylphenol		GC-4 LC-1,5,6,12 TLC-1 PC-1
3,5-Di-*tert*.-butylphenol		GC-4 LC-1
Dibutylphthalate	—1,2-Benzenedicarboxylic acid dibutylester	
3,5-Di-*tert*.-butylpyrocatechol	—3,5-Di-*tert*.-butyl-1, 2-dihydroxybenzene	
3,6-Di-*tert*.-butylpyrocatechol	—3,6-Di-*tert*.-butyl-1, 2-dihydroxybenzene	
4,6-Di-*tert*.-butylresorcinol	—4,6-Do-*tert*.-butyl-1, 3-dihydroxybenzene	
1,4-Dicaffeoylquinic acid		TLC-26
Dichloroacetic acid		LC-17
2,5-Dichloro-3-aminobenzoic acid	—3-Amino-2,5-dichlorobenzoic acid	
2,4-Dichlorobenzoic acid		GC-34
3,4-Dichlorocatechol	—3,4-Dichloro-1,2-dihydroxybenzene	
3,5-Dichlorocatechol	—3,5-Dichloro-1,2-dihydroxybenzene	
3,6-Dichlorocatechol	—3,6-Dichloro-1,2-dihydroxybenzene	
3,4-Dichloro-1,2-dihydroxybenzene		TLC-6
3,5-Dichloro-1,2-dihydroxybenzene		TLC-6
3,6-Dichloro-1,2-dihydroxybenzene		TLC-6
2,6′-Dichloro-4,4′-dihydroxybiphenyl		TLC-7
3,3′-Dichloro-4,4′-dihydroxybiphenyl		TLC-7
4,4′-Dichloro-3,3′-dihydroxybiphenyl		TLC-7
2′,5′-Dichloro-2-hydroxybiphenyl		TLC-7
2′,5′-Dichloro-3-hydroxybiphenyl		TLC-7
2′,5′-Dichloro-6-hydroxybiphenyl		TLC-7
3,4′-Dichloro-4-hydroxybiphenyl		TLC-7
3,5-Dichloro-2-hydroxybiphenyl		TLC-7
3,5-Dichloro-4-hydroxybiphenyl		TLC-7

Compound	Synonym	Number of tables and/or figures
4,4′-Dichloro-3-hydroxybiphenyl		TLC-7
4′,6-Dichloro-4-hydroxybiphenyl		TLC-7
3,6-Dichloro-2-methoxybenzoic acid (Dicamba)		LC-25
Dichlorophene	—2,2′-Methylene *bis*(4-chlorophenol)	
2,3-Dichlorophenol		GC-11
		LC-2
		TLC-1,3
2,4-Dichlorophenol		GC-3,9,10,11
		LC-2,6,13
		TLC-1,3
2,5-Dichlorophenol		GC-11,19
		LC-2,6
		TLC-1,3
2,6-Dichlorophenol		GC-11,19
		LC-2,6
		TLC-1,3,4
3,4-Dichlorophenol		GC-11
		TLC-1,3,14
3,5-Dichlorophenol		GC-10,11
		LC-6
		TLC-3
2,4-Dichlorophenoxyacetic acid (2,4-D)		GC-19
		LC-19,25
2,6-Dichlorophenoxyacetic acid		GC-19
2-(2′,4′-Dichlorophenoxy)propionic acid (Dichloprop)		LC-25
cis-5-(2,2-Dicyanoethyl)-2-phenyl-2,3-dihydro-3,4-furandicarboxylic acid dimethylester		TLC-36
trans-5-(2,2-Dicyanoethyl)-2-phenyl-2,3-dihydro-3,4-furandicarboxylic acid dimethylester		TLC-36
Dicyclohexylphthalate	—1,2-Benzenedicarboxylic acid dicyclohexylester	
Didecylphthalate	—1,2-Benzenedicarboxylic acid didecylester	
Didymic acid		TLC-38
Diffractaric acid		TLC-38
3,6-Dideoxy-*arabino*-hexonic acid		GC-48
3,6-Dideoxy-*arabino*-hexonic acid 1,4-lactone		GC-48
3,5-Dideoxy-*erythro*-pentonic acid		GC-48
3,5-Dideoxy-*erythro*-pentonic acid 1,4-lactone		GC-48
3,4-Dideoxyhexaric acid		GC-48
3,4-Dideoxyhexonic acid		GC-48
2,4-Dideoxy-3-*C*-methylpentaric acid		GC-48
2,4-Dideoxy-3-*C*-methylpentoric acid		GC-48
2,3-Dideoxypentaric acid		GC-48
2,4-Dideoxypentaric acid		GC-48
3,4-Dideoxypentoric acid		GC-48
2,6-Dideoxy-*ribo*-hexonic acid		GC-48
3,6-Dideoxy-*ribo*-hexonic acid		GC-48
2,6-Dideoxy-*ribo*-hexonic acid 1,4-lactone		GC-48
3,5-Dideoxy-*threo*-pentonic acid		GC-48
3,5-Dideoxy-*threo*-pentonic acid 1,4-lactone		GC-48
Di-*n*-dodecylphthalate	—1,2-Benzenedicarboxylic acid di-*n*-dodecylester	
2-Diethylaminoethanol		TLC-30
Diethylethylphthalate	—1,2-Benzenedicarboxylic acid diethylethylester	
Di-2-ethylhexylphthalate	—1,2-Benzenedicarboxylic acid di-2-ethylhexylester	
2,4-Diethylphenol		GC-5

Compound	Synonym	Number of tables and/or figures
2,5-Diethylphenol		GC-5
2,6-Diethylphenol		TLC-4
		PC-4
3,4-Diethylphenol		GC-5
Diethylphthalate	—1,2-Benzenedicarboxylic acid diethylester	
α,α′-Diethyl-4,4′-stilbendiol		TLC-1,12
Diethylterephthalate	—1,4-Benzenedicarboxylic acid diethylester	
Diheptylphthalate	—1,2-Benzenedicarboxylic acid diheptylester	
Di-*n*-hexylphthalate	—1,2-Benzenedicarboxylic acid di-*n*-hexylester	
Di-*n*-hexylpropylphthalate	—1,2-Benzenedicarboxylic acid di-*n*-hexylpropylester	
Dihydrodiol-diphenylhydantoin		GC-16
Dihydro-dihydroxy-2-methoxy-4-propylbenzene		GC-14
3,4-Dihydro-9,10-dimethoxycarbonyl-4-methyl-1-phenyl-1,4-ethano-1H-2-benzopyran		GC-45
2,4-Dihydroxybenzaldehyde		LC-14
		TLC-26
3,4-Dihydroxybenzaldehyde		TLC-26
1,2-Dihydroxybenzene		GC-3,8,9,17,19,36
		LC-2,3,6,8,12,14
		TLC-1,3,5,6,8,12 14,15,26
		PC-2
1,3-Dihydroxybenzene		GC-8,9,19
		LC-2,3,6,8,12,14,27
		TLC-1,3,5,8,12,26
		PC-2
1,4-Dihydroxybenzene		GC-8,9
		LC-2,6,8,12,14,27
		TLC-1,3,8
		PC-2
1,2-Dihydroxybenzene-disulfate		LC-27
1,3-Dihydroxybenzene-disulfate		LC-27
1,4-Dihydroxybenzene-disulfate		LC-27
2,3-Dihydroxybenzoic acid		GC-37
		LC-21
		TLC-3,26,29
2,4-Dihydroxybenzoic acid		GC-37
		LC-15,21,23
		TLC-15,22,26,29
2,5-Dihydroxybenzoic acid		GC-17,37
		LC-14,20,21,30
		TLC-15,26,29
2,6-Dihydroxybenzoic acid		GC-17,37
		TLC-15,26,29
3,4-Dihydroxybenzoic acid		GC-17,37,38
		LC-20,21
		TLC-11,24,26,27
		PC-8
3,5-Dihydroxybenzoic acid		GC-37
		LC-21
		TLC-26
2,2′-Dihydroxybenzophenone		TLC-9
2,4-Dihydroxybenzophenone		TLC-9
4,4′-Dihydroxybenzophenone		TLC-9

Compound	Synonym	Number of tables and/or figures
3′,4′-Dihydroxy-6,7-benzotropolone		LC-10
2,2′-Dihydroxybiphenyl		GC-12
		LC-6,12
		TLC-7
2,5-Dihydroxybiphenyl		GC-12
		TLC-7,11
3,3′-Dihydroxybiphenyl		GC-12
		TLC-7
3,4-Dihydroxybiphenyl		GC-12
		TLC-7
4,4′-Dihydroxybiphenyl		GC-12
		LC-4,6
		TLC-7,11
2,3-Dihydroxybutanoic acid E		PC-6
2,3-Dihydroxybutanoic acid T		PC-6
Dihydroxycaffeic acid (3,4-Dihydroxy-dihydrocinnamic acid)		GC-39
3,4-Dihydroxycinnamic acid		GC-17,36
		LC-20,24,30,31
		TLC-15,26
		PC-7
cis-3,4-Dihydroxycinnamic acid		GC-35
		TLC-24
trans-3,4-Dihydroxycinnamic acid		GC-35
		TLC-24
3,4-Dihydroxycinnamic acid-tartaric acid-glucose ester		LC-24
2,8-Dihydroxycoumarin		GC-17
3,4-Dihydroxycyclohexanecarboxylic acid		TLC-24
trans-2,4′-Dihydroxy-3,3′-dimethoxy-5-(3-hydroxypropyl)stilbenzene		GC-14
2,3-Dihydroxy-1,5-dimethylbenzene		GC-8
		LC-12
2,4-Dihydroxy-1,3-dimethylbenzene		LC-12
2,5-Dihydroxy-1,4-dimethylbenzene		LC-12
4,4′-Dihydroxy-3,3′-dimethylbiphenyl		LC-4
2,2-Di-(2′-hydroxy-3′,5′-dimethylphenyl)propane		LC-4
2,2-Di-(4′-hydroxy-3′,5′-dimethylphenyl)propane		LC-4
Di-(2-hydroxy-3,5-dimethylphenyl)sulfide		LC-4
Di-(4-hydroxy-3,5-dimethylphenyl)sulfide		LC-4
2,3-Dihydroxy-1,5-di-*tert*.-octylbenzene		LC-12
2,2′-Dihydroxydiphenylmethane		GC-14
		TLC-9
2,4′-Dihydroxydiphenylmethane		GC-14
4,4′-Dihydroxydiphenylmethane		GC-14
		TLC-9
2,5-Dihydroxyhippuric acid		TLC-29
2,3-Dihydroxyisopropylbenzene		LC-12
2,5-Dihydroxyisopropylbenzene		LC-12
3,4-Dihydroxyisopropylbenzene		LC-12
6β,7β-Dihydroxykaurenoic acid		TLC-35
3,4-Dihydroxymandelic acid		GC-39
		LC-21
4′,5-Dihydroxy-7-methoxyflavone	—Sakuranetin	
1,4-Dihydroxymethylbenzene		GC-8
2,3-Dihydroxymethylbenzene		GC-8
		LC-12
2,4-Dihydroxymethylbenzene		GC-8
2,5-Dihydroxymethylbenzene		LC-12,14
2,6-Dihydroxymethylbenzene		GC-8
3,4-Dihydroxymethylbenzene		GC-3,8
		LC-3,12

Compound	Synonym	Number of tables and/or figures
3,5-Dihydroxymethylbenzene (Orsinol)		TLC-3,33
2,3-Dihydroxy-2-methylbutanoic acid 2RS-3RS		PC-6
2,3-Dihydroxy-2-methylbutanoic acid 2RS-3SR		PC-6
2,3-Dihydroxy-3-methylbutanoic acid		GC-F2
2,3-Dihydroxy-4-methylbutanoic acid		PC-6
4,6-Dihydroxymethyl-2,3-dimethylphenol		LC-1
3,4-Dihydroxy-5-methyl-*tert*.-octylbenzene		LC-12
2,2-Di-(4′-hydroxy-3′-methylphenyl)butane		LC-4
1,1-Di-(4′-hydroxy-3′-methylphenyl)cyclohexane		LC-4
4,6-Dihydroxymethyl-2,3,5-trimethylphenol		LC-1
1,3-Dihydroxynaphthalene		LC-14
2,7-Dihydroxynaphthalene		LC-14
4,5-Dihydroxy-2,7-naphthalenedisulfonic acid		LC-15
1,4-Dihydroxy-2-naphthoic acid		TLC-19
1,2-Dihydroxy-4-*tert*.-octylbenzene		LC-12
3,4-Dihydroxy-octylbenzene		LC-12
1,3-Dihydroxy-5-pentylbenzene		GC-15
3,4-Dihydroxy-*tert*.-pentylbenzene		LC-12
2,5-Dihydroxyphenylacetic acid		LC-21 TLC-24
3,4-Dihydroxyphenylacetic acid		GC-36,38 LC-21 TLC-29
3-(3,4-Dihydroxyphenyl)alanine (DOPA)		TLC-26
1,1-Di-(4′-hydroxyphenyl)butane		LC-4
2,2-Di-(4′-hydroxyphenyl)butane		LC-4
2,4-Dihydroxyphenyl-diphenylhydantoin		GC-16
3,4-Dihydroxyphenyl-diphenylhydantoin		GC-16
3,4′-Dihydroxyphenyl-diphenylhydantoin		GC-16
4,4′-Dihydroxyphenyl-diphenylhydantoin		GC-16
5-(3′,4′-Dihydroxyphenyl)-5-phenylhydantoin		GC-16
2,2-Di-(4′-hydroxyphenyl)propane		LC-4
3,4-Dihydroxyphenylpropionic acid (Dihydrocaffeic acid)		GC-36
2,3-Dihydroxypropanoic acid	—Glyceric acid	
2,3-Dihydroxy-*n*-propylbenzene		GC-8
C-(2,3-Dihydroxypropyl)glyceraric acid		GC-48
2,4′-Dihydroxystilbene		GC-14
4,4′-Dihydroxystilbene		TLC-11
2,2′-Dihydroxy-3,3′,5,5′-tetramethylbiphenyl		LC-4
4,4′-Dihydroxy-3,3′,5,5′-tetramethylbiphenyl		LC-4
2,5-Dihydroxytoluene	—2,5-Dihydroxy-methylbenzene	
3,4-Dihydroxytoluene	—3,4-Dihydroxy-methylbenzene	
2,6-Di-iodo-3,5-dimethylphenol		TLC-4
2,6-Di-iodophenol		TLC-4
Diisobutylphthalate	—1,2-Benzenedicarboxylic acid diisobutylester	
Diisodecylphthalate	—1,2-Benzenedicarboxylic acid diisodecylester	
2,6-Diisopropylphenol		GC-4,5 LC-1,5,6
Diisopropylphthalate	—1,2-Benzenedicarboxylic acid diisopropylester	
3,4-Dimethoxybenzaldehyde		TLC-26
1,2-Dimethoxybenzene		GC-6
2,6-Dimethoxybenzene-1-*N,N′*-diethylamine		GC-44
Dimethoxybenzenetetracarboxylic acid		GC-41
2,6-Dimethoxybenzene-1-trimethylsilylester		GC-44
3,4-Dimethoxybenzoic acid		GC-36
2,4-Dimethoxycinnamaldehyde		TLC-15
1,2-Dimethoxy-4-methylbenzene		GC-6

Compound	Synonym	Number of tables and/or figures
2,6-Dimethoxyphenol		GC-9 LC-14
3,4-Dimethoxyphenylacetic acid		LC-20
1,2-Dimethoxy-4-*cis*-propenylbenzene		GC-6
1,2-Dimethoxy-4-*trans*-propenylbenzene		GC-6
1,2-Dimethoxy -4-propylbenzene		GC-6
2,6-Dimethoxy-4-(2,4,6-trimethoxyphenoxy)-biphenyl		TLC-13
4-Dimethylaminocinnamic acid		PC-9
4-Dimethylamino-2′-propylcinnamic acid		PC-9
Dimethylanacardate		GC-13 TLC-10
2,2-Dimethylbutanedioic acid		GC-48
meso-2,3-Dimethylbutanedioic acid		GC-48
2,6-Dimethyl-4-*tert*.-butylphenol	—4-*tert*.-Butyl-2,6-dimethylphenol	
Dimethylethylphthalate	—1,2-Benzenedicarboxylic acid dimethyl-ethylester	
3,3-Dimethylglutaric acid		GC-22,25,28
2,5-Dimethylhydraquinone	—2,5-Dihydroxy-1,4-dimethylbenzene	
2,6-Dimethylhydraquinone	—2,5-Dihydroxy-1,3-dimethylbenzene	
2,4-Dimethyl-1-hydroxybenzene	—2,4-Dimethylphenol	
1,3-Dimethyl-2-hydroxybenzene	—2,6-Dimethylphenol	
1,2-Dimethyl-4-hydroxybenzene	—3,4-Dimethylphenol	
1,3-Dimethyl-5-(3′-methoxyphenyl)-5-phenylhydantoin		GC-16
1,3-Dimethyl-5-(4′-methoxyphenyl)-5-phenylhydantoin		GC-16
2,2-Dimethyl-3-methylsuccinic acid		LC-31
2,6-Dimethyl-4-nitrophenol		TLC-1
2,6-Dimethyl-4-*tert*.-octylphenol		LC-12
2,3-Dimethylphenol		GC-1,2,3,4,5,9,10 LC-1,12 TLC-1,2,3,5,8 PC-1
2,4-Dimethylphenol		GC-1,2,4,5,9 LC-1,5,12,14 TLC-1,5,8 PC-1
2,5-Dimethylphenol		GC-1,2,3,4,5,9,10 LC-1,2,12,15 TLC-1,4,5,8
2,6-Dimethylphenol		GC-1,2,3,4,6,9,10 LC-1,5,12,14,15 TLC-1,3,4,5,8 PC-1
3,4-Dimethylphenol		GC-1,2,3,4,5,9 LC-1,14 TLC-1,3,8 PC-1
3,5-Dimethylphenol		GC-1,2,3,4,5,9,10 LC-1 TLC-1,3,4,8 PC-1
Dimethylphthalate	—1,2-Benzenedicarboxylic acid dimethylester	
Diplocin		TLC-38
Diploschisteric acid		TLC-38
Dimethylpropanedioic acid		GC-48
4-(1,1-Dimethylpropyl)phenol		TLC-1
3,5-Dimethylpyrocatechol	—2,3-Dihydroxy-1,5-dimethylbenzene	
3,6-Dimethylpyrocatechol	—2,3-Dihydroxy-1,4-dimethylbenzene	
2,4-Dimethylolphenol		GC-14
2,6-Dimethylolphenol		GC-14

Compound	Synonym	Number of tables and/or figures
Dimethylterephthalate	—1,4-Benzenedicarboxylic acid dimethylester	
2,4-Dinitro-6-*sec.*-butylphenol (Dinoseb)	—2-*sec.*-Butyl-4,6-dinitrophenol	
2,4-Dinitro-6-*tert.*-butylphenol (Dinoterb)	—2-*tert.*-Butyl-4,6-dinitrophenol	
2,4-Dinitro-6-methylphenol (DNDC)	—2-Methyl-4,6-dinitrophenol	
2,4-Dinitro-1-naphthol	—1-Hydroxy-2,4-dinitronaphthalene	
2,4-Dinitrophenol		LC-6 TLC-1,3,5 PC-3
2,5-Dinitrophenol		TLC-1,3
2,6-Dinitrophenol		TLC-1,3
Di-*n*-nonylphthalate	—1,2-Benzenedicarboxylic acid di-*n*-nonylester	
2,4-Di-*tert.*-octylphenol		LC-12
2,6-Di-*tert.*-octylphenol		LC-12
Di-*n*-octylphthalate	—1,2-Benzenedicarboxylic acid di-*n*-octylester	
3,5-Di-*tert.*-octylpyrocatechol	—2,3-Dihydroxy-1,5-di-*tert.*-octylbenzene	
Dipentylphthalate	—1,2-Benzenedicarboxylic acid dipentylester	
2,2-Diphenylacetic acid		GC-42
Diphenyl-2-carboxylic acid		TLC-22
Diphenylethylphthalate	—1,2-Benzenedicarboxylic acid diphenylethylester	
Diphenylglycolate-1-(2,2,2)bicyclooctyl		TLC-30
Diphenylglycolate-2-diethylaminoethyl		TLC-30
Diphenylglycolate-*N*-methyl-3-piperidinyl and HCl salt		TLC-30
Diphenylglycolate-*N*-methyl-4-piperidinyl HCl		TLC-30
Diphenylglycolate-3-quinuclidinyl HCl		TLC-30
Diphenylglycolate-tropinyl HCl		TLC-30
2,2-Diphenylglycolic acid (Benzillic acid)		GC-42
Diphenylhydantoin		GC-16
Diphenylphthalate	—1,2-Benzenedicarboxylic acid diphenylester	
2,2′,4,4′-Diphenyltetrol		TLC-1
Dipropylphthalate	—1,2-Benzenedicarboxylic acid dipropylester	
Di-3,3,5-trimethylcyclohexylphthalate	—1,2-Benzenedicarboxylic acid di-3,3,5-trimethylcyclohexylester	
Di-3,5,5-trimethylhexylphthalate	—1,2-Benzenedicarboxylic acid di-3,5,5-trimethylhexylester	
Di-2,2,4-trimethylpentylphthalate	—1,2-Benzenedicarboxylic acid di-2,2,4-trimethylpentylester	
Divaricatic acid		TLC-38
4-*tert.*-Dodecyl-1,2-dihydroxybenzene		LC-12
Dodecylphenol		TLC-2 PC-1
4-*tert.*-Dodecylpyrocatechol	—4-*tert.*-Dodecyl-1,2-dihydroxybenzene	
Echinocarpic acid		TLC-38
Ellagic acid		GC-17 TLC-15
Emodin		TLC-38
Endocrocin		TLC-38
Entochein		TLC-38
Epanorin		TLC-38
Epicatechin		GC-17
L-Epicatechin		LC-21
Epicatechin gallate		GC-17
(-)-Epicatechin gallate		LC-10

Compound	Synonym	Number of tables and/or figures
(-)-Epigallocatechin gallate		LC-10
Epitheaflavic acid		LC-10
3,4-Epoxy-3-methyl-*cis*-4-phenyl-1(E)-buten-1-carboxylic acid methylester		TLC-36
3,4-Epoxy-3-methyl-*trans*-4-phenyl-1(E)-buten-1-carboxylic acid methylester		TLC-36
3,4-Epoxy-*cis*-4-phenyl-1(E)-buten-1-carboxylic acid methylester		TLC-36
3,4-Epoxy-*trans*-4-phenyl-1(E)-buten-1-carboxylic acid methylester		
Ergosterol		TLC-38
Eriodictyol		TLC-14
Eriodictyol-3′-*O*-*β*-D-glucoside		TLC-14
Eriodictyol-3′,5-*O*-*β*-D-glucoside		TLC-14
Eriodictyol-4′,5-*O*-*β*-D-glucoside		TLC-14
Erythraric acid		GC-48
Erythrin		TLC-38
Erythro-Hex-2-econo-1,4-lactone		GC-48
D-Erythronic acid		GC-27,36,48
Erythronic acid lactone		GC-43
Erythronic acid 1,4-lactone		GC-48
Erythronolactone		GC-27
C-(*Erythro*-1,2,3-Trihydroxypropyl) glyceraric acid		GC-48
Esculin		LC-10
Esculin hydrate		TLC-15
4-Ethoxy butanoic acid		PC-6
2-Ethoxyethoxyacetic acid		GC-28
2-(4′-Ethoxy-3′-nitrophenyl)-2-phenylpropane		LC-7
3-Ethylamino-4-methylphenol		TLC-1
3-Ethylanisol	—3-Ethylmethoxybenzene	
2-Ethyl-4-*tert.*-butylphenol	—4-*tert.*-Butyl-2-ethylphenol	
2-Ethyl-6-*tert.*-butylphenol	—2-*tert.*-Butyl-6-ethylphenol	
α-Ethyl-3-chlorocinnamic acid	—3-Chloro-2′-ethylcinnamic acid	
α-Ethyl-4-chlorocinnamic acid	—4-Chloro-2′-ethylcinnamic acid	
2′-Ethyl-cinnamic acid		PC-9
3-Ethyl-1,2-dihydroxybenzene		GC-8
4-Ethyl-1,2-dihydroxybenzene		LC-12
4-Ethyl-1,3-dihydroxybenzene		GC-8 LC-12
Ethyl-1,4-dihydroxybenzene		GC-8
2-Ethyl-4,6-dihydroxymethylphenol		LC-1
4-Ethyl-1,2-dimethoxybenzene		GC-6
2′-Ethyl-4-dimethylaminocinnamic acid		PC-9
2-Ethyl-4,5-dimethylphenol		GC-5
C-Ethylglyceraric acid		GC-48
2-Ethylhydracrylic acid		GC-F1
1-Ethyl-2-hydroxybenzene	—2-Ethylphenol	
1-Ethyl-3-hydroxybenzene	—3-Ethylphenol	
1-Ethyl-4-hydroxybenzene	—4-Ethylphenol	
2-Ethyl-4-hydroxymethylphenol		LC-1
2-Ethyl-6-hydroxymethylphenol		LC-1
Ethylmalonic acid		GC-F1
2-Ethylmethoxybenzene		GC-6
3-Ethylmethoxybenzene		GC-6 TLC-1
4-Ethylmethoxybenzene		GC-6
2′-Ethyl-4-methoxycinnamic acid		PC-9
2′-Ethyl-4-methylcinnamic acid		PC-9
2-Ethyl-4-methylphenol		GC-5
2-Ethyl-5-methylphenol		GC-5
2-Ethyl-6-methylphenol		GC-5
3-Ethyl-5-methylphenol		GC-5 TLC-1

Compound	Synonym	Number of tables and/or figures
		PC-1
4-Ethyl-2-methylphenol		GC-5
		PC-1
4-Ethyl-3-methylphenol		GC-5
		PC-1
3-Ethyl-6-methylphenol		GC-5
2′-Ethyl-3-nitrocinnamic acid		PC-9
2′-Ethyl-4-nitrocinnamic acid		PC-9
Ethylorsellinate		TLC-38
2-Ethylphenol		GC-1,2,3,4,5,9
		LC-1,11,12,14
		TLC-1,4,5,12
		PC-1
3-Ethylphenol		GC-1,2,4,5,9
		LC-1,11,14
		TLC-1,4
		PC-1
4-Ethylphenol		GC-1,2,3,4,5,9,10
		LC-1,5,11,12,14
		TLC-1,5
		PC-1
Ethylpropanedioic acid		GC-48
2′-Ethyl-4-isopropylcinnamic acid		PC-9
2-Ethyl-5-*n*-propylphenol		GC-5
3-Ethylpyrocatechol	—4-Ethyl-1,2-dihydroxybenzene	
4-Ethylpyrocatechol	—4-Ethyl-1,2-dihydroxybenzene	
Eugenol (see also 4-Allyl-2-methoxyphenol)		TLC-1,14
Evernic acid		TLC-33,38
Fenoprop (2-(2,4,5-triChloro-phenoxy)propionic acid)		LC-25
Ferulic acid	—4-Hydroxy-3-methoxycinnamic acid	
cis-Ferulic acid	—*cis*-4-Hydroxy-3-methoxycinnamic acid	
trans-Ferrulic acid	—*trans*-4-Hydroxy-3-methoxycinnamic acid	
3-Feruloylquinic acid		TLC-26
*bis*Flavanol A		LC-10
2-Fluorophenol		TLC-1,4
3-Fluorophenol		TLC-1,4
4-Fluorophenol		TLC-1
5-Fluoropyridine-3-carboxylic acid		LC-20
Formic acid		GC-24
		LC-14,18,28
		TLC-17,22
Fragilin		TLC-32,38
Fucodiphlorethol-A-decaacetate(2,2′,4′,6,6′-pentaacetoxy-4-(2,6-diacetoxy-4-(2,4,6-triacetoxyphenoxy)-phenoxy)-biphenyl)		TLC-13
Fucotriphloretholdodecaacetate(2,2′,4′,6,6′-pentaacetoxy-4-(2,6-diacetoxy-4-(2,4,6-triacetoxyphenoxy-phenoxy)-phenoxy)-biphenyl)		TLC-13
Fucophlorethol-A-octaacetate(2,2′,4′,6,6′-pentaacetoxy-4-(2,4,6-tri-acetoxy-phenoxy)-biphenyl)		TLC-13
Fumaramic acid		PC-4
Fumaric acid (*trans*-Butenedioic acid)		GC-22,23,24,25,28 36,48
		LC-17,18,28
		TLC-17,18,19,20,22 25,28
		PC-4,7
Fumarprotocetraric acid		TLC-33,38
2,5-Furandicarboxylic acid		GC-32
Galactaric acid		GC-48
Galatonic acid		GC-48
D-Galactonic acid lactone		GC-43

Compound	Synonym	Number of tables and/or figures
Galactonic acid 1,4-lactone		GC-48
2-Galaturonic acid		GC-27
2-D-Galaturonic acid		LC-28
3-Galaturonic acid		GC-27
Galbinic acid		TLC-38
Gallic acid	—3,4,5-Trihydroxybenzoic acid	
Gangaleoidin		TLC-38
Garcina lactone		PC-4
Gentisic acid	—2,5,-Dihydroxybenzoic acid	
Genuine aurintricarboxylic acid		TLC-33
Gibbrellin A1,A3,A4,A5,A7,A9,A13,A14,A20		TLC-35
Gingerol		TLC-14
Gingeryl		GC-15
		TLC-14
Glomelliferic acid		TLC-38
Glucaric acid		GC-27,48
Glucaric acid 1,4-lactone		GC-27
Gluconic acid		GC-36,48
		LC-28
D-Gluconic acid lactone		GC-43
Gluconic acid 1,4-lactone		GC-36,48
Gluconic acid 1,5-lactone		GC-36,48
Glucoric acid		GC-36
Glucuronic acid		LC-28
L-4-Glutamyl hydrazide		PC-5
1-(5-Glutamyl)-methylhydrazide		PC-5
4-*N*-Glutamyl-4-*N*-methyl-*N'*-glyoxylic acid hydrazone		PC-5
Glutarate semialdehyde		PC-4
Glutaric acid (Pentanedioic acid)		GC-22,25,26,36,48
		LC-17,18,28
		TLC-18,19,21,22,25
cis-Glyceraldehyde		GC-36
trans-Glyceraldehyde		GC-36
Glyceraric acid (Tartronic acid)		GC-48
Glyceric acid		GC-29
		LC-28
		TLC-18,20,25
		PC-6
D-Glycero-D-gulo-heptonic acid lactone		GC-43
Glyceronic acid		GC-48
α-Glycerophosphoric acid		GC-36
β-Glycerophosphoric acid		GC-36
Glycolic acid (Hydroxyethanoic acid, Hydroxyacetic acid)		GC-23,24,25,32,36,F2
		LC-17,28
		TLC-18,20,22,25
		PC-6
Glycorate-3-quinuclidinyl		TLC-30
Glyoxalic acid	—Glyoxylic acid	
Glyoxylic acid (Oxoethanoic acid, Oxoacetic acid)		GC-22,23,25,30,F2
		LC-18,28,29
		TLC-22,23,25
		PC-5
Grandidentatin		TLC-32
Grayanic acid		TLC-38
Guaiacol	—2-Methoxyphenol	
Gulonic acid 1,4-lactone		GC-48
Gyrophoric acid		TLC-38
Haemathamnolic acid		TLC-38

Compound	Synonym	Number of tables and/or figures
Haemoventonic acid		TLC-38
Heptanedicic acid (Pimelic acid)		GC-26,42,48,F1
		TLC-31
2,3′,3″,4,5′,5″,6-Heptamethoxy-*p*-terphenyl		TLC-13
Hesperidin		GC-17
Hexachlorophene	—2,2′-Methylene *bis* (3,4,6-trichlorophenol)	
Hexadecylphenol		TLC-2
Hexahydrohippuric acid		TLC-24
3,3′,4′,5,5′,7-Hexahydroxyflavone		GC-17
		TLC-15
Hexanedioic acid		GC-24,25,26,32,36,42,F1
		LC-18,28
		TLC-18,19,22,25,31
4-Hexanoyloxybutanoic acid		PC-6
Hexuronic acid		GC-36
4-Hexyl-1,3-dihydroxybenzene		TLC-1
Hexylheptylphthalate	—1,2-Benzenedicarboxylic acid hexyl-heptylester	
4-*tert*.-Hexyl-2-methylphenol		LC-5
2-*n*-Hexylphenol		GC-5
4-*n*-Hexylphenol		GC-5
4-Hexylresorsinol	—4-Hexyl-1,3-dihydroxybenzene	
Hippuric acid (2-Benzoylaminoethanoic acid)		GC-23,32,36,39,40,F1
		LC-32
		TLC-1,18,19,24
Homogentisic acid	—2,5-Dihydroxyphenylacetic acid	
Homopyrocatechol	—3,4-Dihydroxymethylbenzene	
Homosekikaic acid		TLC-38
Homovanillic acid	—4-Hydroxy-3-methoxyphenylacetic acid	
Hydrocinnamic acid		LC-31
Hydroquinone	—1,4-Dihydroxybenzene	
2-Hydroxyacetanilide		TLC-11
3-Hydroxyacetanilide		TLC-11
4-Hydroxyacetanilide		TLC-11
Hydroxyacetic acid	—Glycolic acid	
2-Hydroxyacetophenone		GC-10
2(2-Hydroxyacetoxy)acetic acid		PC-6
2-Hydroxyaniline		TLC-11
3-Hydroxyaniline		TLC-11
4-Hydroxyaniline		TLC-11
2-Hydroxyanisol	—2-Hydroxymethoxybenzene	
4-Hydroxyanisol	—4-Hydroxymethoxybenzene	
3-Hydroxyanthranilic acid	—2-Amino-3-hydroxybenzoic acid	
4-Hydroxy-azobenzene		TLC-3
2-Hydroxybenzaldehyde		LC-3,14
		TLC-1,8,12,26
3-Hydroxybenzaldehyde		LC-14
		TLC-12,26
4-Hydroxybenzaldehyde		LC-14,31
		TLC-26
5-Hydroxy-1,2,3,4-benzenetetracarboxylic acid		GC-21
Hydroxy-1,3,5-benzenetricarboxylic acid		GC-21
2(2′-Hydroxyphenyl)decane		GC-7
2(4′-Hydroxyphenyl)-decane		GC-7
3(2′-Hydroxyphenyl)-decane		GC-7
3(4′-Hydroxyphenyl)-decane		GC-7
4(2′-Hydroxyphenyl)-decane		GC-7
4(4′-Hydroxyphenyl)-decane		GC-7
5(2′-Hydroxyphenyl)-decane		GC-7

Compound	Synonym	Number of tables and/or figures
5(4′-Hydroxyphenyl)-decane		GC-7
3-Hydroxy-1,2-benzenedicarboxylic acid		GC-21
4-Hydroxy-1,2-benzenedicarboxylic acid		GC-21
2-Hydroxy-1,3-benzenedicarboxylic acid		GC-21
4-Hydroxy-1,3-benzenedicarboxylic acid		GC-21
5-Hydroxy-1,3-benzenedicarboxylic acid		GC-21
		PC-8
Hydroxy-1,4-benzenedicarboxylic acid		GC21
1(2′-Hydroxyphenyl)-nonane		GC-7
1(4′-Hydroxyphenyl)-nonane		GC-7
2(2′-Hydroxyphenyl)-nonane		GC-7
2(4′-Hydroxyphenyl)-nonane		GC-7
3(2′-Hydroxyphenyl)-nonane		GC-7
3(4′-Hydroxyphenyl)-nonane		GC-7
4(2′-Hydroxyphenyl)-nonane		GC-7
4(4′-Hydroxyphenyl)-nonane		GC-7
5(2′-Hydroxyphenyl)-nonane		GC-7
5(4′-Hydroxyphenyl)-nonane		GC-7
1(2′-Hydroxyphenyl)-octane		GC-7
1(3′-Hydroxyphenyl)-octane		GC-7
1(4′-Hydroxyphenyl)-octane		GC-7
2(2′-Hydroxyphenyl)-octane		GC-7
2(4′-Hydroxyphenyl)-octane		GC-7
3(2′-Hydroxyphenyl)-octane		GC-7
3(4′-Hydroxyphenyl)-octane		GC-7
4(2′-Hydroxyphenyl)-octane		GC-7
4(4′-Hydroxyphenyl)-octane		GC-7
5-Hydroxy-1,2,3,4-benzenetetra-carboxylic acid		GC-21
6(2′-Hydroxyphenyl)-undecane		GC-7
2-Hydroxybenzoic acid		GC-17,21,25,34,38
		LC-2,3,14,15,21,22,23,30
		TLC-5,11,12,22,25,26,27, 29
		PC-2,7,8
3-Hydroxybenzoic acid		GC-21,24,34,36
		LC-2,14,19,21,22,30
		TLC-11,15,25,29
4-Hydroxybenzoic acid		GC-17,21,34,36,37,38
		LC-2,14,19,20,21,22, 30,31
		TLC-1,11,15,22,26,27,29
		PC-8
4-Hydroxybenzoic acid propylester		TLC-1
2-Hydroxybenzophenone		TLC-9
4-Hydroxybenzophenone		TLC-9
1-(4′-Hydroxybenzyl)-4-hydroxybenzene		LC-4
2-Hydroxybiphenyl		GC-12,19
		LC-6,8,12
		TLC-1,4,7,11
3-Hydroxybiphenyl		GC-12
		TLC-4,7
4-Hydroxybiphenyl		GC-12
		LC-12
		TLC-1,7,11
2-Hydroxybutanoic acid-(2-Hydroxybutyric acid)		GC-36,F1
		TLC-22,25
		PC-6
3-Hydroxybutanoic acid (3-Hydroxybutyric acid)		GC-F1
		LC-18,28

Compound	Synonym	Number of tables and/or figures
		TLC-18,19,22,26
		PC-6
4-Hydroxybutanoic acid (4-Hydroxybutyric acid)		GC-25
		LC-17
2-Hydroxybutanoyl-2-hydroxybutanoic acid		GC-48
4-Hydroxy-*tert.*-butylbenzene	—4-*tert.*-Butylphenol	
2-Hydroxybutyric acid	—2-Hydroxybutanoic acid	
β-Hydroxybutyric acid	—3-Hydroxybutanoic acid	
4-Hydroxybutyric acid	—4-Hydroxybutanoic acid	
2-Hydroxychlorobenzene	—2-Chlorophenol	
3-Hydroxychlorobenzene	—3-Chlorophenol	
4-Hydroxychlorobenzene	—4-Chlorophenol	
Hydroxycinnamic acid		GC-24
cis-Hydroxycinnamic acid		GC-35
trans-Hydroxycinnamic acid		GC-35
2-Hydroxycinnamic acid		GC-17
		TLC-15,26
3-Hydroxycinnamic acid		TLC-26
4-Hydroxycinnamic acid		GC-17,37
		LC-20,21,31
		TLC-15,26,27
4-Hydroxycinnamic acid-tartaric acid-glucose ester		LC-24
4-Hydroxycoumarin		TLC-11
7-Hydroxycoumarin		GC-17
		TLC-11
2-Hydroxycyclohexanecarboxylic acid		GC-42
		TLC-31
cis-2-Hydroxycyclohexanecarboxylic acid		GC-30
trans-2-Hydroxycyclohexanecarboxylic acid		GC-30
cis-3-Hydroxycyclohexanecarboxylic acid		GC-30
trans-3-Hydroxycyclohexanecarboxylic acid		GC-30
4-Hydroxycyclohexanecarboxylic acid		GC-36
2(1-Hydroxycyclohexane)acetic acid		PC-6
4(1-Hydroxycyclohexane)-2-butanoic acid		PC-6
2(1-Hydroxycyclohexane)-2-ethylacetic acid		PC-6
4-Hydroxy-3,5-dimethoxybenzaldehyde		GC-17
		TLC-1,26
4-Hydroxy-3,5-dimethoxybenzoic acid (Syringic acid)		GC-17,24,37
		LC-6,20,21,31
		TLC-15,26
4-Hydroxy-3,5-dimethoxycinnamic acid (Sinapic acid)		GC-17,24,37
		LC-21
		TLC-15,26,27
cis-4-Hydroxy-3,5-dimethoxycinnamic acid		GC-35
		TLC-24
trans-4-Hydroxy-3,5-dimethoxycinnamic acid		GC-35
		TLC-24
1-Hydroxy-2,4-dimethylbenzene	—2,4-Dimethylphenol	
2-Hydroxy-1,3-dimethylbenzene	—2,6,-Dimethylphenol	
1-Hydroxy-1,2-dimethylbenzene	—3,4-Dimethylphenol	
1-(2′-Hydroxy-3′,5′-dimethylbenzyl)-l2-hydroxy-3,5-dimethylbenzene		LC-4
1-(4′-Hydroxy-3′,5′-dimethylbenzyl)-4-hydroxy-3,5-dimethylbenzene		LC-4
2,4-*bis*(4-Hydroxy-α,α-dimethylbenzyl)phenol		LC-9
1-Hydroxy-2,4-dinitronaphthalene		TLC-1
		PC-3
4′-Hydroxy-2,4-dinitro-diphenylamine		TLC-3
4′-Hydroxy-2,4-dinitro-diphenylamine-3′-sulfonic acid		TLC-3
2-(4′-Hydroxy-3′,5′-dinitrophenyl)-2-(4″-nitrophenyl)propane		LC-7
2-(4′-Hydroxy-3′,5′-dinitrophenyl)-2-phenylpropane		LC-7

Compound	Synonym	Number of tables and/or figures
3-Hydroxydiphenylamine		TLC-3
4-Hydroxydiphenylamine		TLC-3
2-Hydroxydiphenylmethane		TLC-9
4-Hydroxydiphenylmethane		TLC-9
(Hydroxyethanoyl)hydroxyethanoic acid		GC-48
2-Hydroxy-ethylbenzene	—2-Ethylphenol	
4-Hydroxy-ethylbenzene	—4-Ethylphenol	
C-(2-Hydroxyethyl)glyceraric acid		GC-48
4-Hydroxy-β-D-glucopyranoside		TLC-15
2-Hydroxyglutaric acid		GC-36
6-Hydroxyhexanoic acid		GC-42
		TLC-31
2-Hydroxyhippuric acid (Salicyluric acid)		GC-36
		LC-21,30
		TLC-29
3-Hydroxyhippuric acid		GC-32,36
4-Hydroxyhippuric acid		GC-32,36
5-Hydroxy-3-indoleacetic acid		GC-36,38
		LC-21
		TLC-29
2-Hydroxyisobutanoic acid (2-Hydroxyisobutyric acid)		GC-F1
3-Hydroxyisobutanoic acid		GC-32,36,F1
α-Hydroxyisocaproic acid	—2-Hydroxyisohexanoic acid	
2-Hydroxyisohexanoic acid		GC-31
2-Hydroxyisophthalic acid	—2-Hydroxy-1,3-benzenedicarboxylic acid	
4-Hydroxyisophthalic acid	—4-Hydroxy-1,3-benzenedicarboxylic acid	
5-Hydroxyisophthalic acid	—5-Hydroxy-1,3-benzenedicarboxylic acid	
1-Hydroxy-2-isopropylbenzene	—2-Isopropylphenol	
4-Hydroxy-isopropylbenzene	—4-Isopropylphenol	
4-Hydroxymandelic acid		GC-36,38,F1
		LC-21
		TLC-29
3-Hydroxytrimellitic acid(3-Hydroxybenzene-1,2,4-tri carboxylic acid)		GC-21
5-Hydroxytrimellitic acid(5-Hydroxybenzene-1,2,4-tri carboxylic acid)		GC-21
Hydroxytrimesic acid		GC-21
3-Hydroxy-4-methoxybenzaldehyde		LC-14
		TLC-26
4-Hydroxy-3-methoxybenzaldehyde (Vanillin)		GC-9,17
		LC-2,14,31
		TLC-1,14,26
		PC-2
4-Hydroxy-3-methoxybenzoic acid (Vanillic acid)		GC-17,24,36,37,38,F1
7-Hydroxy-6-methoxycoumarin (Scopoletin)		GC-17
		LC-10
		TLC-15,27
3-Hydroxy-4-methoxymandelic acid		GC-37
		LC-22
		TLC-24
		PC-7
4-Hydroxy-3-methoxymandelic acid (Vanillylmandelic acid)		GC-36,37,38,39,F1
		LC-20,21,22,30
		TLC-24,29
		PC-7
4-Hydroxy-3-methoxyphenylacetaldehyde		LC-14
3-Hydroxy-4-methoxyphenylacetic acid		GC-38
4-Hydroxy-3-methoxyphenylacetic acid 63(Homovanillic acid)		GC-36,38,39,F1

Compound	Synonym	Number of tables and/or figures
		LC-20,21
		TLC-29
4-Hydroxy-3-methoxyphenyllactic acid		LC-21
5-(4-Hydroxy-3-methoxyphenyl)-5-phenylhydantoin		GC-16
4-Hydroxy-3-methoxyphenylpropionic acid		LC-21
4-Hydroxy-3-methoxyphenylpyruvic acid		LC-21
4-Hydroxy-5-methyl-1,3-benzenedicarboxylic acid		GC-21
2-Hydroxy-3-methylbenzoic acid		LC-23
2-Hydroxy-4-methylbenzoic acid		LC-23
3-Hydroxy-2-methylbutanoic acid		GC-F1
4-Hydroxymethyl-2,3-dimethylphenol		LC-1
6-Hydroxymethyl-2,3-dimethylphenol		LC-1
5-Hydroxymethylfuroic acid		GC-32,36
5-Hydroxy-2-methylfuroic acid		GC-F1
3-Hydroxy-3-methylglutaric acid		GC-36
		TLC-18
4-Hydroxy-5-methylisophthalic acid		4-Hydroxy-5-methyl-1,3-benzene dicarboxylic acid
6-Hydroxy-2′-methyl-2-naphthaleneacetic acid		LC-24
2-*C*-(Hydroxymethyl)pentonic acid		GC-48
2-Hydroxy-2-methylpropanoic acid		PC-6
3-Hydroxymethylstyryl-arylether		GC-14
4-Hydroxymethyl-2,3,5-trimethylphenol		LC-1
6-Hydroxymethyl-2,3,5-trimethylphenol		LC-1
4-Hydroxymethyl-2,3,6-trimethylphenol		LC-1
2-Hydroxy-3-methyl-*n*-valeric acid		GC-31
1-Hydroxynaphthalene (α-naphthol)		GC-3,10,12
		LC-6,9,12
		TLC-1,2,3,5,11
		PC-1,2
2-Hydroxy-1-naphthaldehyde		LC-9
		TLC-3
2-Hydroxy naphthalene (β-naphthol)		GC-10
		LC-6,12
		TLC-1,3,5,11
		PC-2
1-Hydroxy-2-naphthalenesulfonic acid		LC-15
1-Hydroxy-3-naphthalenesulfonic acid		LC-15
1-Hydroxy-4-naphthalenesulfonic acid		LC-15,30
1-Hydroxy-5-naphthalenesulfonic acid		LC-15
1-Hydroxy-3,6-naphthalenedisulfonic acid		LC-15
1-Hydroxy-3,8-naphthalenedisulfonic acid		LC-15
2-Hydroxy-6-naphthalenesulfonic acid		LC-15
2-Hydroxy-3,6-naphthalenedisulfonic acid		LC-15
6-Hydroxy-1-naphthalenesulfonic acid		LC-30
2-Hydroxy-1-naphthoic acid		LC-22
1-Hydroxy-2-naphthoic acid		LC-22
3-Hydroxy-2-naphthoic acid		LC-22
		PC-8
2-Hydroxynitrobenzene	—2-Nitrophenol	
3-Hydroxynitrobenzene	—3-Nitrophenol	
4-Hydroxynitrobenzene	—4-Nitrophenol	
2-Hydroxy-3-nitrobenzoic acid		LC-23
2-Hydroxy-4-nitrobenzoic acid		LC-23
2-Hydroxy-5-nitrobenzoic acid		LC-23
2-(4′-Hydroxy-3′-nitrophenyl)-2-phenylpropane		LC-7
1-Hydroxy-2-nitronaphthalene		PC-2
2-Hydroxy-1-nitrosonaphthalene		PC-2
2-Hydroxy-3-octyltoluene	—2-Methyl-6-octylphenol	

Compound	Synonym	Number of tables and/or figures
DL-*trans*–3-Hydroxy-5-oxo-2,3-tetrahydrofuran dicarboxylate (DL-Hibiscus acid lactone)		
DL-*trans*-3-Hydroxy-5-oxo-2,3-tetrahydrofuran dicarboxylic acid		TLC-37
4-Hydroxyphenoxyacetic acid		TLC-22
2-Hydroxyphenylacetic acid		GC-32,36,38,39,F1
		LC-20,21
		TLC-15
3-Hydroxyphenylacetic acid		GC-32,36,38,39,F1
		LC-20,21
4-Hydroxyphenylacetic acid		GC-23,28,32,36,38,39, 40,F1
		LC-20,21
		TLC-15,29
4-(2′-Hydroxyphenyl)butanoic acid		TLC-24
trans-4-(2′-Hydroxyphenyl)but-3-enoic acid		TLC-24
2-Hydroxyphenyl-diphenylhydantoin		GC-16
3-Hydroxyphenyl-diphenylhydantoin		GC-16
4-Hydroxyphenyl-diphenylhydantoin		GC-16
3-Hydroxyphenylhydracrylic acid		GC-36
4-Hydroxyphenylhydracrylic acid		GC-36
3-(3-Hydroxyphenyl)hydracrylic acid		GC-32
6-(2′-Hydroxyphenyl)hydroxyhexanoic acid		TLC-24
4-Hydroxyphenyllactic acid		GC-36,F1
		LC-21
5-(4′-Hydroxyphenyl)-5-phenylhydantoin		GC-16
2-(4′-Hydroxyphenyl)-2-phenylpropane		LC-7
3-Hydroxyphenylpropionic acid		GC-36,39
		LC-21
3-Hydroxy-2-phenylpropionic acid (Tropic acid)		GC-42
		LC-31
4-Hydroxyphenylpropionic acid		GC-23,39
		LC-21
4-Hydroxyphenylpyruvic acid		GC-23,32
		LC-21,24
4,4′-Hydroxyphenyl-2,2,4-trimethylchroman		LC-9
4-Hydroxy-*o*-phthalic acid	—4-Hydroxy-1,2-benzenedicarboxylic acid	
2-Hydroxypropane-1,2,3-tricarboxylic acid	—Citric acid	
2-Hydroxypropanoic acid	—Lactic acid	
3-Hydroxypropanoic acid (β-Lactic acid)		PC-6
2-Hydroxypropanoyl-2-hydroxypropanoic acid		GC-48
3-Hydroxypropanoyl-3-hydroxypropanoic acid		GC-48
1-Hydroxypropane-1,2,3-tricarboxylic acid		GC-48
C-(3-Hydroxypropyl)glyceraric acid		GC-48
2-Hydroxypyridine-5-carboxylic acid		LC-31
3-Hydroxyquinaldine		TLC-37
8-Hydroxyquinoline		PC-2
4-Hydroxysalicylic acid	—2,4-Dihydroxybenzoic acid	
4-Hydroxystilbene		TLC-11
2-Hydroxy-5-sulfobenzoic acid (3-Carboxy-4-hydroxy-benzenesulphonic acid)		LC-23
		PC-8
Hydroxyterephthalic acid	—Hydroxy-1,4-benzenedicarboxylic acid	
3-Hydroxy-tetrahydro-6-pyridazinone-3-carboxylic acid		PC-5
2-Hydroxytoluene	—2-Methylphenol	
3-Hydroxytoluene	—3-Methylphenol	
4-Hydroxytoluene	—4-Methylphenol	
1-Hydroxy-2,4,5-trimethylbenzene	—2,4,5-Trimethylphenol	
2(1-Hydroxy-3,3,5-trimethylcyclohexyl)acetic acid		PC-6

Compound	Synonym	Number of tables and/or figures
2-Hydroxyvaleric acid		TLC-22
2-Hydroxyisovaleric acid		GC-31
3-Hydroxyisovaleric acid		GC-32,36,F1
Hypoprotocetraric acid		TLC-38
Hypothamnolic acid		TLC-38
Idaric acid		GC-48
Idonic acid		GC-48
Idonic acid 1,4-lactone		GC-48
Imbricaric acid		TLC-38
Iminodiacetic acid		GC-24
4-Indanol		GC-5
5-Indanol		GC-5
Indole-3-acetic acid (Indoleacetic acid)		GC-25,32,36,F1
		LC-21,32
		TLC-24,29
Indolebutanoic acid		LC-21
3-Indolecarboxaldehyde		LC-24
Indolepropionic acid		LC-21
3-Indolepyruvic acid		LC-24
2-Iodobenzoic acid		GC-34
3-Iodobenzoic acid		GC-34
4-Iodobenzoic acid		GC-34
2-Iodophenol		TLC-4
3-Iodophenol		TLC-4
4-Iodophenol		TLC-12
2-Isobutylphenol		GC-5
4-Isobutylphenol		GC-5
Isochlorogenic acid		GC-17
Isocitric acid		GC-22,25,27,28,36
		LC-18,28
		TLC-22
DL-4-Isocitricamide		PC-4
Isocitric lactone		PC-4
Isoeugenol	—2-Methoxy-4-propenylphenol	
Isohomovanillic acid		GC-28
Isonicotinic acid		LC-20,31
Isopalustric acid		GC-46
Isophthalic acid	1,3-Benzenedicarboxylic acid	
Isopimaric acid		GC-46
4-Isopropylcinnamic acid		PC-9
2-Isopropylhydroquinone	—2,5-Dihydroxy-isopropylbenzene	
4-Isopropyl-1,2-dihydroxybenzene	—3,5-Dihydroxy-isopropylbenzene	
2-Isopropyl-5-methylphenol	—3-Methyl-6-isopropylphenol	
2-Isopropyl-6-methylphenol	—2-Methyl-6-isopropylphenol	
2-Isopropylphenol		GC-4,5
		LC-1,6,11,12
		TLC-1,4,5
		PC-1
3-Isopropylphenol		GC-5
		LC-11
4-Isopropylphenol		GC-5
		LC-11,12
		TLC-5
		PC-1
4(4′-Isopropylphenyl)-3-butanoic acid		PC-9
3-Isopropylpyrocatechol	—2,3-Dihydroxyisopropylbenzene	
4-Isopropylpyrocatechol	—3,4-Dihydroxyisopropylbenzene	
2,4′-Isopropylidenediphenol		LC-9

Compound	Synonym	Number of tables and/or figures
4,4′-Isopropylidenediphenol		LC-9
Isosteviol		TLC-35
Isovaleric acid		GC-25
Itaconic acid		GC-24
		TLC-20
Kaempferol (3,4′,5,7-Tetrahydroxyflavone)		LC-21
Kaempferol-3-rhamnoglucoside		GC-17
α-Ketoadipic acid	—2-Oxoadipic acid	
α-Ketobutyric acid	—2-Oxobutanoic acid	
α-Ketocaproic acid	—2-Oxohexanoic acid	
α-Ketoisocaproic acid	—2-Oxoisohexanoic acid	
α-Ketocaprylic acid	—2-Oxooctanoic acid	
α-Keto-δ-*N*-carbamidovaleric acid	—2-Oxo-5-*N*-carbamidovaleric acid	
α-Keto-γ-ethiolbutyric acid	—2-Oxo-4-ethiolbutanoic acid	
α-Ketoglutaric acid	—2-Oxoglutaric acid	
α-Ketoglutaric acid-γ-ethylether	—2-Oxoglutaric acid-4-ethylether	
α-Keto-β-dimethylbutyric acid	—2-Oxo-3-dimethylbutanoic acid	
α-Keto-γ-(methylthio)butyric acid	—2-Oxo-4-(methylthio)butanoic acid	
α-Keto-β-methyl-*n*-valeric acid	—2-Oxo-3-methyl-*n*-valeric acid	
α-Ketooctanoic acid	—2-Oxooctanoic acid	
α-Ketovaleric acid	—2-Oxovaleric acid	
α-Ketoisovaleric acid	—2-Oxoisovaleric acid	
Lactic acid (2-Hydroxypropanoic acid)		GC-23,24,25,27,28,29, 36,F1,F2
		LC-17,18,28
		TLC-17,18,20,22,25
		PC-4,6,7
Lecanoic acid		TLC-33,38
Leprapinic acid		TLC-38
Lepraric acid		TLC-38
Leucocyanidin		GC-17
Levopimaric acid		GC-48
Levulinic acid		LC-28
		TLC-23
Lichesterinic acid		TLC-38
Lichexanthone		TLC-38
Lividic acid		TLC-38
Lobaric acid		TLC-38
Loxodin		TLC-38
Luteolin-7-glucoside		GC-17
Lyxonic acid		GC-48
D-Lyxonic acid lactone		GC-43
Lyxonic acid 1,4-lactone		GC-48
Maleamic acid		PC-4
Maleic acid (*cis*-Butenedioic acid)		GC-24,25,48
		LC-28
		TLC-20,25,28
		PC-4
Malic acid (2-Hydroxybutanedioic acid)		GC-22,23,24,25,27,28 36,48
		LC-16,17,18,28
		TLC-17,18,20,22,25
		PC-4,7
Malonic acid (Propanedioic acid)		GC-22,24,25,26
		LC-17,18,28
		TLC-18,20,22,25
Malvidin		GC-17
Malvidin chloride		TLC-15
Malvin		GC-17
Mandelic acid		GC-25,32
		LC-20,32
		TLC-15,25,30

Compound	Synonym	Number of tables and/or figures
Mannaric acid		GC-48
L-Mannonic acid lactone		GC-43
Mannonic acid 1,4-lactone		GC-48
Mannonic acid 1,5-lactone		GC-48
Merochlorophaeic acid		TLC-38
Mesaconic acid		GC-25,F1 LC-18
Meso-2,3-dimethylbutanedioic acid		GC-48
Mesotartaric acid		GC-27
Mesoxalic acid	—Oxopropanedioic acid	
Methacrylic acid		LC-28
3-Methoxybenzaldehyde		TLC-26
4-Methoxybenzaldehyde		TLC-26
Methoxybenzene		GC-4,6 TLC-1
3-Methoxy-1,2-benzenedicarboxylic acid		GC-21
4-Methoxy-1,2-benzenedicarboxylic acid		GC-21
2-Methoxy-1,3-benzenedicarboxylic acid		GC-21
4-Methoxy-1,3-benzenedicarboxylic acid		GC-21
5-Methoxy-1,3-benzenedicarboxylic acid		GC-21
Methoxy-1,4-benzenedicarboxylic acid		GC-21
3-Methoxy-1,2,4-benzenetricarboxylic acid		GC-21
5-Methoxy-1,2,4-benzenetricarboxylic acid		GC-21
Methoxy-1,3,5-benzenetricarboxylic acid		GC-21
5-Methoxy-1,2,3,4-benzenetetracarboxylic acid		GC-21
2-Methoxybenzoic acid		GC-21,33
3-Methoxybenzoic acid		GC-21,33
4-Methoxybenzoic acid		GC-21,33
5-*C*-(2-Methoxycarbonyl)-vinyl)-2*C*-phenyltetrahydro-3γ,4C-furandicarboxylic acid dimethylester		TLC-36
5-*C*-(2-(Methoxycarbonyl)-vinyl)-2t-phenyl-tetrahydro-3γ,4C-furandicarboxylic acid dimethylester		TLC-36
2-Methoxycinnamaldehyde		TLC-26
4-Methoxycinnamic acid		PC-9
3-Methoxytrimellitic acid	—3-Methoxy-1,2,4-benzenetricarboxylic acid	
5-Methoxytrimellitic acid	—5-Methoxy-1,2,4-benzenetricarboxylic acid	
Methoxytrimeric acid	—Methoxy-1,3,5-benzenetricarboxylic acid	
2-Methoxymethylbenzene		GC-6 TLC-1
3-Methoxymethylbenzene		GC-6 TLC-1
4-Methoxymethylbenzene		GC-6 TLC-1
2-Methoxy-1,3-dimethylbenzene		GC-6
2-Methoxy-1,4-dimethylbenzene		GC-6
2-Methoxy-1,5-dimethylbenzene		GC-6
2-Methoxy-1,6-dimethylbenzene		GC-6
3-Methoxy-1,5-dimethylbenzene		GC-6
3-Methoxy-1,6-dimethylbenzene		GC-6
4-Methoxy-5-methyl-1,3-benzenedicarboxylic acid		GC-21
4-Methoxy-2′-methylcinnamic acid		PC-9
6-Methoxy-2′-methyl-2-naphthaleneacetic acid		LC-24
4-Methoxy-5-methylisophthalic acid	—4-Methoxy-5-methyl-1,3-benzene-dicarboxylic acid	
2-Methoxy-4-methylphenol		GC-5,10 TLC-1
6-Methoxy-2-naphthaleneacetic acid		LC-24
2-(4-Methoxy-3,5-dinitrophenyl)-2-(4′-nitrophenyl)propane		LC-7
2-(4-Methoxy-3,5-dinitrophenyl)-2-phenylpropane		LC-7
2-Methoxyphenol		GC-5,9,10

Compound	Synonym	Number of tables and/or figures
		LC-2,6,14
		TLC-1,3,11,24,26
3-Methoxyphenol		GC-5,9
		LC-14
4-Methoxyphenol		GC-3,5,9
		LC-3,14
		TLC-2,11,12
4-Methoxyphenylacetic acid		LC-20
3-Methoxy-*o*-phthalic acid	—3-Methoxy-1,2-benzenedicarboxylic acid	
4-Methoxy-*o*-phthalic acid	—4-Methoxy-1,2-benzenedicarboxylic acid	
2-Methoxyisophthalic acid	—2-Methoxy-1,3-benzenedicarboxylic acid	
4-Methoxyisophthalic acid	—4-Methoxy-1,3-benzenedicarboxylic acid	
5-Methoxyisophthalic acid	—5-Methoxy-1,3-benzenedicarboxylic acid	
5-Methoxyprehnitric acid		GC-21
4-Methoxypropenylbenzene		GC-26
2-Methoxy-4-propenylphenol		TLC-1,14
2-Methoxy-propylbenzene		GC-6
3-Methoxy-propylbenzene		GC-6
4-Methoxy-propylbenzene		GC-6
3-Methoxy-2′-propylcinnamic acid		PC-9
4-Methoxy-2′-propylcinnamic acid		PC-9
2-Methoxy-4-propylphenol		GC-10
		LC-6,14
Methoxyterephthalic acid	—Methoxy-1,4-benzenedicarboxylic acid	
2-Methylacetoacetic acid		GC-F1
3-Methyladipic acid		GC-36,F1
Methylanacardate		GC-13
		TLC-10
o-Methylanisol	—2-Methylmethoxybenzene	
m-Methylanisol	—3-Methylmethoxybenzene	
p-Methylanisol	—4-Methylmethoxybenzene	
2′-*O*-methylanziaic acid		TLC-38
3-*O*-Methylarabinaric acid		GC-48
2-*C*-Methylarabinonic acid		GC-48
2-*C*-Methylarabinonic acid 1,4-lactone		GC-48
4-Methylbenzenesulphonic acid		LC-15
Methylbenzilate		TLC-30
2-Methylbenzoic acid		GC-25,33,34
3-Methylbenzoic acid		GC-33,34
4-Methylbenzoic acid		GC-33,34
		LC-15,30
4-Methylbenzoylformic acid		GC-34
Methylbutanedionic acid		GC-48
cis-Methylbutanedioic acid		GC-48
2-Methyl-5-*tert*.-butylhydroquinone	—4-*tert*.-Butyl-2,5-dihydroxy-methylbenzene	
2-Methyl-6-*tert*.-butylhydroquinone	—3-*tert*.-Butyl-2,5-dihydroxy-methylbenzene	
3-Methyl-5-*tert*.-butylpyrocatechol	—3-*tert*.-Butyl-2,5-dihydroxy-methylbenzene	
3-Methyl-6-*tert*.-butylpyrocatechol	—4-*tert*.-Butyl-2,3-dihydroxy-methylbenzene	
2-Methylcardanol dimethylester		GC-13
		TLC-10
2-Methylcardol		GC-13
		TLC-10
2-Methylcardol diacetate		GC-13
		TLC-10
2-Methylcardol diene		TLC-10
2-Methylcardol monoene		TLC-10
2-Methylcardol triene		TLC-10
5-Methyl-3-(2-chlorophenyl)isoxazol-4-*N,N′*-diethylamide		GC-44
5-Methyl-3-(2,6-dichlorophenyl)isoxazol-4-*N,N′*-diethylamide		GC-44

Compound	Synonym	Number of tables and/or figures
5-Methyl-3-(2-chlorophenyl)isoxazol-4-trimethylsilylester		GC-44
5-Methyl-3-(2,6-dichlorophenyl)isoxazol-4-trimethylsilylester		GC-44
Methyl-3,5-dichlorolecanorate		TLC-38
2′-Methylcinnamic acid		PC-9
4-*O*-Methylcryptochlorophaeic acid		TLC-38
Methylcyclopentylphenylglycolate		TLC-30
O-Methylerythraric acid		GC-48
2-*C*-Methylerythronic acid		GC-48
2-*C*-Methylerythronic acid 1,4-lactone		GC-48
Methylcyclopentylphenylglycolate		TLC-30
O-Methylerythraric acid		GC-48
2-*C*-Methylerythronic acid		GC-48
2-*C*-Methylerythronic acid 1,4-lactone		GC-48
Methylfumaric acid		GC-32
3-Methylgarcina lactone		PC-4
4-*O*-Methylglucaric acid (3-*O*-Methylgularic acid)		LC-25
O-(4-*O*-Methyl-α-D-glucopyranosyl-uronic acid)-(1-2)-D-β-D-xylopyranosyl-(1-4)-D-xylonic acid		LC-25
O-(4-*O*-Methyl-α-D-glucopyranosyl-uronic acid)-(1-2)-*O*-α-D-xylopyranosyl-(1-4)-*O*-β-D-xylopyranosyl-(1-4)-D-xylonic acid		LC-25
2-*O*-(4-*O*-Methyl-α-D-glucopyranosyl-uronic acid)-D-xylonic acid		LC-25
C-Methylglyceraric acid		GC-48
2-*C*-Methylglyceronic acid		GC-48
Methylglyoxal-2,4-dinitrophenylhydrazone		TLC-37
Methylglyoxylate-2-acetyl-2-benzylhydrazone		TLC-34
Methylglyoxylate-2-acetyl-2-methylhydrazone		TLC-34
Methylglyoxylate-2-benzylhydrazone		TLC-34
Methylglyoxylate-2-benzyl-4-phenyl semicarbazone		TLC-34
Methylglyoxylate-2-benzyl semicarbazone		TLC-34
Methylglyoxylate-2-ethoxycarbonyl-methylsemicarbazone		TLC-34
Methylglyoxylate-2-methylhydrazone		TLC-34
Methylglyoxylate-2-methyl-4-phenylsemicarbazone		TLC-34
Methylglyoxylate-2-methylsemicarbazone		TLC-34
Methylglyoxylate*(E)*-4-phenylsemicarbazone		TLC-34
Methylglyoxylate*(Z)*-4-phenylsemicarbazone		TLC-34
Methylglyoxylate*(Z)*semicarbazone		TLC-34
3-*O*-Methylgularic acid		GC-48
3-*O*-Methylgulonic acid		GC-48
Methylgyrophorate		TLC-38
4-*O*-Methylgyrophoric acid		TLC-38
Methylhexylparaconic acid		TLC-28
Methylhippuric acid		TLC-24
1-Methyl-3-hydroxy-4-isopropylbenzene	—5-Methyl-2-isopropylphenol	
N-Methyliminodiacetic acid		GC-24
5-Methyl-4-indanol		GC-5
Methylmalonic acid (2-Methylpropanedioic acid)		GC-23,25,36 LC-18,28 TLC-17,18,19
3-Methylmercaptophenol		TLC-1
4-Methylmercaptophenol		TLC-1
2′-Methyl-4-dimethylaminocinnamic acid		PC-9
2′-Methyl-4-methylcinnamic acid		PC-9
Methyl-2,2′-*O*-di-*O*-methylevernate		TLC-38
Methyl-2′,4-di-*O*-methylhypoprotocetrarate		TLC-38
3-Methyl-4-methylmercaptophenol		TLC-1
Methyl-4-*O*-methylphysodate		TLC-38
Methyl-2,2′,4-tri-*O*-methyldivaricatate		TLC-38
2′-Methyl-3-nitrocinnamic acid		PC-9
2′-Methyl-4-nitrocinnamic acid		PC-9
2-Methyl-5-nitrophenol		TLC-1

Compound	Synonym	Number of tables and/or figures
3-Methyl-4-nitrophenol		TLC-1
3-Methyl-6-nitrophenol		TLC-1
4-Methyl-2-nitrophenol		TLC-1
4-Methyl-3-nitrophenol		TLC-1
2-Methyl-4,6-dinitrophenol		LC-25
		TLC-1
3-Methyl-5-*tert*.-octylpyrocatechol	—3,4-Dihydroxy-5-methyl-*tert*.-octylbenzene	
4-Methyl-2-*tert*.-octylphenol		LC-12
2-Methylpentanedioic acid		GC-48
3-Methylpentanedioic acid		GC-48
3,3-Dimethylpentanedioic acid		GC-48
2-*O*-Methylperlatolic acid		TLC-38
2′-*O*-Methylperlatolic acid		TLC-38
2-Methylphenol		GC-1,2,3,4,5,9,10,18,19, 32,36
		LC-1,6,11,12,14,15
		TLC-1,2,3,4,5,8,11,12
		PC-1,2
3-Methylphenol		GC-1,2,3,4,5,9,10,17, 18,36
		LC-1,6,11,12,14,15
		TLC-1,2,3,4,8,11,12
		PC-1,2
4-Methylphenol		GC-1,2,3,4,5,9,10,18, 34,36
		LC-1,5,6,11,12,14,15
		TLC-1,2,5,8,11,12
		PC-1,2
5-Methyl-3-phenylisoxazole-4-*N,N*′-diethylamide		GC-44
5-Methyl-3-phenylisoxazole-4-trimethylsilylester		GC-44
4(4-Methylphenyl)-3-butanoic acid		PC-9
4-*O*-Methylphysodic acid		TLC-38
N-Methyl-3-piperidinol and HCl salt		TLC-30
N-Methyl-4-piperidiol		TLC-30
Methylpropanedioic acid		G-48
2′-Methyl-4-isopropylcinnamic acid		PC-9
4-Methyl-2′-propylcinnamic acid		PC-9
2-Methyl-4-*n*-propylphenol		GC-5
2-Methyl-4-isopropylphenol		GC-5
2-Methyl-5-isopropylphenol (Carvacrol)		TLC-1
2-Methyl-6-*n*-propylphenol		GC-5
		LC-5
2-Methyl-6-isopropylphenol		GC-5
		TLC-1
3-Methyl-4-*n*-propylphenol		GC-5
3-Methyl-5-isopropylphenol		GC-5
		TLC-4
		PC-1
3-Methyl-6-propylphenol		GC-5
3-Methyl-6-isopropylphenol		GC-5,10,19
		TLC-4,5,14
4-Methyl-2-*n*-propylphenol		GC-5
2-Methylpyrocatechol	—2,3-Dihydroxy-methylbenzene	
4-Methylpyrocatechol	—3,4-Dihydroxy-methylbenzene	
2-Methyl-4-*tert*.-octylphenol		LC-12
		PC-1
2-Methyl-6-octylphenol		TLC-2
2-Methyl-6-*tert*.-octylphenol		LC-12
		PC-1
2-*C*-Methylribonic acid		GC-48

Compound	Synonym	Number of tables and/or figures
2-*C*-Methylribonic acid 1,4-lactone		GC-48
3-Methylsalicylic acid	—2-Hydroxy-3-methylbenzoic acid	
4-Methylsalicylic acid	—2-Hydroxy-4-methylbenzoic acid	
Methylsuccinic acid		GC-44,F1
O-Methylthrearic acid		GC-48
2-*O*-Methylxylaric acid		GC-48
2,2′-Methylene-bis-(4-chlorophenol)-(Dichlorophene)		TLC-1,4
2,2′-Methylene-bis-(3,4,6-trichlorophenol)(Hexachlorophene)		TLC-1,4
Methylenebutanoic acid		GC-48
2-Methyolphenol		GC-14
4-Methyolphenol		GC-14
Mevalonic acid		TLC-18
Microphyllinic acid		TLC-38
Miriquidic acid		TLC-38
Myricetin (3,3′,4′,5,5′,7-Hexahydroxyflavone)		TLC-15
Myricitrin		GC-17
Myristic acid (Tetradecanoic acid)		GC-36
Myristicin		TLC-14
2-Naphthalenecarboxylic acid		TLC-22
1,5-Naphthalenediol		TLC-3
1-Naphthol (α-Naphthol)	—1-Hydroxynaphthalene	
2-Naphthol (β-Naphthol)	—2-Hydroxynaphthalene	
1-Naphthol-2-sulfonic acid	—1-Hydroxy-2-naphthalenesulfonic acid	
1-Naphthol-3-sulfonic acid	—1-Hydroxy-3-naphthalenesulfonic acid	
1-Naphthol-4-sulfonic acid	—1-Hydroxy-4-naphthalenesulfonic acid	
1-Naphthol-5-sulfonic acid	—1-Hydroxy-5-naphthalenesulfonic acid	
1-Naphthol-3,6-disulfonic acid	—1-Hydroxy-3,6-naphthalenedisulfonic acid	
1-Naphthol-3,8-disulfonic acid	—1-Hydroxy-3,8-naphthalenedisulfonic acid	
2-Naphthol-6-sulfonic acid	—2-Hydroxy-6-naphthalenesulfonic acid	
2-Naphthol-3,6-disulfonic acid	—2-Hydroxy-3,6-naphthalenedisulfonic acid	
1-Naphthylmethylcarbamate		LC-9
2-Naphthylmethylcarbamate		LC-9
Naringenin		GC-17
Naringin		GC-17
Neoabietic acid		GC-46
Neochlorogenic acid		GC-17
		LC-10
Nephroarctin		TLC-38
Nicotinic acid		GC-25
		LC-20,30,31
		TLC-25
Nigracin		TLC-32
Nitriloacetic acid		GC-24
Nitrilotriacetic acid		GC-24
3-Nitro-1,2-benzenedicarboxylic acid		PC-8
4-Nitrobenzenesulfonic acid		LC-15
2-Nitrobenzoic acid		GC-25,34
		PC-8
3-Nitrobenzoic acid		LC-14,19
		PC-8
4-Nitrobenzoic acid		GC-25
		LC-15,19
		PC-8
3-Nitrocinnamic acid		PC-9
4-Nitrocinnamic acid		PC-9
2-Nitrophenol		GC-9,10,11
		LC-2,15
		TLC-1,2,3,5,11,12
3-Nitrophenol		GC-9,11
		LC-2,6,14

Compound	Synonym	Number of tables and/or figures
		TLC-1,2,3,11,12
4-Nitrophenol		GC-9,10,11,19
		LC-2,6,14,15
		TLC-1,2,3,5,11,12
4-(*p*-Nitrophenyl)-azo-resorcinol	—4′-Nitrophenyl-3,5-dihydroxyazobenzene	
3-Nitrophthalic acid	—3-Nitro-1,2-benzenedica boxylic acid	
3-Nitro-2′-propylcinnamic acid		PC-9
4-Nitro-2′-propylcinnamic acid		PC-9
3-Nitrosalicylic acid	—2-Hydroxy-3-nitrobenzoic acid	
3-*tert*.-Octylpyrocatechol	—2,3-Dihydroxy-*tert*.-0ctylbenzene	
4-*tert*.-Octylpyrocatechol	—3,4-Dihydroxyoctylbenzene	
Oenin		GC-17
Olinetoric acid		TLC-38
Orotic acid		LC-31
Orsellinic acid (2,4-Dihydroxy-6-methylbenzoic acid)		TLC-33
Orsinol	—3,5-Dihydroxymethylbenzene	
Oxalacetic acid	—Oxaloacetic acid	
2-Oxalacetic acid		GC-45
Oxalic acid (Ethanedionic acid)		GC-22,23,24,25,36
		LC-17,18,28
		TLC-22,25
Oxaloacetic acid (2-Oxobutanedioic acid)		GC-22,23,28,30,45,F2
		LC-17,18
		TLC-22
2-Oxoadipic acid (2-Oxohexanedioic acid)		GC-31
		TLC-37
α-Oxobutyric acid	—2-Oxobutanoic acid	
2-Oxobutanoic acid		GC-30,31
		LC-29
		TLC-19,23
		PC-5
3-Oxobutanoic acid		TLC-17,23
2-Oxoisocaproic acid	—2-Oxoisohexanoic acid	
2-Oxo-5-*N*-carbamidovaleric acid		PC-5
2-Oxo-4-ethiolbutanoic acid		PC-5
2-Oxoglutaric acid (2-Oxopentanedioic acid)		GC-22,23,28,30,31,32,36,F1,F2
		LC-17,18,28,29
		TLC-18,25,37
		PC-4,5
2-Oxoglutaric acid-4-ethylether		TLC-22,23
		PC-5
2-Oxohexanoic acid		LC-29
		TLC-23
2-Oxoisohexanoic acid		GC-31,F2
		TLC-22,23
		PC-5
2-Oxoisovaleric acid		GC-31
		TLC-23
		PC-5
2-Oxo-3-dimethylbutanoic acid		GC-30
		TLC-23
2-Oxo-3-methylbutyric acid	—2-Oxo-3-methylbutanoic acid	
2-Oxo-4-(methylthio)butanoic acid		GC-31
2-Oxo-3-methyl-*n*-valeric acid		GC-30,31
		PC-5
2-Oxo-4-methylvaleric acid		GC-30
2-Oxononanoic acid		LC-29

Compound	Synonym	Number of tables and/or figures
2-Oxooctanoic acid		GC-31
		LC-29
Oxopropanedioic acid		PC-5
2-Oxovaleric acid		GC-31,32
		LC-29
		TLC-22,23
Paludosic acid		TLC-38
Palustric acid		GC-46
Pannaric acid		TLC-38
Pannarin		TLC-38
Paradol		GC-15
		TLC-14
Paradyl		GC-15
		TLC-14
Parietin		TLC-38
Pelargonidin		GC-17
Pelargonin		GC-17
Pentachloromethoxybenzene		GC-19
Pentachlorophenol		GC-10,11
		LC-13,26
		TLC-1,5,8
3,3′,4′,5,7-Pentahydroxyflavone(Quercetin)		GC-17
		LC-21
		TLC-15
2,3′,4,5′,6-Pentamethoxybiphenyl		TLC-13
Pentamethylphenol		GC-5
2-*n*-Pentylphenol		GC-5
4-*n*-Pentylphenol		GC-5
4-*tert.*-Pentylphenol		GC-4,5
		LC-1,5
		TLC-1
Peonin		GC-17
Perlatolic acid		TLC-38
Phenol		GC-1,2,3,4,5,9,10, 17,19,34,36
		LC-1,9,11,12,14,15, 19,21,27
		TLC-1,2,3,4,5,8,11,12
		PC-1,2
Pheoxyacetic acid		TLC-25
Phenylacetic acid		GC-9,23,25,32,36,39, 40,42
		LC-20,31
		TLC-24,25
Phenylacetylglutamic acid		TLC-18
4-Phenyl-1*(E)*,3*(E)*-butadien-1,3-dicarboxylic acid dimethylester		TLC-36
4-Phenyl-1*(E)*,3*(Z)*-butadien-1,3-dicarboxylic acid dimethylester		TLC-36
4-Phenylbutanoic acid		GC-28,40
		TLC-24
4-Phenyl-2-butenoic acid		TLC-24
		GC-40
4-Phenyl-3-butenoic acid		GC-40
		TLC-24
4-Phenyl-4-butenoic acid		GC-40
Phenyl-1,1-di-(3′-*tert.*-butyl-2′-hydroxy-5′-methylphenyl)-methane		LC-4
Phenyl-1,1-di-(5′-*tert.*-butyl-4′-hydroxy-3′-methylphenyl)-methane		LC-4
2-Phenylchromone		GC-17
Phenylglyoxylic acid		GC-25
		LC-32

Compound	Synonym	Number of tables and/or figures
Phenyllactic acid		GC-32,36
2-Phenylphenol	—2-Hydroxybiphenyl	
3-Phenylphenol	—3-Hydroxybiphenyl	
4-Phenylphenol	—4-Hydroxybiphenyl	
2-Phenyl-*n*-propanol		GC-15
2-Phenylisopropanol		GC-15
3-Phenylpropionic acid(Phenylpropionic acid)		GC-25,39
		LC-32
		TLC-24
Phenylpyruvic acid		GC-31,32,36
		LC-24
		PC-5
3′-Phenylpyruvic acid		LC-31
Phenylsulfate		LC-27
Phloretin		TLC-15
Phloridzin		TLC-15
Phloroglucine acetate		TLC-13
Phloroglucinol	—1,3,5-Trihydroxybenzene	
Phospho-enolpyruvic acid		PC-7
3-Phosphoglyceric acid		PC-7
Phthalaldehydecarboxylic acid		GC-41
Phthalic acid	—1,2-Benzenedicarboxylic acid	
Physodalic acid		TLC-38
Physodic acid		TLC-38
Picein (*p*-Hydroxyacetophenone-D-blucoside)		TLC-32
Picric acid	—2,4,6-Trinitrophenol	
Picrolichenic acid		TLC-38
Pimaric acid		GC-46
Pinastric acid		TLC-38
Placodiolic acid		TLC-38
Planaic acid		TLC-38
Polyporic acid		TLC-38
Populin		TLC-32
Porphyrilic acid		TLC-38
Prehnitic acid (1,2,3,4- Benzenetetracarboxylic acid)		GC-21
Propane-1,2,3-tricarboxylic acid		GC-36,48
		TLC-17,18
trans-Propene-1,2,3-tricarboxylic acid		GC-48
Propenoic acid		LC-17,28
Propionic acid		LC-18,28
		TLC-17,22
4-Propioxy-butanoic acid		PC-6
2′-Propyl-3-bromocinnamic acid	—3-Bromo-2′-propylcinnamic acid	
2′-Propyl-4-bromocinnamic acid	—4-Bromo-2′-propylcinnamic acid	
2′-Propyl-3-chlorocinnamic acid	—3-Chloro-2′-propylcinnamic acid	
2′-Propyl-4-chlorocinnamic acid	—4-Chloro-2′-propylcinnamic acid	
2′-Propylcinnamic acid		PC-9
2′-Propyl-3-iodocinnamic acid		PC-9
2′-Propyl-4-iodocinnamic acid		PC-9
2′-Propyl-3-methoxycinnamic acid	—3-Methoxy-2′-propylcinnamic acid	
2′-Propyl-4-methoxycinnamic acid	—4-Methoxy-2′-propylcinnamic acid	
2′-Propyl-4-methylcinnamic acid	—4-Methyl-2′-propylcinnamic acid	
2′-Propyl-3-nitrocinnamic acid	—3-Nitro-2′-propylcinnamic acid	
2′-Propyl-4-nitrocinnamic acid	—4-Nitro-2′-propylcinnamic acid	
2-*n*-Propylphenol		GC-4,5
		LC-1,11
		TLC-4
3-*n*-Propylphenol		GC-5

Compound	Synonym	Number of tables and/or figures
4-Propylphenol		GC-4,5 LC-1,5,11 TLC-1
2′-Propyl-4-isopropylcinnamic acid		PC-9
Psoromic acid		TLC-38
Protocatechuic acid	—3,4-Dihydroxybenzoic acid	
Protocetraric acid		TLC-38
Protolichesterinic acid		TLC-38
Pseudoplacodiodic acid		TLC-38
Pulvinic acid		TLC-38
Pulvinic dilactone		TLC-38
Pyrazinoic acid		GC-47
Pyridinealdehydetricarboxylic acid		GC-41
1-Pyridyl-2-azo-naphthol		PC-2
Pyrocatechuic acid	—2,3-Dihydroxybenzoic acid	
Pyrocatechol	—1,2-Dihydroxybenzene	
Pyrogallol	—1,2,3-Trihydroxybenzene	
Pyroglutamic acid		GC-32,36 LC-28 TLC-18
Pyrrolidonecarboxylic acid		LC-18 TLC-18
L-Pyrrolidine-5-carboxylic acid		LC-28
Pyruvic acid (2-Oxopropanoic acid)		GC-22,23,30,31,32, 36,45 LC-17,18,24,28,29 TLC-17,19,22,23,25,37 PC-4,5
Quercetin	—3,3′,4′,5,7-Pentahydroxyflavone	
Quercitrin		GC-17
Quinaldic acid		GC-25
Quinic acid	—1,2,3,4-Tetrahydroxycyclo-hexanecarboxylic acid	
Quinol		TLC-26 PC-2
Quinolindicarboxylic acid		GC-41
Quinoxalindicarboxylic acid		GC-41
3-Quinuclidiol		TLC-30
Rangiformic acid		TLC-38
Resorcinol	—1,3-Dihydroxybenzene	
β-Resorcinol (β-1,3-Dihydroxybenzene)		GC-36
Resveratrol		TLC-33
Rhizocarpic acid		TLC-38
Rhodocladonic acid		TLC-38
Rhodophyscin		TLC-38
Ribaric acid		GC-48
Ribonic acid		GC-27,48
D-Ribonic acid lactone		GC-43
Ribonic acid-1,4-lactone		GC-36,48
Ribonolactone		GC-27
Ritalinic acid		LC-31
Roccellic acid		TLC-38
Rutin (3,3′,4′,5,7-Pentahydroxyflavone-3-rutinoside)		GC-17 LC-10,21 TLC-15
Safrole		TLC-14
Sakuranetin (4′,5-Dihydroxy-7-methoxyflavone)		TLC-15
Sakuranin		TLC-15

Compound	Synonym	Number of tables and/or figures
Salazinic acid		TLC-38
Salicin (α,2-Dihydroxytoluene)		TLC-32
Salicortin		TLC-32
Salicylaldehyde	—2-Hydroxybenzaldehyde	
Salicylic acid	—2-Hydroxybenzoic acid	
Salicyloylsalicin		TLC-32
Salicyloylsalicin-2-*O*-benzoate		TLC-32
Salicyloylsalicin-6-*O*-benzoate		TLC-32
Salicyluric acid	—2-Hydroxyhippuric acid	
Salidroside		TLC-32
Salireposide		TLC-32
Sandaracopimaric acid		GC-46
Schizopeltic acid		TLC-38
Scopoletin	—7-Hydroxy-6-methoxycoumarin	
Scopolin		LC-10
Scrobiculin		TLC-38
Sebacic acid	—Decandioic acid	
Sekikaic acid		TLC-38
Sepacic acid		GC-24
Shikimate-3-phosphate		TLC-24
Shikimic acid		GC-24,48
		TLC-24
		PC-7
Shogaol		GC-15
		TLC-14
Sinapic acid	—4-Hydroxy-3,5-dimethoxycinnamic acid	
cis-Sinapic acid	—*cis*-4-Hydroxy-3,5-dimethoxycinnamic acid	
trans-Sinapic acid	—*trans*-4-Hydroxy-3,5μ Dimethoxcinnamic acid	
Succinic acid (Butanedioic acid)		GC-22,23,24,25,26,28,36,40,44,48,F1
		LC-17,18,28
		TLC-17,18,19,22,25
		PC-4,7
2-Succinylbenzoate		TLC-19
5-Sulfo-1,3-benzenedicarboxylic acid(3,5-Dicarboxy-benzenesulfonic acid)		PC-8
2-Sulfobenzoic acid(2-Carboxy-benzenesulfonic acid)		PC-8
5-Sulfosalicylic acid	—2-Hydroxy-5-sulfobenzoic acid	
Syringaldehyde	—4-Hydroxy-3,5-dimethoxybenzaldehyde	
Syringic acid	—4-Hydroxy-3,5-dimethoxybenzoic acid	
Syringol		TLC-26
Talonic acid		GC-48
D-Talonic acid lactone		GC-43
Talonic acid 1,4-lactone		GC-43
Tannic acid		LC-10
Tartaric acid		GC-24,25,36
		LC-28
		TLC-18,20,25
Taxifolin		GC-17
		TLC-15
Tenuiorin		TLC-38
Terebic acid		TLC-28
Terephthalic acid	—1,4-Benzenedicarboxylic acid	
3,3′,5,5′-Tetra-*tert*.-butyl-2,2′-dihydroxybiphenyl		LC-4
3,3′,5,5′-Tetra-*tert*.-butyl-4,4′-dihydroxybiphenyl		LC-4
Tetrachlorocatechol	—3,4,5,6-Tetrachloro-1,2-dihydroxybenzene	

Compound	Synonym	Number of tables and/or figures
3,4,5,6-Tetrachloro-2-(2′,3′,4′,5′,6′-pentachlorophenoxy)phenol		GC-11
3,4,5,6-Tetrachloro-1,2-dihydroxybenzene		TLC-6
3,3′,5,5′-Tetrachloro-2,6′-dihydroxybiphenyl		TLC-7
3,3′,5,5′-Tetrachloro-4,4′-dihydroxybiphenyl		TLC-7
2,3,4,5-Tetrachloro-methoxybenzene		GC-19
2,3,4,6-Tetrachloro-methoxybenzene		GC-19
2,3,4,5-Tetrachlorophenol		GC-11
		LC-2
2,3,4,6-Tetrachlorophenol		GC-11
		TLC-1
2,3,5,6-Tetrachlorophenol		GC-11
		TLC-4
3,4,5,6-Tetrahydrohippuric acid		TLC-24
1,4,5,6-Tetrahydro-6-pyridazinone-3-carboxylic acid		PC-5
1,2,5,8-Tetrahydroxy-anthraquinone		PC-2
1,2,3,5-Tetrahydroxybenzene-2,5-disulfate		LC-27
1,2,3,4-Tetrahydroxycyclohexanecarboxylic acid (Quinic acid)		GC-17,22,48
		LC-18
3′,4′,5,7-Tetrahydroxyflavone (Kaempfrol)		GC-17
		LC-21
		TLC-27
4-(1,1,3,3-Tetramethylbutyl)phenol		TLC-1
		PC-1
2,3,4,5-Tetramethylphenol		GC-5
2,3,4,6-Tetramethylphenol		GC-5
2,3,5,6-Tetramethylphenol		GC-4,5
		LC-1,5
Thamnolic acid		TLC-38
Theaflavin		LC-10
Theaflavin monogallate		LC-10
Theaflavin digallate		LC-10
Thioglycolic acid		GC-25
Thiophanic acid		TLC-38
Thiophaninic acid		TLC-38
Threaric acid		GC-48
threo-hex-2-enono-1,4-lactone		GC-48
C-(*threo*-1,2,3-trihydroxypropyl)glyceraric acid		GC-48
Threonic acid		GC-36,48
L-Threonic acid lactone		GC-43
Threonic acid 1,4-lactone		GC-48
Threono-1,4-lactone		GC-36
Thymol	—3-Methyl-6-isopropyl-1-phenol	
Tocol		LC-26
α-Tocopherol		LC-26
β-Tocopherol		LC-26
γ-Tocopherol		LC-26
4-Toluenesulphonic acid	—4-Methylbenzenesulfonic acid	
2-Toluic acid	—2-Methylbenzoic acid	
3-Toluic acid	—3-Methylbenzoic acid	
4-Toluic acid	—4-Methylbenzoic acid	
Tremulacin		TLC-32
Tremuloidin		TLC-32
Triandrin		TLC-32
2,4,6-Tribromophenol		LC-13
Tricarballylic acid		GC-22,28
		TLC-20
Trichloroacetic acid		GC-25
		LC-17
3,4,5-Trichlorocatechol	—3,4,5-Trichloro-1,2-dihydroxybenzene	

Compound	Synonym	Number of tables and/or figures
3,4,5-Trichloro-1,2-dihydroxybenzene		TLC-6
2,3′,6′-Trichloro-4-hydroxybiphenyl		TLC-7
3,3′,5-Trichloro-4,4′-dihydroxybiphenyl		TLC-7
3,4′,5-Trichloro-4-hydroxybiphenyl		TLC-7
2,4,5-Trichloro-methyoxybenzene		GC-19
2,3,4-Trichlorophenol		GC-11
2,3,5-Trichlorophenol		GC-11
2,3,6-Trichlorophenol		GC-11
2,4,5-Trichlorophenol		TLC-1,4 GC-11,19 LC-2
2,4,6-Trichlorophenol		TLC-1,5 GC-10,11,19 LC-2,6,13
3,4,5-Trichlorophenol		TLC-1 GC-11 LC-2
2,4,5-Trichlorophenoxyacetic acid (2,4,5-T)		GC-19 LC-19,25
2,4,5-Trichlorophenoxypropionic acid		GC-19
Trichoside		TLC-32
2,3,4-Trideoxyhexaric acid		GC-48
3-Trifluoromethylphenol		LC-9
2′,3′,4′-Trihydroxyacetophenol		TLC-1
1,2,3-Trihydroxybenzene		GC-9 LC-10,12,21 TLC-1,3,5,8,26 PC-2
1,2,4-Trihydroxybenzene		LC-6,21 TLC-1,8
1,3,5-Trihydroxybenzene		GC-17 LC-2,6,14,21,27 TLC-1,4,5,8,26 PC-2
1,3,5-Trihydroxybenzene-monosulfate		LC-27
1,3,5-Trihydroxybenzene-disulfate		LC-27
1,3,5-Trihydroxybenzene-trisulfate		LC-27
1,3,5-Trihydroxybenzoic acid		LC-3,12
2,4,6-Trihydroxybenzoic acid		LC-12,21 TLC-29
3,4,5-Trihydroxybenzoic acid		GC-17 LC-14,20,21,26 TLC-3,15,22,26,27 PC-7
2,4,4′-Trihydroxybenzophenone		TLC-1
2,4,5-Trihydroxybutyrophenone		LC-26
3,4,5-Trihydroxy-1-cyclohexane-1-carboxylic acid		GC-17
4′,5,7-Trihydroxyflavone		TLC-15,27
3′,5,7-Trihydroxy-4′-methoxyflavone		GC-17 GC-21
Trimellitic acid	—1,2,4-Benzenetricarboxylic acid	
Trimesic acid (1,3,5-Benzenetricarboxylic acid)		
3,4,5-Trimethoxybenzaldehyde		TLC-26
3,4,5-Trimethoxybenzoic acid		GC-36
3,4,5-Trimethoxyphenylacetic acid		GC-41 LC-20
2,4,6-Trimethylbenzoic acid		GC-42
2,3,4-Trimethylphenol		GC-5 TLC-5 PC-1

Compound	Synonym	Number of tables and/or figures
2,3,5-Trimethylphenol		GC-1,2,4,5
		LC-1
		TLC-1
2,3,6-Trimethylphenol		GC-1,4,5
		LC-1
2,4,5-Trimethylphenol		GC-4,5
		LC-1,5,14
		TLC-1,5
2,4,6-Trimethylphenol		GC-1,2,4,5,9
		LC-1,12
		TLC-1,5
3,4,5-Trimethylphenol		GC-4,5,9
		LC-1
		TLC-1
Trimethylpyruvic acid		PC-5
2,4,6-Trimethylolphenol		GC-14
2,4,6-Trinitrophenol		LC-14
		TLC-1,3,25
		PC-2
1,3,5-Tris(hydroxymethyl)benzenetriacetate		GC-41
1,3,5-Tris(hydroxymethyl)cyclohexanetriacetate		GC-41
Tropic acid	—3-Hydroxy-2-phenylpropionic acid	
Tropine		TLC-30
Uric acid		GC-36
		LC-32
Ursolic acid		TLC-38
Usnic acid		TLC-33,38
Valeric acid		TLC-22,25
Vanilacetic acid		GC-40
Vanilglycolic acid		GC-40
Vanillic acid	—4-Hydroxy-3-methoxybenzoic acid	
Vanillin	—4-Hydroxy-3-methoxybenzaldehyde	
3-Vanillylhydraacrylic acid		GC-F1
Vanilmandelic acid	—4-Hydroxy-3-methoxymandelic acid	
Variolaric acid		TLC-38
Veratric acid	—3,4-Dimethoxybenzoic acid	
Vicanicin		TLC-38
Vimalin		TLC-32
Virensic acid		TLC-38
Vitexin		GC-17
		TLC-15
Vulpinic acid		TLC-38
Xylaric acid		GC-48
o-Xylenol (2,3-Xylenol)	—2,3-Dimethylphenol	
2,4-Xylenol	—2,4-Dimethylphenol	
2,5-Xylenol	—2,5-Dimethylphenol	
2,6-Xylenol	—2,6-Dimethylphenol	
3,4-Xylenol	—3,4-Dimethylphenol	
3,5-Xylenol	—3,5-Dimethylphenol	
Xylonic acid		GC-48
L-Xylonic acid lactone		GC-48
Xylonic acid 1,4-lactone(Xylono-1,4-lactone)		GC-36,48
Zeorin		TLC-38
Zingerone		TLC-14

INDEX

A

B

C

D

E

F

G

H

I

K

L

M

N